AF904594

BIODIVERSITY LOSS IN A CHANGING PLANET

Edited by **Oscar Grillo** and **Gianfranco Venora**

Biodiversity Loss in a Changing Planet
Edited by Oscar Grillo and Gianfranco Venora

Published by InTech

Janeza Trdine 9, 51000 Rijeka, Croatia

Copyright © 2011 InTech

All chapters are distributed under the Creative Commons Attribution 3.0 license, which allows users to download, copy and build upon published articles even for commercial purposes, as long as the author and publisher are properly credited, which ensures maximum dissemination and a wider impact of our publications. After this work has been published by InTech, authors have the right to republish it, in whole or part, in any publication of which they are the author, and to make other personal use of the work. Any republication, referencing or personal use of the work must explicitly identify the original source.

As for readers, this license allows users to download, copy and build upon published chapters even for commercial purposes, as long as the author and publisher are properly credited, which ensures maximum dissemination and a wider impact of our publications.

Notice

Statements and opinions expressed in the chapters are these of the individual contributors and not necessarily those of the editors or publisher. No responsibility is accepted for the accuracy of information contained in the published chapters. The publisher assumes no responsibility for any damage or injury to persons or property arising out of the use of any materials, instructions, methods or ideas contained in the book.

Publishing Process Manager Dragana Manestar

Technical Editor Teodora Smiljanic

Cover Designer InTech Design Team

Additional hard copies can be obtained from orders@intechopen.com

Biodiversity Loss in a Changing Planet, Edited by Oscar Grillo and Gianfranco Venora
p. cm.
ISBN 978-953-307-707-9

We are IntechOpen, the world's leading publisher of Open Access books
Built by scientists, for scientists

4,000+
Open access books available

116,000+
International authors and editors

120M+
Downloads

151
Countries delivered to

Our authors are among the
Top 1%
most cited scientists

12.2%
Contributors from top 500 universities

BOOK CITATION INDEX
CLARIVATE ANALYTICS
INDEXED

WEB OF SCIENCE™

Selection of our books indexed in the Book Citation Index in Web of Science™ Core Collection (BKCI)

Interested in publishing with us?
Contact book.department@intechopen.com

Numbers displayed above are based on latest data collected.
For more information visit www.intechopen.com

For EU product safety concerns:
IN TECH d.o.o., Prolaz Marije Krucifikse Kozulić 3, 51000 Rijeka, Croatia,
info@intechopen.com or visit our website at intechopen.com.

Meet the editors

Dr. Oscar Grillo was born in Catania (Sicily) in 1977, he is a food technologist with an international PhD in applied and environmental botany. Since 2003 he has been working as researcher at the Stazione Sperimentale di Granicoltura per la Sicilia, a governmental institute of agronomic research, mainly working with computer vision applied to food matrices and plant structures, above all seeds, and in particular studying wheat and the related leguminous. Currently, he is working also at the Sardinian Germplasm Bank of the Biodiversity Conservation Centre of the University of Cagliari on projects devoted to seed characterization and identification by image analysis. Results of his work have been published in many peer-reviewed journals papers and international conference papers. Referee for a few peer-reviewed journals, many times he has been invited as teacher/lecturer/speaker by some universities and research centres in Spain and Italy. He has trained many MSc and PhD students, who have made their own contributions to the agronomical and botanical research.

Gianfranco Venora is a biologist, born in Caltagirone (Sicily) in 1958, where he lives and works. He took his University degree in 1981, and since 1982 he has been working as researcher at the Stazione Sperimentale di Granicoltura per la Sicilia. His working expertise is mainly about durum wheat and leguminous breeding. At the beginning of 1990, after some years of experience on karyotyping cropped and wild species of agronomical importance, he was fascinated by computer vision applied to food matrices and plant structures, above all seeds, studying wheat and the related leguminous. He was recently nominated as professor to the research doctorate on Applied and Environmental Botany of the University of Cagliari.
Many peer-reviewed journals published his papers and he was invited as speaker in many international conferences, and as teacher/lecturer by some universities and research centres in Germany, Netherlands, Bulgaria, Spain, Czech Republic and Italy. Gianfranco Venora is currently referee for about 15 peer-reviewed journals and tutor of many MSc and PhD students.

Contents

Preface

The deep biophysical alterations that are currently happening on our planet, are the natural consequence to the climate changes which the biosphere is subjected to. The loss of biodiversity is one of the macroscopic effects of this modification. Although this certainly isn't the first time the Earth undergoes such evolutions, mankind today should have the opportunity to make this change little less critical, considering, above all, that probably one of the most important causes of the actual loss of biodiversity is closely related to human development.

Every ecosystem is a complex organization of carefully mixed life forms, a dynamic and particularly sensible system. Consequently, their progressive decline may accelerate climate change and vice versa, influencing flora and fauna composition and distribution, resulting in the loss of biodiversity.

Mainly focused on climate change effects, this book includes very interesting case studies about biodiversity evaluations and provisions in several different ecosystems, such as tropical forests and coral reefs, analysing the current life condition of many life forms, from shellfish to reptiles and amphibians, and covering diverse biogeographic zones, from Europe to Oceania, Africa to Asia, as well as from Pacific Ocean to Indian Ocean.

Oscar Grillo
Stazione Sperimentale di Granicoltura per la Sicilia, Caltagirone
Biodiversity Conservation Centre, University of Cagliari
Italy

Gianfranco Venora
Stazione Sperimentale di Granicoltura per la Sicilia, Caltagirone,
Italy

1

Climate Change: Wildfire Impact

Mirza Dautbasic[1], Genci Hoxhaj[2], Florin Ioras[3], Ioan Vasile Abrudan[4] and Jega Ratnasingam[5]
[1]Sarajevo University
[2]Ministry of Environment, Forest and Water Administration
[3]Buckinghamshire New University
[4]Transilvania University
[5]Putra University
[1]Bosnia Herzegovina
[2]Albania
[3]United Kingdom
[4]Romania
[5]Malaysia

1. Introduction

The European forests harbour biological wealth of international importance (circa 6,000 species are of conservation importance according to IUCN). Changes to come in climate are challenging science, governments, and local communities in order to sustain the health of its ecosystems, which will, in turn, also help protect the quality of life.

European climate system are supported by various factors such as soils, topography, available plant species. Some of these factors are contributing to both natural ecosystems and their fire regimes. Long-term patterns of temperature and precipitation determine the moisture available to grow the vegetation that fuels wildfires (Stephenson, 1998). Climatic inconsistency on inter-annual and shorter scales governs the flammability of these fuels (Westerling, 2003; Heyerdahl et al., 2001). Flammability and fire frequency in turn affect the amount and continuity of available fuels. Therefore, long-term trends in climate can have profound implications for the location, frequency, extent, and severity of wildfires and for the character of the ecosystems that support them (Westerling, 2006a). Human determined climatic change may, over a relatively short time period (< 100 years), give rise to climates outside anything experienced in Europe, since the establishment of an industrial civilization, currently sustaining a population that has increased approximately 270% since 1850. Changes in wildfire regimes driven by climate change are likely to impact ecosystem services that European citizens rely on, including carbon sequestration; water quality and quantity; air quality; wildlife habitat; and recreational facilities. In addition to climate change, the continued growth of continent's population and the spatial pattern of development that accompanies that growth are consequently affecting wildfire regimes through their impact on the availability and continuity of fuels and the availability of ignitions.

South East Europe ecosystems are a vast mosaic of different habitat types. The biodiversity patterns we encounter today are a result of millions of years of climatic and geologic change.

Over years, populations of their native biota expanded and contracted in range - some at local scales, others at hemispheric scales, some up and others down slopes - to find and adapt to the local conditions that allowed them to persist to this day. During drier periods, for example, some species seek out the refuge of mountaintops that provided the conditions necessary for survival; on contrary during wetter periods, those species may have moved from those refuges to re-sort across the landscape that is now found in Europe.

What this dynamism demonstrates us is that change occurs at various temporal and spatial scales, and that while today's climate may be our baseline, our climate has not been and will not be static. It also highlights how critical connectivity is in our landscape: the extraordinary biological richness is to a great degree a product of species being able to shift in their range and adapt to changing climatic conditions. If that landscape connectivity is lost, or if the climate changes overtakes the ability of species to respond, or if populations are already reduced or stressed by other factors, species may be unable to survive through the climate changes to come.

In the case of many species and ecological processes, the effect of past and future land use change may induce significant stresses, that left unmanaged could see species to extinction. Some of these land use impacts may have a more significant impact than a changing climate. The challenge for South East Europe is to describe out the anticipated effects of past and future land use change from those of climate change - so that we can better plan our strategies to protect ecosystem health and conserve the native biodiversity for future generations.

This chapter endeavours to investigate what impact has the climate change, with specific reference to wildfire, on biodiversity and ecological processes in South East Europe and is presenting some considerations on how species native to the region will have to adapt.

2. Climate change

A changing climate will interact with other drivers in pertaining ways and generate feedback cycles with significant consequences. The effects of habitat fragmentation on native species may be dependent on intra- and inter-annual variation in rainfall (Morrison, 2000); so changes in rainfall and development patterns may deepen impacts. Increasing fires, in combination with increasing nitrogen deposition as a result of ash deposition on soil, may facilitate invasive of non-native weeds that in turn increase fire risk. Decreasing water supplies due to human pressure may have negative effects on native plants and animals, like species found in rivers. Meanwhile, increased irrigation run-off from non-porous soil in an urbanized watershed can fundamentally alter hydrological regimes in other ways (White et al, 2002).

These threats may lead to population pressure for native species, and possibly lead to extinction. The urbanization stress on southern part of South East Europe has increased recently, and most of the direct impacts to resources have occurred in the recent past. This means that the indirect effects have yet to be seen. Once these changes have occurred, it is expect that in some areas of South East Europe (eg Croatia, Bulgaria) it will only accelerate. Compounding the ecological impacts of land use change is perhaps an unprecedentedly rapid change in climate. The "climatic envelopes" species need (the locations where the temperature, moisture and other environmental conditions are suitable for persistence) will shift. For many species, a changing climate is not the problem, per se. The problem is the pace of the change: the envelope may shift faster than species are able to follow. For some species, the envelope may shift to areas already changed to human land use. Human impacts may have undermined the resilience of some species to adapt to the change (e.g., by lowering their overall population). Human land uses also may have disconnected the ecological connectivity in the landscape that would provide the movement corridor from the current to the future range.

This degree of alteration of ecological processes and jeopardise of native species will complete the transformation of the entire region to a "managed ecosystem"(Ioras, 2009). This reality will require that the local politicians articulate what the wanted future condition is for the area in question. Only with an informed thorough assessment of the current and future challenges confronting native species, and a clear articulation of ecological and socio-economic goals, we will be able to manage South East Europe native species and systems through the transformation ahead.

2.1 Climate and forest wildfire

2.1.1 Moisture, fuel availability, and fuel flammability

Climate increases wildfire risks primarily through its effects on moisture availability. Wet conditions during the growing season promote fuel—especially fine fuel—production via the growth of vegetation, while dry conditions during and prior to the fire season increase the flammability of the live and dead vegetation that fuels wildfires (Swetnam and Betancourt 1990, 1998; Veblen et al. 1999, 2000; Donnegan 2001). Moisture availability is determined by both precipitation and temperature. Warmer temperatures can reduce moisture availability via an increased potential for evapo-transpiration (evaporation from soils and surface water, and from vegetation), a reduced snowpack, and an earlier snowmelt. Snowpack at high altitude is an important mean of making water available as runoff in late spring and early summer (Sheffield et al. 2004), and a reduced snowpack and earlier snowmelt potentially lead to a longer, drier summer fire season in many mountain forests (Westerling, 2006b).

For wildfire risks in most Eastern European forests, inter-annual variability in precipitation and temperature appear to be determinant on forest wildfire through their short-term effects on fuel flammability, as opposed to their longer-term affects on fuel production. One way of illustration this is with the use of average Palmer Drought Severity Index (PDSI). The Palmer Drought Severity Index (PDSI) was developed by Palmer (1965) based on monthly temperature and precipitation data as well as the soil-water holding capacity at that location to represent the severity of dry and wet spells over the U.S. The global PDSI data (Dai et al., 2004) consist of the monthly surface air temperature (Jones and Moberg 2003) and precipitation (Dai et al., 1998; Chen et al., 2002) over global land areas from 1870 to 2006. These date is represented as PDSI values in 2.5°x 2.5° global grids.

The time series of the PDSI variations are determined by the mean values from all grid data from the selected area. The mean values are computed by means of the robust Danish method (Kegel, 1987). This method allows to detect and isolate outliers and to obtain accurate and reliable solution for the mean values. The global PDSI variations for the period 1870-2006 are between +1 in the beginning and -2 in 2002. The Palmer classification of drought conditions is in terms of minus numbers: between 0.49 and -0.49 - near normal conditions; -0.5 to -0.99 -incipient dry spell; -1.0 to -1.99 - mild drought; -2.0 to -2.99 - moderate drought; -3.0 to -3.99 - severe drought; and -4.0 or less - extreme drought. The positive values are similar about the wet conditions.

The PDSI variations over the South-East Europe are determined for area between longitude 10°30' E and latitude 32.5°50' N (Fig.1). This area consists of 44 grids of the global PDSI data.

The maximal errors are below 0.08 and the mean value of the all PDSI points is 0.02 (Fig.2). The PDSI variations over the South-East Europe from Fig.3 show several severe wet and dry events.

Fig. 1. Area of South-East Europe between longitude 10°-30° E and latitude 32°.5-50° N.

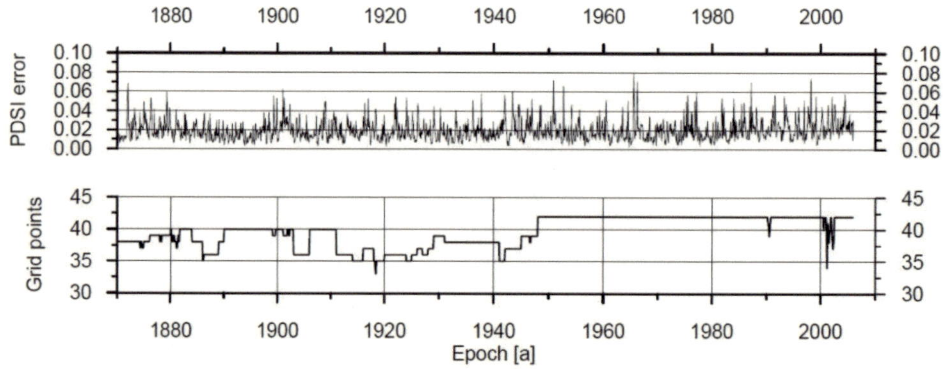

Fig. 2. Number of the grid points and errors of PDSI for South-East Europe (source Chapanov and Gambis, 2010).

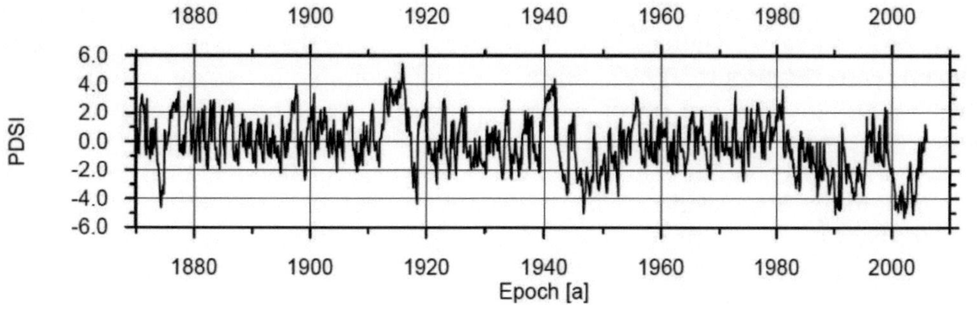

Fig. 3. Variations of the PDSI for South-East Europe (source Chapanov and Gambis, 2010).

Positive values of the index represent wet conditions, and negative values represent dry conditions. This is used here as an indicator of the moisture available for the growth and wetting of fuels.

This analysis included all fires over 400ha -large wildfires threshold (Running, 2006) that have burned since 1970, and account for the majority of large forest wildfires in South East Europe. The fires have been aggregated for each country using the European Forest Institute Database on Forest Disturbances in Europe (Table 1).

Country/Decade	1970-1979	1980-1989	1990-1999
Albania	0	0	9
Austria	1	0	0
Bosnia	0	0	12
Bulgaria	0	2	29
Croatia	10	37	66
Czech	8	4	6
Cyprus	12	10	4
Greece	60	33	46
Hungary	5	5	15
Italy	58	50	102
Macedonia	0	0	25
Moldova	0	0	0
Romania	0	0	6
Slovakia	13	1	3
Slovenia	0	0	13
Yugoslavia	12	17	2

Table 1. Number of forest fire that affected an area over 400ha in South East Europe between 1970-2000.

Note: 0 means no reported data

In the South, the frequency of large wildfires peaks in Italy and Greece, in the East in Bulgaria and Croatia often ignited by lightning strikes before the summer rains wet the fuels (Swetnam and Betancourt, 1998). Since the lightning ignitions are associated with subsequent precipitation, it is possible that the monthly drought index may tend to appear to be somewhat wetter than conditions were at the time of ignition.
In the two northern countries - Slovak and Check Republic-conditions also tended to be drier than normal in the 70s: extended drought increased the risk of large forest wildfires in these wetter northern forests for fires above 1700 meters in elevation, the importance of surplus moisture in the preceding year was greatest for the southern countries. According to Swetnam and Betancourt (1998) moisture availability in predecessor growing seasons was important for fire risks in open conifer forests as fine fuels play an important role in providing a continuous fuel cover for spreading wildfires, but not in mixed conifer forests. Looking at the western part of South East Europe more generally, the moisture necessary to support denser forest cover tends to increase with latitude and elevation. Consequently, the shift in forest fire incidence as one moves from the forests of the SW to those of the NE is broadly consistent with a decreasing importance of fine fuel availability—and an increasing importance of fuel flammability— as limiting factors for wildfire as moisture availability increases on average.

2.1.2 Forest wildfire and the timing of spring

There has been a remarkable increase in the incidence of large forest wildfire in some of the countries in the South East Europe since the early 1980s (Table 2). Understanding the factors behind such increase in forest wildfire activity is key to understanding the recent trends and inter-annual variability in forest wildfire. According to Westerling et al. (2006b) the length of the average season completely free of snow cover is highly sensitive to variability in regional temperature, increasing approximately 30 percent in the latest third of snowmelt years and this has a positive effect on wildfire incidence. In years with an early spring snowmelt, spring and early summer temperatures were higher than average, winter precipitation was below average, the dry soil moistures typical of summer in the region came sooner and were more intense, and vegetation was drier (Westerling et al., 2006b).

Country	Time period	Average number of fires	Average area burned, ha
Albania	1981-2000	667	21456
Bulgaria	1978-1990	95	572
	1991-2000	318	11242
Croatia	1990-1997	259	10000
Cyprus	1991-1999	20	777
Greece	1990-2000	4502	55988
Romania	1990-1997	102	355
Slovenia	1991-1996	89	643

Table 2. Fire statistical data of the SE Europe. Source: GFMC.

The statistics presented here are for only those wildfires greater than 400ha that burned primarily in forests, of which there were 676 in South East Europe since 1970. This region has experienced a number of large wildfires that ignited spread to and burned substantial forested area (Table 2). The consequences of an early spring for the fire season are profound.

Comparing fire seasons for the earliest versus the latest third of years by snowmelt date, the length of the wildfire season (defined here as the time between the first report of a large fire ignition and last report of a large fire controlled) was 45 days (71 percent) longer for the earliest third than for the latest third. Sixty-six percent of large fires in South East Europe occur in early snowmelt years, while only nine percent occur in late snowmelt years. Large wildfires in early snowmelt years, on average, burn 25 days (124 percent) longer than in late snowmelt years. As a consequence, both the incidence of large fires and the costs of suppressing them are highly sensitive to spring and summer temperatures. Both large fire frequency and suppression expenditure appear to increase with spring and summer average temperature in a highly non-linear fashion. In the case of Albania, Bosnia Herzegovina and Romania (Hoxhaj, 2005; Alexandru et al, 2007; Ciobanu and Ioras ed, 2007) suppression expenditure in particular appears to undergo a shift near 15°C during of 2007 (Figure 4 and 5). Year 2007 was used as reference year due to the significant increase of wildfire (Figure 6) and also this year was known to have had a heat wave. Temperatures taken separately above and below that threshold are not significantly correlated with expenditures, but the mean and variance of expenditures increase dramatically above it.

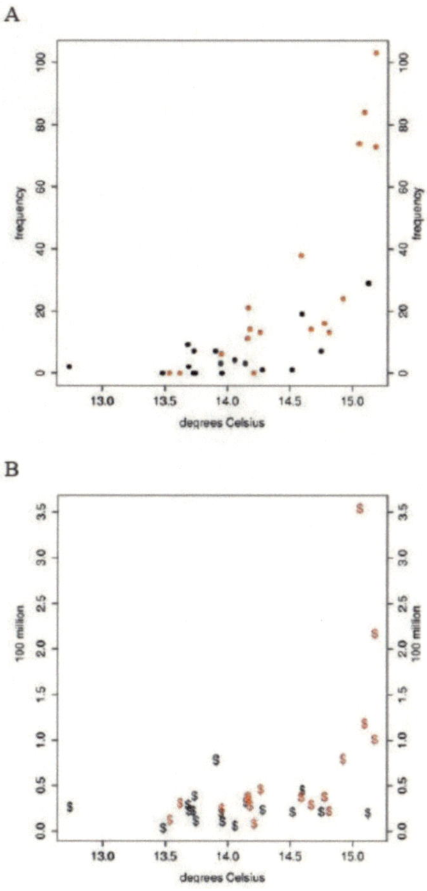

Fig. 4. The annual number of large forest fires in Albania, Bosnia and Romania versus average March - August temperature in 2007.

Fig. 5. The forest fire spread in the South East Europe on 25 July 2007 as seen by the Terra Satellite (Source GFMC).

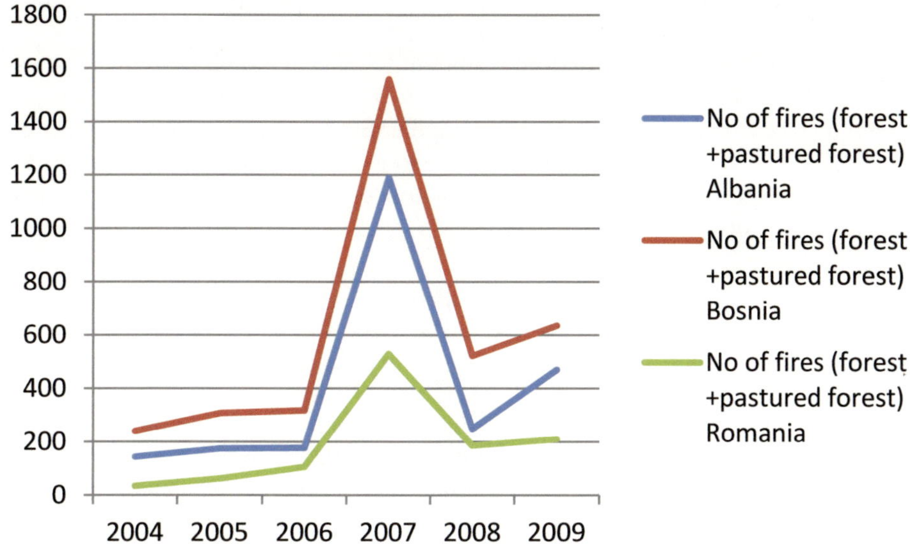

Fig. 6. Forest fire numbers (to include forested pastures) in Albania, Bosnia Herzegovina and Romania between 2004 and 2009.

3. Land use patterns

Looking across South East Europe it is obvious that land uses changes have determined significant, often cascading impacts to biodiversity and ecosystems - and more recently it was witnessed how these have threatened the quality of life for the human residents as well. Ecological impacts of land use have been well documented through pioneering research on habitat fragmentation. Fragmentation can affect communities from the "bottom up". Suarez et al, (1998) research on habitat fragmentation, showed how when non-native species invade, and native ant species disappear other species up the food chain will soon also disappear because they have lost the native species that are their main food resource (Chen et al., 2011). Such "ecosystem decay" leading to loss of biodiversity may take decades to complete following the fragmentation. The cahoots between climate change and habitat fragmentation is the most threatening aspect of climate change for biodiversity, and is a central challenge facing conservation (Ioras, 2006).

3.1 Increasing population

As the human population grows, there will be increased competition for resources (like space and water) with plants and animals. Demand for housing will displace rural land uses like farming that can provide important habitat for some native species. With increased development we will witness more introduction, establishment, and invasion of habitat altering non-native species. More people will demand more opportunity for recreation - yet low intensity recreational uses like hiking when is done is an intensive way can damage fragile environments (Ioras, 1997).

Increased demand for resources, goods, and services will increase demand for transport infrastructure (roads, power lines, pipelines, etc.) which may fragment otherwise intact landscapes and provide an entry point for non-native species such as weeds. More people means increased susceptibility to fire ignitions - and subsequently more restrictions on fire management for ecological outcomes. More people will also increase the potential for human-wildlife conflicts in the remaining wildlands (e.g. interactions with predators like bears, wolfs; biodiversity impacts from efforts to control insect-borne disease vectors). Hence, even distant human land uses can damage natural resources. Pollution, for example - whether it is represented by airborne toxins when wildfires burn, or nitrogen, ozone from urban areas, or wastewater that fouls beaches and other coastal areas - will pose great challenges for the health of the ecosystems.

3.2 Interaction of climate, land use, and wildfire

Fire in the recent years has become a key ecological process in South East Europe. Many plant species display adaptations that are finely tuned to a particular frequency and intensity of fire. Some plants may re-sprout from roots following fire. The seeds of other plants may require heat or chemicals from smoke to germinate. Some animals may be especially suited to invade recently burned areas; others may only succeed in habitats that have not burned for a relatively long time. In some cases species that are highly adapted to - even reliant on - fire can also be put at risk by fire. If fire behaviour is changed by human activities such that it is outside of its natural range of variation, it can have great significant adverse impact on native species. For example Pinus heldreichii H. Christ requires fire to reproduce, but if fires recur too frequently (i.e., before the trees have a chance to mature to reproductive age) fire can kill the young trees and break that finely-tuned life cycle. Its areal covers Albania, Bosnia Herzegovina, Bulgaria, Greece, Macedonia and Serbia (Critchfield et al 1966).

Due to human activities the fire behaviour of the entire region have greatly altered - fires generally occur too frequently in the coastal areas and too infrequently in the higher elevation forests. Fires set during wind conditions can have enormous ecological consequences (see the fire that engulfed Dubrovnik coast during of summer 2007); for some highly restricted species, an individual fire could lead to extinction. Future land use and climate changes will only exacerbate the alteration fire regimes in South East Europe. These have consequences not only on biodiversity conservation but there are also important implications for public safety, the quality of our air and water, and the economy.

Some parts of Croatia, Bulgaria already have the most severe wildfire conditions in the region, and the situation is only likely to worsen with climate change—meaning dangerous consequences for both humans and biological diversity. South East Europe's coastal area exceptional combination of fire-prone, shrubby vegetation and extreme fire weather means that fires here are not only going to become very frequent, but occasionally huge and extremely intense. The combination of a changing climate and an expanding human population threatens to increase both the number and the average size of wildfires even more. Increasing fire frequency--or ever shortening intervals between repeated fires at any particular location--poses the greatest threat to the region's coastal natural communities (except perhaps in high altitude forests), whereas increasing incidence of the largest, most intense fires poses the greatest threat to human communities.

A region's fire regime is defined by the number, timing, size, frequency, and intensity of wildfires, which are in turn largely determined by weather and vegetation. Vegetation on the region's coastal plains and foothills—where humans are most concentrated—is dominated by shrub species that burn hot and fast, and that renew themselves in the aftermath of fire (so long as inter-fire intervals are sufficiently long to allow individual plants to mature and reproduce by resprouting or setting seed between fires). In the

Mediterranean climate, this coastal sage and chaparral vegetation rapidly grows fine new twigs and leaves during the moist winters. This new growth then dries to a highly flammable state during the arid summer-fall season. Consequently, most fires burn during summer when fine, dry fuels become abundant, whereas the greatest total acreage burns in fall, when the largest fires are driven by winds.

Different climate change models yield somewhat different predictions about the frequency, timing, and severity of future region wind conditions, leading to uncertainty about just how fire regimes may change in the future. However, preliminary analyses for the period 2002-2006 suggest that wind conditions may significantly increase earlier in the fire season (especially end of July- start of September) while they may decrease somewhat later in the season (especially towards the end of September). This predicted change to earlier winds occurrences would likely increase the frequency of huge fires as severe fire weather would coincide more closely with the period of most frequent fire ignitions (Fig. 7).

Of course, fires also require an ignition source. Fires started naturally, by lightening strikes, are actually quite rare during the most dangerous autumn fire weather—when the hot, dry sea winds blow. Nowadays, however, the vast majority of ignitions are caused by humans or their inventions; and even without climate change, the number of fires in southern part of

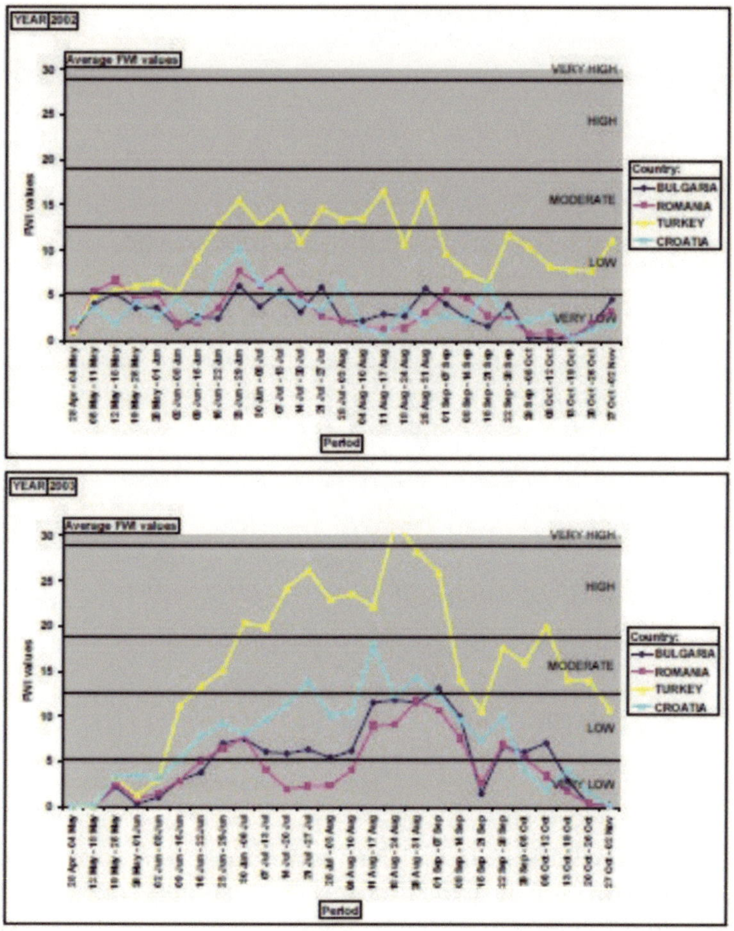

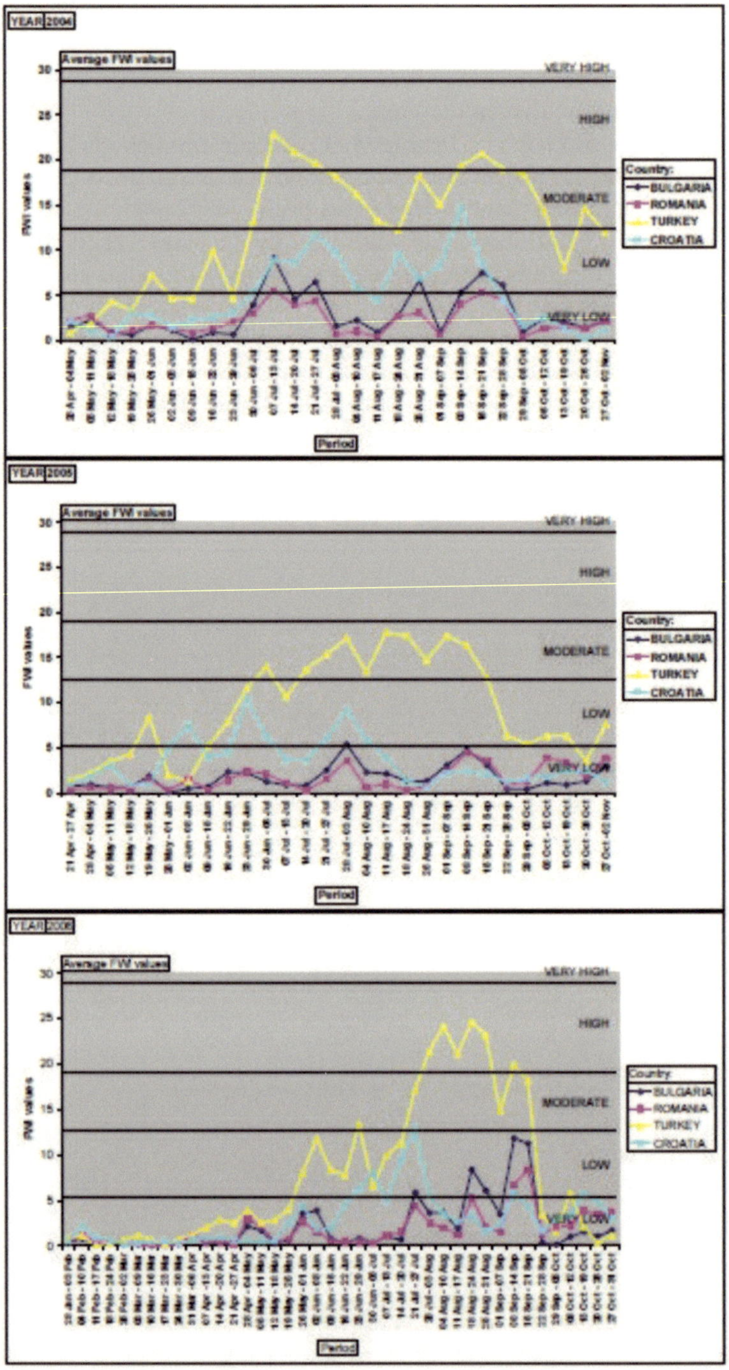

Fig. 7. Fire risk trends (Fire Weather Index -FWI) between 2002 and 2006 in Bulgaria, Croatia, Romania and Turkey. Source EFFIS "Forest Fire in Europe 2006" (http://effis.jrc.ec.europa.eu/reports/fire-reports/doc/2/raw)

the region has been steadily increasing in direct proportion to human population (www.effis.jrc.ec.europa.eu/reports/category/40/fire-reports). This increase in ignitions, especially if coupled with a longer fire-weather season, creates more opportunities for fires to start when conditions are most extreme, however, huge firestorms such as those during 2003 and 2007 are not new phenomena in this region. Studies of charcoal layers deposited on the sea floor near the Cyclades Islands indicate that such major fire events have recurred on average every 20 to 60 years, or roughly two to five times per century over the past 12 to 13 centuries (Bryne et al., 1977). These huge firestorms inevitably occur following very wet years, at the beginning of drought periods (Mensing et al., 1999). How these inter-annual wet-dry cycles may change with changing climate is as yet unclear.

Due to the combined forces of changing climate, increasing fire ignitions, and invasive weedy species, fires are likely to burn ever more frequently in a positive feedback loop. Studies have shown that frequent fires over short time intervals increase invasions by weedy annual plants into native communities. These weedy invaders then set seed, die, and dry out earlier than the natives, thereby starting the fire season even earlier and increasing chances of another fire. These weedy annuals, referred to as "flash fuels" by firefighters, also ignite more readily and burn more rapidly than native perennial plants, thus creating a more favourable environment for themselves at the expense of the natives, which evolved under longer fire-return intervals. The potential for these interactions between climate change, weedy invasions, and changing fire regimes paints a grim picture for South East Europe's biological diversity and watershed quality, as vast stands of rich biodiverse and soil-holding shrub communities are replaced by biologically sparse, shallow-rooted, fire-perpetuating weeds.

4. Specific challenges of climate change in South and Eastern Europe

The impact of climate change promises to be more visible in southern part of the region because there is such a great diversity of plants and animals. Every species has unique requirements for persistence. This means that species will respond differently to the same climatic change. The range of a species is determined by external conditions like temperature, but also by conditions like interactions with other species.

Thus, native species will face novel environmental conditions - and will have precious little time to adjust. Even if the changes in climate are gradual, it has been recognized that the changes will be steep. Species with limited ability to move will have an especially difficult time keeping pace as Chen et al. (2011) reported that the distributions of species have recently shifted to higher elevations at a median rate of 11.0 meters per decade, and to higher latitudes at a median rate of 16.9 kilometres per decade. Some species may even require assistance moving to new regions.

Of greatest concern for local scientist, however, is that even with a gradual change there may be "tipping points" in the system, whereby ecological complexities interact and there is a dramatic "step change" in the system. These may include massive scale die-back of forests due to abnormal drought conditions, conversion of scrub habitat to non-native grassland with a few too frequent fires, and the scouring of watersheds, excessive erosion, and alteration of geomorphology of region's streams and rivers, with rain after catastrophic fire. Such fundamental conversion of the region's ecosystems could be abrupt and irreversible. It is not currently known where such thresholds in the system might be.

4.1 Climate change and forest ecosystem

In South East Europe, as in many other places in the world, the distribution of plant and animal populations will not be able to suddenly shift northward or to higher elevations because the potential habitat has been claimed by development, invaded by non-native species, or has unsuitable soils or other physical limitations (Parmesan, 2006).

Extended drought can stress individual trees, increase their susceptibility to insect attack, and result in widespread forest decline. Plant species respond differently and entire species may die off when drought occurs in an area that already has predictable seasonal droughts. Stressed trees have less resistance to insects, such as bark beetles. More indirectly, warmer winter temperatures as predicted for the region's future can increase insect survival and population levels. Drought and abnormally warm years that began in the 1980s have resulted in unprecedented pest outbreaks and tree dieback in southern part of the region (Logan et al, 2003).

Extended drought can also increase the severity of wildfires when they are ignited. The 2003 and 2007 wildfire events in South East Europe were shaped by extended drought that reduced fuel moisture of trees, the sea borned winds and high temperatures, and the ignition in shrubs - maquis type of vegetation that burned "uphill" into the forests.

Forests may not regenerate to historical species composition, when wildfires burn with higher intensity than tree species are adapted to. For example Franklin et al. (2006) surveyed areas in Cuyamaca Rancho State Park, USA, during the first two post-fire growing seasons following the Cedar Fire, and found that most conifers were killed by the high-intensity fire and that pine seedlings have not re-established. Oaks and ceanothus species now dominate the forest.

Forest-dependent fish and wildlife species may be lost in the indirect effects of climate change, drought, and wildfire. For example the Sweetwater Creek State park native trout and stickleback populations in Atlanta were totally eliminated in the Cedar Fire in 2003, and the last native trout population is threatened in Pauma Creek by sediments filling pools (after wildfires and rainstorms).

To understand the impact of climate change particular focus has to be given to shrubland communities that support a diversity of sensitive plant and animal species in the region. To begin to understand how changing climate conditions might affect these natural communities, a climate sensitivity analyses for coastal sage scrub and maquis vegetation and for plant and animal species found in these shrublands is needed (Preston et al, 2008).

To assess sensitivity of species and vegetation types to climate, the model that uses varied temperature and precipitation compared with current climate conditions could be employed. These values fall within the range of various climate forecasts for the region, although the emerging consensus is that the region will become more arid (IPPC, 2007). In response to increasing temperature and reduced precipitation, each vegetation type moves to higher elevations where current conditions are cooler and there is greater precipitation compared with locations where these shrublands / maquis vegetation occur.

In Europe some work was done on modelling habitat shifts due to climate change, however, the most conclusive one took place in the USA. For example analyses was conducted for five different coastal sage scrub shrub species in the USA; California sagebrush (*Artemisia californica*), brittlebush (*Encelia farinos*), flat-topped buckwheat (*Eriogonum fasciculatum*), laurel sumac (*Malosma laurina*), and white sage (*Salvia apiana*). The model developed by The Center for Conservation Biology (CCB) at the University of California, Riverside also modelled two annual host plants, California plantain (*Plantago erecta*) and white snapdragon

(*Antirrhinum coulterianum*) for the endangered butterfly, Quino Checkerspot (*Euphydryas editha quino*). All plant species, except brittlebush, flat-topped buckwheat and white snapdragon showed similar sensitivities as coastal sage scrub and chaparral to altered climate conditions. These three exceptions showed higher levels of potential habitat remaining at elevated temperatures, particularly flat-topped buckwheat.

The CCB also used modelling for the USA-endangered Quino Checkerspot butterfly and threatened California Gnatcatcher (*Polioptila californica*) (Preston et al, 2008). Other models included associations between species and compared predictions under altered climate conditions with models that did not. The CCB found that when vegetation, shrub or host plant species were included in the animal models, potential habitat for the butterfly and songbird was significantly reduced at altered climate conditions. Such models could be used to predict distribution changes with climate change.

Climate change and the pressures associated with human pressure can each lead to large changes in biodiversity. While the ecological effects of each of these stressors are increasingly being documented, the complex effects of climate change, harvesting pressure and urbanization on ecosystems remain inadequately understood. Yet such effects are likely to be extremely important in regions such as southern part of South East Europe.

Exploitation of intertidal and subtidal species as well as runoff and nutrient loading into coastal waters continues to increase as a result of rapid population growth at a time when the species involved are also being subjected to large scale changes in the environment driven by global warming. It is not unreasonable to presume that harvesting can undermine the resiliency of species to climate change. For example, historic data show that body size of molluscs plays an important role in determining which species are likely to shift their geographic distributions in response to climate change (Roy et al, 2001). Yet body sizes of many intertidal species have decreased substantially over the last century as a result of human harvesting of these species. Furthermore such size declines can result in major changes in growth rates, reproductive outputs and life histories of species and can even lead to changes in the compositions of ecological communities (Roy et al, 2003). How such changes in the biology of the species involved affects their resiliency to global warming is still poorly understood but the potential for feedback effects certainly exists.

5. Conclusion

The South Eastern Europe must brace for change. Even without the climate changes to come, native plants and animals, and the ecosystems on which the region rely, will be severely affected in the decades ahead. Climate change will only accelerate - and perhaps dramatically - changes already afoot in natural community composition and distribution. Some species may disappear as their habitat shifts to outside of the region; the range of others may expand to include the region. Species with limited dispersal ability will be most likely tested. Some of region's native species may be wholly reliant on how the region community mobilizes to ease them through the transitions to come.

The most important strategy to increase the likelihood of natural systems to adapt to the new climate regime is to maintain the connectivity between conservation reserve networks of core area representing the diversity of communities in the region for ecological cohesion of the landscape.

A favourable condition for biodiversity and ecosystems in the year to come is continued functionality of ecosystem processes; this would "save the evolutionary stage" and so perhaps allow the greatest complement of native species to persist. While the current configuration and composition of general vegetation communities will surely be different, it is desirable the communities to be characterized predominantly by species native to the region.

Forest wildfire in the South East Europe is strongly influenced by spring and summer temperatures and by cumulative precipitation. The effect of temperature on wildfire risks is related to the timing of spring, and increases with latitude and elevation. The greatest effects of higher temperatures on forest wildfire in recent decades have been seen in the southern countries - Croatia, Greece, Italy- and a handful of fire seasons account for the majority of large forest wildfires. A seasonal climate forecast for spring and summer temperatures would thus be of value in anticipating the severity and expense of the forest wildfire season in much of the South East Europe, and would be of particular value in Albania, Bosnia, Croatia.

6. Acknowledgement

This paper is based on work conducted by some of the authors within the project entitled "MSc technology-enhanced Forest Fire Fighting Learning" (project number 510184-LLP-1-2010-1-UK-ERASMUS-ECDC) financed by the EACEA Agency of European Union.

7. References

Alexandru, V., Ioras, F., Stihi, C. & Horvath, B. (2007). European forests and fires frequency in these forests. In Proceedings "Lucrările sesiunii ştiinţifice Pădurea şi Dezvoltarea Durabilă" Braşov, Romania, 2006, pp. 531-534.

Byrne, R.; Michaelsen, J. & Soutar, A. (1977). Fossil charcoal as a measure of wildfire frequency in Southern California: a preliminary analysis. In H.A. Mooney and C.E. Conrad (eds.). In *Proceedings of the Symposium on Environmental Consequences of Fuel Management in Mediterranean Ecosystems*. USDA Forest Service, General Technical Report WO-3, pp 361-367.

Chapanov, Y. & Gambis, D. (2010). Drought cycles over South-East Europe for the period 1870-2005 and their connection with solar activity. In *Proceedings BALWOIS 2010 - Ohrid, Republic of Macedonia - 25, 29 May 2010.*

Chen, M.; Xie, P.; Janowiak, J.E. & Arkin, P.A. (2002). Global land precipitation: a 50-yr monthly analysis based on gauge observations. *J. Hydrometeorol.*, 3, pp. 249-266.

Chen, C., Hill, J.K., Ohlemuler, R., Roy, D.B. & Thomas, C.D., (2011). Rapid Range Shifts of Species Associated with High Levels of Climate Warming. Science, Vol. 333 no. 6045, pp. 1024-1026.

Ciobanu, V. & Ioras, F (ed) (2007). Forest Fires, Transilvania University Publishing House.

Critchfield, W. B. & Little, E. L. (1966): Geographic distribution of the pines of the world. *USDA Forest Service Miscellaneous Publication* 991.

Dai, A.; Trenberth, K. E. & T. Karl, (1998). Global variations in droughts and wet spells: 1900-1995. *Geophys. Res. Lett.*, 25, pp. 3367-3370.

Donnegan, J.A.; Veblen, T.T. & Sibold, S.S. (2001). Climatic and human influences on fire history in Pike National Forest, central Colorado. *Canadian Journal of Forest Research* 31, pp 1527-1539.

Franklin, J.; Spears-Lebrun, L. A.; Deutschman, D. H. & Marsden, K. (2006). Impact of a high-intensity fire on mixed evergreen and mixed conifer forests in the Peninsular Ranges of southern California, USA. *Forest Ecology and Management Volume 235,* Issues 1-3, pp 18-29.

Heyerdahl, E.K; Brubaker, L.B. & Agee, J.K. (2001). Factors controlling spatial variation in historical fire regimes: A multiscale example from the interior West, USA. *Ecology* 82(3) pp 660-678.

Hoxhaj, G. (2005). Forest Fires in Albania. Regional Balkan Wildland Fire Network/Global Wildland Fire Network International Technical and Scientific Consultation "Forest Fire Management in the Balkan Region", 4-5 April 2005, Ohrid, Republic of Macedonia.

Ioras F (1997) Forest management techniques for sustainable management of the forests with tourism designation, Proceedings of the XI World Forestry Congress, Anatalya Turkey

Ioras, F. (2006). Assessing sustainable development in the context of sustainable forest management and climate change mitigation. In Proceedings "Lucrarile sesiuni ştiinţifice Padurea şi Dezvoltarea Durabila", Braşov, Romania, 2005, pp. 595-600.

Ioras, F. (2009). Climate change as a threat to biodiversity and ecological processes in Eastern Europe. In Proceedings "Lucrarile sesiunii ştiinţifice Padurea şi Dezvoltarea Durabila" Braşov, Romania, 2008, pp 257-264.

IPCC (Intergovernmental Panel on Climate Change), (2007). Climate change: The Physical Science Basis. Summary for Policymakers. Contribution of Working Group I to the *Fourth Assessment Report of the Intergovernmental Panel on Climate Change.*

Jones, P.D. & Moberg, A. (2003). Hemispheric and large-scale surface air temperature variations: An extensive revision and an update to 2001. *J. Climate,* 16, pp. 206-223.

Kegel, J. (1987). Zur Lokalizierung grober Datenfehler mit Hilfe robuster Ausgleichungsfervahren, *Vermessungstechnik,* 35, No. 10, Berlin.

Logan, J.; Regniere, J. & Powell, J. A. (2003). Assessing the Impacts of Global Warming on Forest Pest Dynamics. *Frontiers in Ecology and the Environment.* 1: pp 130-137.

Mensing, S.A.; Michaelsen, J. & Byrne, R. (1999). A 560-year record of Santa Ana fires reconstructed from charcoal deposited in the Santa Barbara Basin, *California Quaternary Research* 51: pp 295-305.

Morrison, S. A. (2000). *Demography of a fragmentation-sensitive songbird: Edge and ENSO effects.* Ph.D. Dissertation. Dartmouth College.

Parmesan, C. (2006). Ecological and evolutionary responses to recent climate change; *Annu Rev. Ecol. Syst.* 37: pp 637-669.

Palmer, W. C. (1965). Meteorological Drought. *Res. Paper No.45,* U.S. Weather Bureau, Washington, 58pp.

Preston, K.L.; Rotenberry, J.T.; Redak, R. & Allen, M.F. (2008). Habitat shifts of endangered species under altered climate conditions: Importance of biotic interactions. *Global Change Biology.*

Roy, K.; Jablonski, D. & Valentine, J.W. (2001). Climate change, species range limits and body size in marine bivalves. *Ecology Letters* 4: pp 366-370.

Roy, K.; Collins, A.G.; Becker, B.J.; Begovic, E. & Engle, J.M. (2003). Anthropogenic impacts and historical decline in body size of rocky intertidal gastropods in southern California. *Ecology Letters* 6: pp 205-211.

Running, S.W. (2006). Is Global Warming Causing More, Larger Wildfires? *Science* 313, pp. 927.

Suarez, A. V.; Bolger, D. T. & Case, T. J. (1998). Effects of fragmentation and invasion on native ant communities in coastal southern California. *Ecology* 79: pp. 2041-2056.

Sheffield, J.; Goteti, G.; Wen, F. & Wood, E.F. (2004). A simulated soil moisture based drought analysis for the United States. *Journal of Geophysical Research* 109:D24108.doi:10.1029/2004JD005.82.

Swetnam, T.W. & Betancourt, J.L. (1990). Fire-southern oscillation relations in the Southwestern United States. *Science* 249:pp 1017-1020.

Swetnam, T.W. & Betancourt, J.L. (1998). Mesoscale disturbance and ecological response to decadal climatic variability in the American Southwest. *Journal of Climate* 11:pp 3128-3147.

Veblen, T.T.; Kitzberger, T. & Donnegan, J. (2000). Climatic and human influences on fire regimes in ponderosa pine forests in the Colorado Front Range. *Ecological Applications* 10:pp 1178-1195.

Veblen, T.T.; Kitzberger, T.; Villalba, R. & Donnegan, J. (1999). Fire history in northern Patagonia: The roles of humans and climatic variation. *Ecological Monographs* 69:pp 47-67.

Westerling, A.L.; Brown, T.J.; Gershunov, A.; Cayan, D.R. & Dettinger, M.D. (2003). Climate and wildfire in the Western United States. *Bulletin of the American Meteorological Society* 84(5):pp 595-604.

Westerling, A.L. (2006a). *Climate and Forest Wildfire in the Western United States.* In Economics of Forest Disturbance, T. Holmes, Ed., Springer.

Westerling, A.L.; Hidalgo, H.G.; Cayan, D.R. & Swetnam, T.W. (2006b.) Warming and earlier spring increases Western U.S. forest wildfire activity. *Science* 313:pp 940-943.

White, M. D. & Greer, K. A. (2002). The effects of watershed urbanization on stream hydrological characteristics and riparian vegetation of Los Peñasquitos Creek, California. Conservation Biology Institute. (16.12.2010) Available from http://consbio.org/cbi/projects/ pp. 557-558.

2

Assessing Loss of Biodiversity in Europe Through Remote Sensing: The Necessity of New Methodologies

Susana Martinez Sanchez[1], Pablo Ramil Rego[1],
Boris Hinojo Sanchez[1] and Emilio Chuvieco Salinero[2]
[1]GI-1934 TB Botany and Biogeography Lab., IBADER,
Campus of Lugo, University of Santiago de Compostela
[2]Department of Geography – University of Alcalá
Spain

1. Introduction

There is a global consensus on the idea of the present loss of biodiversity is intimately linked with human development, and that the conservation and sustainable use of present biological diversity is paramount to current and future generations of all life on Earth (Duro et al. 2007).

The United Nation Convention on Biological Diversity (CBD, http://www. biodiv.org, last accessed May 2011) lays down that countries are responsible for conserving their biological diversity and for using their biological resources in a sustainable manner. It expands until 2020 with the global *Strategy Plan for Biodiversity 2011-2020 and the Aichi biodiversity targets* (http://www.cbd.int/2011-2020, last accessed May 2011) to promote effective implementation of the CDB and to stem biodiversity loss by 2020. It compels the contracting countries to develop scientific and technical capacities to provide the appropriate measures in order to prevent and halt the pace of biodiversity loss all around the world.

It was during 90s and 2000s when scientific community became conscious that habitat destruction is the most prominent driver of biodiversity loss (Dirzo and Raven 2003)and together with degradation and fragmentation represent the most important factors leading to worldwide species decline and extinction (Chhabra et al. 2006; Soule and Terborgh 1999).

To improve the current conservation efforts and draw new strategies around the commitments under the CBD, it is crucial that our progress is monitored (Pereira and Cooper 2006). Biodiversity monitoring should be focused on trends in the abundance and distribution of populations and habitat extent (Balmford et al. 2005) and be carried out at different scales, regional and global and even local (Pereira and Cooper 2006).

There are several biophysical features influence species distributions, population sizes and ranges like land cover, primary productivity, temporal vegetation dynamics, disturbance events or climate (Hansen et al. 2004). All of them could be used as biophysical predictors of biodiversity at different scales. Remote sensing has been shown to be effective in some extent to measure and mapping those indicators and it has become a powerful tool for ecological studies because it allows monitoring over significant areas (Kerr and Ostrovsky 2003).

Remote sensing technologies contribute to biodiversity monitoring both direct and indirectly and they have been intensively improved in the last two decades, especially since the beginning of 2000s when the very high spatial resolution sensors were launched. Medium spatial resolution images from sensors on satellites make especially available information related to biophysical factors. In this sense, Landsat TM and ETM+ sensors are widely used in ecological investigations and applications because they have several advantages)(Cohen and Goward 2004): 30m of spatial resolution that facilitates characterization of land cover and land cover change; measurements acquired in all major portions of the solar electromagnetic spectrum (visible, near-infrared, shortwave-infrared); more than 30 years of Earth imaging and a temporal resolution of 16 days makes possible a complete analysis of the dynamic of the ecosystem; moderate cost (actually, all Landsat data from the USGS_ U.S. Geological Survey_ archive are free since the end of 2008). Anyway, to some extent, indirect measures that rely on remote sensing of biophysical parameters are, especially for national-level analyses, not enough accurate when the aim is the analysis of some aspects of biodiversity.
In a direct way, hyperspatial and hyperspectral sensors potentially supply land elements like individual organisms, species assemblages or ecological communities. Finally, LIDAR and RADAR technologies make possible to map vegetation structure (Lefsky et al. 2002; Zhao et al. 2011). Direct measures of biodiversity are becoming feasible with this kind of sensors although processes are still expensive and time-consuming, at least to regional levels.
Then, through RS it is possible to estimate in some extent habitat loss and fragmentation and trends in natural populations. At global and regional level the keystone is how translating remote-sensing data products into real and accurate knowledge of habitats and species distributions and richness. Subsequently, at present it is recognized that remote sensing technologies are especially crucial for conservation-related science (Kerr and Ostrovsky 2003) but they are still challenging. In addition, what is finally missing are global and regional standards for developing methodologies so systematic monitoring can be carried out (Strand et al. 2007).

2. Land-cover versus habitat data

Land cover is the observed physical cover of the Earth's surface (bare rock, broadleaved forests, etc…) (Eurostat 2001). Land cover data are usually derived by using multispectral remotely sensed data and statistical clustering methods. Remotely-sensed land cover data have been used at different scales (local, regional and global) as: i) input variables in biosphere –atmosphere models simulating exchanges of energy and water between land surface and the atmosphere and in terrestrial ecosystem models simulating carbon dynamics at global scales; ii) input variables in terrestrial vegetation change assessments; iii) proxies of biodiversity distribution (DeFries 2008; Hansen et al. 2004; Thogmartin et al. 2004).
On the other hand, *habitat* is a three-dimensional spatial entity that comprises at least one interface between air, water and ground spaces, it includes both the physical environment and the communities of plants and animals that occupy it, it is a fractal entity in that its definition depends on the scale at which it is considered" (Blondel 1979). *Natural habitats* means terrestrial or aquatic areas distinguished by geographic, abiotic and biotic features, whether entirely natural or semi-natural (EU Habitats Directive, 92/43/EC). The identification of one habitat implies a holistic perspective and involves not only the expression of the vegetation (land cover) but also the species and other biophysical parameters like topography, aspect, soil characteristics, climate or water quality.

Blondel's definition of habitat was adopted in different habitat classifications at European level like CORINE Biotopes (Devillers et al. 1991), Classification of Paleartic Habitats (Devillers et al. 1992), the database PHYSIS of the Institut Royal des Sciences Naturelles de Belgique, or in the EUNIS program - *Habitat* of the European Environment Agency (EEA).

We can understand habitat monitoring as the repeated recording of the condition of habitats, habitat types or ecosystems of interest to identify or measure deviations from a fixed standard, target state or previous status (Hellawell 1991; Lengyel et al. 2008). Habitat monitoring has two attributes (Lengyel et al. 2008; Turner et al. 2003): i) it can cover large geographical areas, then it can be used to evaluate drivers of biodiversity change over different spatial and temporal scales; being specially interesting at regional scales; ii) it provides information on the status of characteristic species because many species are restricted to discrete habitats; then if the link between some key species and discrete habitat types has been previously established, habitat monitoring can be used as a proxy for simultaneous monitoring of several species.

Nature resources management and biological conservation assessments require spatially explicit environmental data that come from remote sensing or derived thematic layers. Most of these studies assume that the selected geospatial data are an effective representation of the ecological target (such as habitat) and provide an appropriate source of information to the objectives (McDermid et al. 2009). For example, biodiversity has been frequently studied indirectly through associations with land cover, which represents that mapping land cover has been often used as a surrogate for habitats (Foody 2008). The suitability of these assumptions is a current scientific concern and a dynamic research issue (Glenn and Ripple 2004; McDermid et al. 2009; Thogmartin et al. 2004).

Some studies (McDermid et al. 2009) have evaluated the suitability of general-purpose land cover classifications and compared to other data sources like vegetation inventory or specific-purpose maps: they show the constraints of general-purpose remote sensing land cover maps for explain wildlife habitat patterns and recommend the use of specific-purpose databases based on remote sensing along with field measurements. Then, traditional or general-purpose land cover maps may not be appropriate proxies of habitats, as we will show after assessing the suitability of the CORINE Land Cover product (European Environment Agency, http://www.eea.europa.eu/publications/COR0-landcover, last accessed May 2011) in Europe on this target (Section 4). Multi-purpose land cover maps meeting the needs of a large number of users but they are not specifically designed to represent the habitat of one/some key specie/-s. Furthermore, land-cover classifications used for wildlife habitat mapping and modeling must be appropriate spatial and thematic resolution to identify reliably the habitats that the target species potentially occupy (Kerr and Ostrovsky 2003).

Thus, the identification of habitats through remote sensing must be suited the characteristics of each habitat type, rather than follow general or standard images processing approaches. For example, binary and one-class classifiers have been used in the implementation of the European Union's Habitat Directive (Boyd et al. 2006; Foody 2008). Moreover, it should be based on *in situ* and ancillary measurements (Kerr and Ostrovsky 2003) and also on ecological expert knowledge that allow finding the relationship between key species and their potential habitats.

There are some ongoing challenges with this issue "*habitat monitoring*": the identification of the habitats as ecological units and not simply as land covers and the assessment and quantification of habitats degradation and fragmentation. Currently, one of the main scientific challenges and one of the big issues are if we are able to identify proper and accurately habitats from remote sensing at landscape level: the mapping and monitoring of the territory in terms of its habitats. We have to say not yet, at least not only with remote sensing technologies and with an adequate budget and an optimal time. We also need ancillary information, ecological expert knowledge, field work and other auxiliary tools like landscape ecology indices.

3. European efforts for habitats mapping and monitoring

There are different scientific and legislative agreements that define habitats in Europe (Groom et al. 2006). The European Union's Habitats Directive since 1992 sets the rules in Europe for developing a coherent ecological network, called *Natura 2000,* which is the centerpiece of the EU nature and biodiversity policy (http://ec.europa.eu/environment/nature/natura2000/index_en.htm, last accessed May 2011). The aim of the network is to assure the long-term survival of Europe's most valuable and threatened species and habitats. It is comprised of Special Areas of Conservation (SAC) designated by Member States under the Habitats Directive, and also incorporates Special Protection Areas (SPA) designed under the Birds Directive (Directive 2009/147/EC)). Habitats Directive describes two kind of habitat, from the viewpoint of their conservation status (See Annex 1 of the Directive): i) *natural habitat types of Community interest,* habitat types in danger of disappearance and whose natural range mainly falls within the territory of the European Union; ii) *priority natural habitat types,* for the conservation of which the Community has particular responsibility (Appendix 1, Table 13). The establishment of this network of protected areas also fulfills a Community obligation under the UN Convention on Biological Diversity.

The characteristics and identification of the different habitat types included in the Habitats Directive were firstly described in the Manuel d'Interprétation des Habitats de l'Union Européenne -EUR 15/2 that has been revised several times from 1999 until the present (http://ec.europa.eu/environment/nature/legislation/habitatsdirective/index_en.htm#interpretation, last accessed May 2011). The manual enhances that habitat interpretation should be flexible and revised especially in regions with fragmented landscapes that has also a high anthropic influence. Consequently many European regions have developed their own handbooks for the interpretation of the habitats at regional level (Ramil Rego et al. 2008) (Italy: http://vnr.unipg.it/habitat/; France: http://natura2000.environnement.gouv.fr/habitats/cahiers.html; last accessed May 2011) and their own methodologies for habitats mapping and monitoring (Izco Sevillano and Ramil Rego 2001; Jackson and McLeod 2000). At present various methodologies are being used with different fieldwork efforts and levels of complexity and, in some cases, with critical limitations for appropriate and accurate monitoring.

The 2020 EU Biodiversity Strategy (http://www.consilium.europa.eu/uedocs/cms_Data/docs/pressdata/en/ec/113591.pdf, last accessed May 2011), urges countries to conserving and restoring nature. Countries are responsible of the habitats and species conservation and they must adopt measures to promote it and report about repercussions of this measures on their conservation status. The target 1 of the strategy lay down "*To halt the deterioration in the status of all species and habitats covered by EU nature legislation and achieve a significant and measurable improvement in their status so that, by2020, compared to current assessments: (i) 100% more habitat assessments and 50% more species assessments under the Habitats Directive show an improved conservation status; and (ii) 50% more species assessments under the Birds Directive show a secure or improved status*". Indirectly it requires the development of methodologies to get this goal in an appropriate and accurate way. Any loss of protected habitats must be compensated for by restoration or new assignations with the same ecological value and surface area.

Through its 17th Article, Habitat Directive forces countries to monitor habitat changes every six years and to assess and report to the European Union on the conservation status of the habitats and wild flora and fauna species of Community interest: the mapping of the distribution area, the trends, the preservation of their structure and functions together with the future perspectives and an overall assessment.

Then, to meet the requirements of global and regional biodiversity targets such as the *Strategy Plan for Biodiversity 2011-2020 and the Aichi biodiversity targets, the 2020 EU Biodiversity Strategy* or the European *Natura 2000* Network, the development of more cost and time effective monitoring strategies are mandatory (Bock et al. 2005).

At the moment, the first habitats reports were submitted in electronic format to the European Environmental Agency (www.eunis.eea.europa.eu, last accessed May 2011) (EEA) until March 2008, through an electronic platform on the Internet. This platform is managed by the EEA and the European Environment Information and Observation Network (EIONET) (http://bd.eionet.europa.eu/, last accessed May 2011). Currently, this information was supplied by 25 of the 27 countries that currently comprise the European Union (all except Bulgaria and Romania).

We have developed a map (Figure 1) about the distribution of habitats of Community interest derived from this information. The data were compiled, refined and standardized in

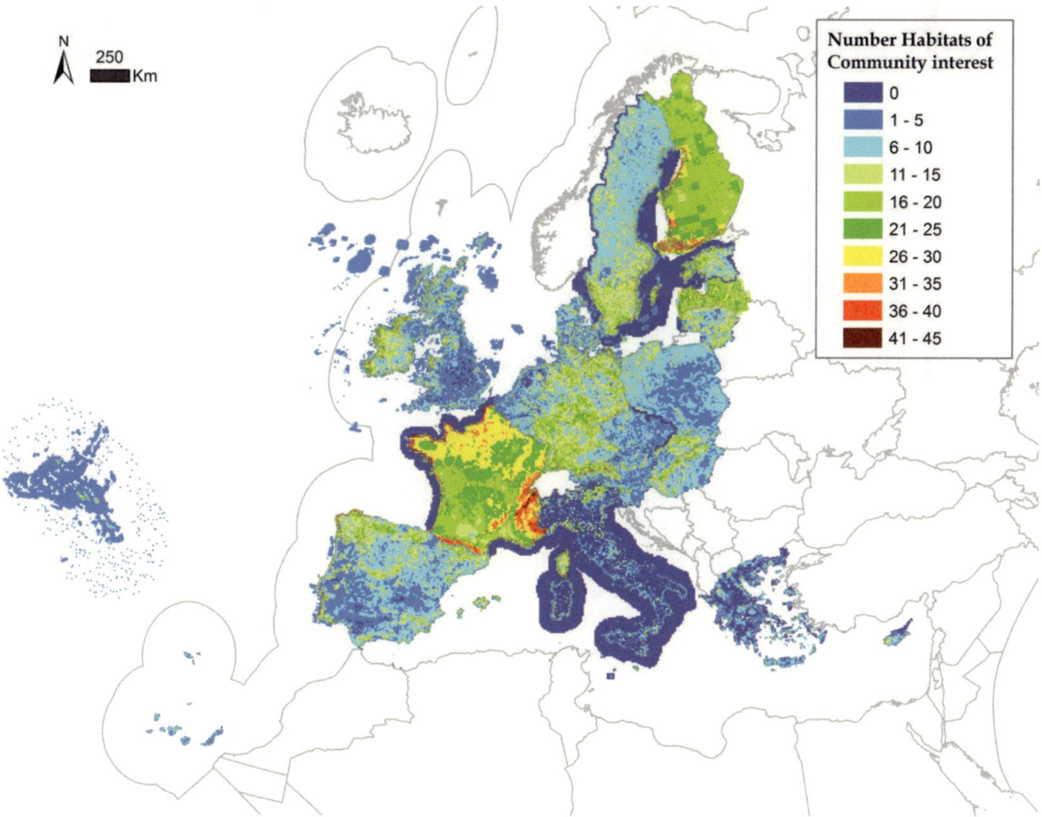

Fig. 1. Distribution of habitats of Community interest in Europe (Source: Developed from EIONET 2011)

a database using the *ETRS 1989 Lambert azimuthal equal-area* projection system (following INSPIRE Directive). All the spatial data of the habitats of Community interest, derived from each country, were harmonized and represented in a 10 km^2 UTM grid (Universal Transverse Mercator Projection) following the recommendations of the European Commission (EuropeanCommission 2006). The output map follows the EUNIS (European Nature Information System) classification system and represents finally all the European habitats of Community interest.
The EUNIS system constitutes a pan-European classification proposed by the EEA (www.eunis.eea.europa.eu, last accessed May 2011). It is developed and managed by the European Topic Centre for Nature Protection and Biodiversity (ETC/NPB in Paris), and covers the whole of the European land and sea area, i.e. the European mainland as far east as the Ural Mountains, including offshore islands (Cyprus; Iceland but not Greenland), and the archipelagos of the European Union Member States (Canary Islands, Madeira and the Azores), Anatolian Turkey and the Caucasus (Davies et al. 2004). It represents a common classification scheme for the whole of European Union, as it is compatible with the units of protection established in the strategy of *Natura 2000*-protected areas. It covers all types of habitats from natural to artificial, from terrestrial to freshwater and marine. EUNIS is also cross-comparable with CORINE Land Cover (Bock et al. 2005; Moss and Davies 2002) (Appendix 1, Table11).

4. The CORINE land cover map as a proxy of biodiversity: difficulties and constraints

The *CORINE land cover* project (EEA, 1999, http://www.eea.europa.eu/publications/COR0-landcover, last accessed May 2011) constitutes the first harmonized European land cover classification system, based on photo-interpretation of Landsat images. The minimum unit for inventory is 25 ha and the minimum width of units is 100m. Only area elements (polygons) are identified. Areas smaller than 25 ha are allowed in the national land cover database as additional thematic layers, but should be generalized in the European database. The CORINE land cover (CLC) nomenclature is hierarchical and distinguishes 44 classes at the third level, 15 classes at the second level and 5 classes at the first level. Third level is mandatory although additional national levels can be mapped but should be aggregated to level 3 for the European data integration. Any unclassified areas appear in the final version of the dataset (See CLC Legend in Appendix 1, Table 12).
Because of general land cover maps may not be suitable proxies of habitat maps, we have analyzed the spatial inconsistencies between a remote-sensed land cover map (CORINE Land Cover 2000) and some selected habitats of the map obtained from EIONET (Figure 1) which represents the spatial distribution of the natural and semi-natural habitats in the European Community. CORINE Land Cover is cross-comparable with habitats of Community interest (Council Directive 92/43/EEC) through the EUNIS system (www.eunis.eea.europa.eu, last accessed May 2011) (Tables 1 and 2). The comparative analysis was done using a 10 km^2 UTM grid which is the base of the EIONET map. Spatial gaps and contradictions that arise between both sources, when land cover maps are used to assess biodiversity status, were evaluated through the analysis of coincidences at the cells of the grid between both databases.
The CORINE Land Cover (CLC) map was analyzed and compared at the third level in the European context and at the fifth level at the scale of Spain (Table 3 and 4). Following

Corine Land Cover codes (3rd level)	Correlation with habitat of Community interest
311	9120 9180* 91E0* 91F0 9230 9330 9340 9380
312	9410
321	6170 6410 6510
322	4020* 4030 5130
323	5130
331	2120 2130*
332	8130 8220
411	7230
412	7110* 7120 7130*
421	1310
423	1140
511	3260 3270
512	3150
521	1150*
522	1130

[*], Priority natural habitat types as Habitats of the Council Directive 92/43/EEC

Table 1. Correspondences between Corine Land Cover classification (3rd level) and habitats of Community interest

Corine Land Cover codes (5th level)														
31110	31120	31150	31210	32100	32110	32111	32111	32112	32120	32121	32122	32210	32312	33110
Correlation with habitat of Community interest														
9330 9340 9380	9120 9180* 9230	91E0* 91F0	9410	6510	6410	6170 6410	6170 6410	6170 6410	6410 6510	6410 6510	6410 6510	4020* 4030 5130	5130	2120 2130*

Corine Land Cover codes (5th level)									
33210	33220	41100	41200	42100	42300	51110	51210	52100	52200
Correlation with habitat of Community interest									
8130	8220	7230	7110* 7120 7130*	1310	1140	3260 3270	3150	1150*	1130

[*], Priority natural habitat types as Habitats of the Council Directive 92/43/EEC

Table 2. Correspondences between Corine Land Cover classification (5th level) and habitats of Community interest

similarities with a gap analysis (Jennings 2000; Scott et al. 1993), spatially explicit correlations was carried out and five thresholds representing the grade of inconsistency between both maps were set: i) less than 10% of coincidences represent a **total gap**; ii) coincidences between 10-30%, **very high gap**; iii) coincidences between 30-50%, **high gap**; iv) coincidences between 50-90%, **moderate gap**; v) coincidences upper 90% represent **no gap**.

At European level (Table 3), and by countries, some relevant habitats showed:

- *Habitat 2130* and 2120* (correspondence with CLC 331 class BEACHES, DUNES AND SAND PLAINS): at European level it shows a moderate gap. By countries: TOTAL GAP in Finland; VERY HIGH GAP in Denmark; HIGH GAP in UK and Sweden. Netherlands, Lithuania y Latvia present a right correspondence.
- *Habitat 1150** (correspondence with CLC 521 class COASTAL LAGOONS): at global level it shows a very high gap. By countries: TOTAL GAP in Cyprus, Slovenia, Finland, Ireland, Latvia, Malta and UK (Null values Cyprus, Finland, Latvia and Malta); VERY HIGH GAP in Denmark, Spain, Estonia, Portugal y Sweden; HIGH GAP in Italy, France, and Germany. Only Lithuania presents a right correspondence. It is important to note and comment on null values in Finland, Cyprus, Slovenia, Latvia and Malta.
- *Habitat 1130* (correspondence with CLC 522 class ESTUARIES): at European level it shows a high gap. By countries: TOTAL GAP in Denmark, Slovenia, Estonia, Finland, Italy, Lithuania and Poland; VERY HIGH GAP in Sweden and Greece; HIGH GAP in Germany, France, Ireland and UK; only Portugal shows a right correspondence. It is important to note and comment on null values in Denmark, Slovenia, Estonia, Lithuania and Poland.
- *Habitat 4020** (correspondence with CLC 322 class MOORS and HEATHLANDS): at European level it shows a high gap. By countries: VERY HIGH GAP in France; MODERATE GAP in Spain and Portugal; right correspondence in UK.
- *Habitat 7110** (correspondence with CLC 412 class PEATBOGS): at European level it shows a high gap. By countries: TOTAL GAP in Slovakia, Slovenia, Spain, France, Hungary, Italy, Poland and Portugal; VERY HIGH GAP in Austria, Netherlands, Czech Republic; HIGH GAP in Belgium, Latvia and UK; right correspondence in Ireland. They are serious inconsistencies (represented by null values) in Portugal, Hungary, Italy and Slovenia.
- *Habitat 7130* (correspondence with CLC 412 class PEATBOGS): at European level it shows a moderate gap. By countries: TOTAL GAP in Spain, France and Portugal (France and Portugal with null values); MODERATE GAP in UK; right correspondence in Sweden and Ireland.
- *Habitat 9180** (correspondence with CLC 311 class BROAD-LEAVED FORESTS): at European level it shows a moderate gap. By countries: VERY HIGH GAP in Finland; HIGH GAP in Austria; right correspondence in Poland, Luxemburg, Hungary, Lithuania, Italy, Greece, France, Estonia, Spain, Slovenia, Slovakia and Belgium.
- *Habitat 91E0** (correspondence with CLC 311 class BROAD-LEAVED FORESTS): at European level it shows a moderate gap. By countries: HIGH GAP in Austria; right correspondence in Poland, Luxembourg, Hungary, Lithuania, Greece, France, Estonia Spain, Slovenia and Slovakia.

At European level, the type of habitats (among the evaluated set) with a worse representation on CLC map are coastal lagoons (1150*), mires and bogs (7110*, 7120, 7230), water courses (3260 and 3270), heaths (4020* and 4030), *Molinia* and lowland hay meadows (6410 and 6510) and siliceous rocky slopes (8220). The different types of broadleaved forests show an acceptable representation, although a very important question is that CLC does not

HAB	BE	CZ	DK	DE	EE	IE	EL	ES	FR	IT	CY	LV	LT	LU	HU	MT	NL	AT	PL	PT	SL	SK	FI	SE	UK
	COUNTRIES																								
1130	57,38	-	0,00	28,99	0,00	34,44	13,64	37,93	41,38	5,00	-	-	0,00	-	-	-	70,59	-	0,00	56,45	0,00	-	1,45	13,19	43,28
1140	78,26	-	42,86	63,57	0,00	62,50	0,00	25,64	94,09	0,00	-	-	-	-	-	-	84,75	-	-	2,63	0,00	-	-	0,00	57,67
1150*	-	-	24,72	44,71	17,39	6,45	55,81	17,24	41,38	48,84	0,00	0,00	100,00	-	-	0,00	-	-	50,00	15,15	0,00	-	0,00	14,72	2,01
1310	21,74	-	-	72,94	4,35	19,83	53,16	30,45	43,57	44,30	25,00	6,25	-	-	-	0,00	57,89	-	0,00	64,79	33,33	-	-	2,78	41,33
2120	79,17	-	-	62,82	66,10	72,90	38,57	73,13	62,09	58,42	-	98,00	100,00	-	-	-	100,00	-	77,27	93,18	-	-	4,92	32,26	49,30
2130*	72,00	-	32,88	60,47	66,10	73,87	-	72,92	60,09	63,64	-	95,83	100,00	-	-	-	92,00	-	62,96	81,91	-	-	5,68	32,26	43,51
3150	22,22	45,83	26,34	41,83	92,31	82,35	52,38	20,74	27,16	39,29	50,00	55,59	100,00	-	58,11	-	73,57	22,42	71,43	28,72	54,55	46,05	68,36	92,37	62,03
3260	7,53	-	-	12,89	7,58	8,96	22,22	7,63	10,83	26,42	-	11,76	18,75	-	16,79	-	33,80	30,65	7,94	6,52	56,14	36,17	12,56	17,97	2,81
3270	41,38	-	-	65,56	-	40,00	-	14,98	11,82	24,24	-	30,93	84,62	-	73,18	-	100,00	68,97	52,35	38,57	56,25	50,00	-	-	-
4020*	-	-	-	-	-	-	-	88,05	19,26	-	-	-	-	-	-	-	-	-	-	61,36	-	-	-	-	100,00
4030	18,94	-	49,01	12,96	28,57	18,19	-	91,12	22,19	16,67	-	-	9,33	0,00	0,00	-	66,45	8,70	0,11	64,79	-	4,55	10,03	3,30	81,51
5130	5,77	-	-	13,92	61,62	31,91	-	100,00	23,81	31,79	-	0,00	9,09	0,00	0,00	-	79,49	55,56	14,29	-	14,71	6,12	-	4,78	81,08
6170	-	-	-	91,07	-	-	88,31	84,89	94,64	97,98	-	-	-	-	-	-	-	87,27	83,33	-	84,85	83,33	-	82,34	91,14
6410	0,55	-	31,31	29,48	65,63	36,39	-	72,44	10,61	70,50	-	14,74	1,62	0,00	41,83	-	3,17	40,58	2,01	64,89	37,25	32,50	0,00	3,15	64,86
6510	1,15	-	-	23,39	58,04	61,90	50,00	78,31	9,93	73,49	-	10,16	2,41	5,71	36,20	-	16,67	35,06	4,76	89,47	45,07	20,32	0,00	1,71	15,93
7110*	47,73	24,36	50,00	61,36	78,74	100,00	-	0,74	1,86	0,00	-	42,00	75,68	-	0,00	-	16,22	13,92	7,01	0,00	0,00	4,17	65,77	81,95	43,46
7120	37,14	40,43	65,63	54,31	85,45	97,01	-		1,78	0,00	-	42,00	76,00	-	-	-	61,54	16,88	6,10	0,00	-	10,00	62,10	84,62	43,46
7130	-	-	-	-	-	96,74	-	6,58	0,00	-	-	-	-	-	-	-	-	-	-	0,00	-	-	-	100,00	57,00
7230	13,33	6,38	42,29	25,52	67,15	31,46	0,00	4,46	10,29	4,62	-	31,51	45,57	0,00	59,33	-	11,76	14,47	27,35	-	18,75	4,84	5,63	16,11	3,31
8130	-	-	-	-	-	-	-	39,59	51,38	34,15	-	-	0,00	-	-	-	-	66,67	-	26,32	-	-	-	-	-
8220	0,00	0,45	-	0,88	0,00	7,46	0,00	21,24	10,52	58,55	6,67	0,00	0,00	-	0,00	-	-	28,79	0,00	18,78	78,95	11,11	2,71	48,85	55,93
9120	59,77	-	-	98,08	-	-	-	99,69	94,98	-	-	-	-	-	-	-	78,26	-	-	-	-	-	-	-	98,80
9180*	95,06	-	-	85,87	100,00	-	100,00	97,96	95,00	93,08	-	83,83	98,97	100,00	100,00	-	-	47,01	93,41	-	96,00	93,66	27,12	88,26	81,51
91E0*	66,81	-	85,53	83,75	97,62	60,73	100,00	90,75	95,00	88,45	-	83,83	97,95	96,77	98,54	-	76,79	44,05	92,31	83,33	100,00	92,93	56,51	70,57	87,95
91F0	85,71	-	-	91,97	100,00	-	58,82	76,92	99,10	86,21	-	83,83	100,00	-	99,68	-	62,50	75,72	96,75	68,42	100,00	99,05	-	100,00	-
9230	-	-	-	-	-	-	-	91,19	94,43	-	-	-	-	-	-	-	-	-	-	87,37	-	-	-	-	-
9330	-	-	-	-	-	-	-	92,33	91,54	91,11	-	-	-	-	-	-	-	-	-	91,62	-	-	-	-	-
9340	-	-	-	-	-	-	65,71	76,74	89,88	86,76	-	-	-	-	-	0,00	-	-	-	93,09	100,00	-	-	-	-
9380	-	-	-	-	-	-	100,00	98,77	96,02	100,00	-	-	-	-	-	-	-	-	-	87,50	-	-	-	-	-
9410	-	-	-	100,00	-	-	100,00	-	95,60	99,21	-	-	-	-	-	-	-	97,12	94,74	-	100,00	100,00	-	-	-

Countries: Belgium [BE], Czech Republic [CZ], Denmark [DK], Germany [DE], Estonia [EE], Ireland [IE], Greece [EL], Spain [ES], France [FR], Italy [IT], Cyprus [CY], Latvia [LV], Lithuania [LT], Luxembourg [LU], Hungary [HU], Malta [MT], Netherlands [NL], Austria [AT], Poland [PL], Portugal [PT], Slovenia [SL], Slovakia [SK], Finland [FI], Sweden [SE], United Kingdom [UK]

HAB: Annex I Habitats of the Council Directive 92/43/EEC of 21 May 1992, on the conservation of natural habitats and of wild fauna and flora (Table 13, Appendix 1)

- **Total gap**: less than 10% of coincidences
- **Moderate gap**: coincidences between 50-90%
- **Very high gap**: coincidences between 10-30 %
- **No gap**: coincidences upper 90%
- High gap: coincidences between 30-50%

Table 3. Spatial correlations between Corine Land Cover cartography (3rd level) and CD 92/43/EEC habitats cartography in the EU Countries (units in percentage)

identify differences between forest compositions, i.e. the correspondence at the third level is with the broad CLC 311 class (broad-leaved forests) (Table1). Similar situation occurs with many other habitats like dunes whose correspondence is with CLC 331 class *Beaches, dunes and sand plains*, or alluvial forests.

This constrains many possibilities in the use of CLC at the third level to monitor biodiversity. For instance, the CLC 311 class (Broad-leaved forests) could correspond to *Eucalyptus globulus* or any native broad-leaved forests as *Quercus robur*, both phenomena with very different implications for biodiversity (Pereira and Cooper 2006).

Also, the finer the nomenclature detail, the worse the spatial correlations are. As the scale is finer and CLC is considered at 5th level (for example at Spain level) results get worse and inconsistencies increase.

At Spain level (Table 4) and using a 10km grid results show:

- TOTAL GAP: water courses (3260), mires and bogs (7110*, 7120, 7130, 7230) and alluvial forests (91E0*).
- VERY HIGH GAP: sandflats and coastal lagoons (1140, 1150*), lakes and water courses (3150, 3270), alpine calcareous grasslands (6170), rocky habitats (8130, 8220) and forests of *Ilex aquifolium* (9380).
- HIGH GAP: estuaries (1130), salt marshes (1310) and dry heaths (4030).
- MODERATE GAP: dunes (2120, 2130*), *Molinia* and lowland hay meadows (6410, 6510), woolands of *Quercus spp*. (9230, 9330, 9340).
- GOOD CORRELATION: sclerophyllous scrubs (5130), atlantic forests (9120, 9180*).

HAB	CLC
1130	37,93
1140	25,64
1150*	18,39
1310	31,82
2120	72,50
2130*	71,88
3150	14,74
3260	8,75
3270	17,51
4020*	60,90
4030	45,53
5130	100,00
6170	22,81
6410	70,00
6510	71,93

HAB	CLC
7110*	0,74
7120	0,00
7130	6,58
7230	4,46
8130	16,30
8220	17,31
9120	98,44
9180*	91,84
91E0*	8,79
9230	77,98
9330	72,79
9340	58,21
9380	19,63

(red)	**Total gap**: less than 10% of coincidences
(orange)	**Very high gap**: coincidences between 10-30 %
(yellow)	**High gap**: coincidences between 30-50%
(green)	**Moderate gap**: coincidences between 50-90%
(blue)	**No gap**: coincidences upper 90%

Table 4. Spatial correlations between Corine Land Cover cartography (5th level) and CD 92/43/EEC habitats cartography in Spain (units in percentage)

It is relevant that at Spain level CLC map shows important inconsistencies with bogs and mires (7110*, 7120, 7130, 7230), water courses (3260), alluvial forests (91E0*) or coastal lagoons (1150*).
Total, very high and high gaps should be considered as important inconsistencies which enhance the limited capacity of the CLC map for representing natural and semi-natural habitats and reveal the inappropriate use of the CLC map as a biodiversity proxy, both at European and regional level. In some cases gaps can be explained because of the CLC methodology which makes not possible to identify habitats with less than 25ha or linear features below 100m in width. Also, discrepancies among countries could be attributed to differences among the skills and expert knowledge of image interpreters.
Then, though theoretically possible (Groom et al. 2006; Hansen et al. 2004), the use of some components of the complex habitat entity, such land covers, as a surrogate parameter of a particular habitat is uncertain and it should be previously evaluated.

5. Different approaches on the habitat identification through RS in the context of Europe

The lack of a simple and direct relationship between habitats and any biophysical feature detected by RS restricts the possibilities for automated image classification processes to habitat identification. In this sense, the current wide range of remote sensing techniques and products have supported many suggestions at different scales and using different approaches. The rationale underlying for all of them is the idea of selecting key variables and algorithms to the identification of the habitat entity, integrating knowledge from ancillary data sources. Some of these approaches are mentioned and briefly described in the next paragraphs. Also we propose a new methodology (based on a previous model proposed in Martinez et al., 2010(Martínez et al. 2010)) which presents some key concepts to be consider in a future standardized process.

5.1 Decision rules implemented through a Geographical Information System (GIS): the example of the European PEENHAB project (Mücher et al. 2004; Mücher et al. 2009)

The overall objective of the European PEENHAB project was to develop a methodology to identify spatially all major habitats in Europe according to the Annex I of the Habitats Directive (231 habitats, (EuropeanCommission 2007). This should result in a European Habitat Map with a spatial scale of 1: 2,5M and a minimum mapping unit of 100km^2 with a minimum width of 2,5km. It was expected that this European Habitat Map was the main data layer in the design of the Pan-European Ecological Network (PEEN), which is widely recognized as an important policy initiative in support of protected *Natura 2000* sites.
PEENHAB proposed a new methodology to allow the spatial identification of individual habitats to European scale, based on specific expert knowledge and the design of decision rules on the basis of their description in Annex I. Habitats were identified by a combination of spatial data layers implemented in a GIS decision rule. The methodology was implemented following five steps: i) the selection of appropriate spatial data sets; ii) the definition of knowledge rules using the descriptions of Annex I habitats; iii) the use of additional ecological expert knowledge; iv) the implementation of the models for the individual habitats; v) validation (Mücher et al. 2009).
The spatial datasets used as ancillary data were: CORINE land cover database, biogeographic regions, distribution maps of individual plant species, digital elevation models, soil databases and other geographic and topographic data.

For example, for the Annex I habitat "Calcareous Beech Forest (code 9150)", first a rule was defined that selects the broadleaf forests from the CORINE land cover database, then a second rule was used to select the beech distribution map from the Atlas Florae Europaeae, and a third rule identified the calcareous soils from the European soil database. The combination of these three filters will form the decision rule that delimitates the spatial extension of calcareous beech forest.

The main advantage of this approach is the suggestion of using specific knowledge, implemented as a GIS decision rule, to identify individual habitat maps as they are described in Annex I of Habitats Directive. The approach use remote sensing data in an indirect way (through the use of CORINE land cover and other input variables) along with other suitable ancillary data. Results are appropriate at European scale in order to set guidelines for the strategic design of the Pan-European Ecological Network.

5.2 Object oriented approaches (Bock et al. 2005; Díaz Varela et al. 2008)

Bock et al. (2005) have proposed an object-oriented approach for EUNIS habitat mapping using remote sensing data at multiple scales with good results. The approach performs well when applied to high resolution satellite data (Landsat 30m) for the production of habitat maps at regional level with coarse thematic resolution; also it performs extremely well when applied to very high spatial resolution data (Quickbird 0,7m) for the production of local scale maps with fine thematic resolution.

The use of a multi-scale segmentation (implemented in the software package eCognition, Definiens, http://www.definiens.com) allow for the accurate classification of habitat types, which occur at different scales: for example, large-scale woodland habitat can be detected at coarser segmentation levels, while small-scale habitats such as woodland corridors can be detected at finer segmentation levels.

The main advantages of the object-oriented approaches to habitats mapping are: i) the ability to integrate ancillary data into the classification processes, related to shape, texture, context, etc.; ii) the option of developing knowledge-based rules in the classification process. Both questions make especially possible the accurate identification of habitats with similar spectral properties. Some results that show the advantage of these issues are (Bock et al. 2005): i) the effective separation of different grassland types like calcareous and mesotrophic grassland habitats to a high degree of accuracy through the use of geological data; ii) the use of multi-temporal remote sensing data to distinguish among arable lands, manage grasslands and semi-natural habitats.

5.3 The use of binary classifications by decision trees (DT) (Boyd et al. 2006; Foody et al. 2007; Franklin et al. 2002; Franklin et al. 2001)

Some studies have shown binary classifications as one of the more appropriate methods to identify habitats in the territory. Binary classifications can be implemented by non-parametric Decision Trees (DT) algorithms. Some of these studies have focused on the mapping of one specific thematic class (Boyd et al. 2006; Foody et al. 2007) hypothesizing that non-parametric algorithm would be more suitable to habitats of conservation interest because of the scarce spatial distribution usually associated to them (the size of the training sample will be smaller). Other studies have combined that kind of techniques (binary classifications by DT) in hybrid approaches (Franklin et al. 2001). The hybrid approaches assumes that parametric algorithms like standard maximum likelihood (ML) are the best option with spectrally different habitats while applies non-parametric algorithms to other complex habitats, in a so-called Integrated Decision Tree Approach (IDTA). The IDTA (Franklin et al. 2001) consist on a process with a simple set of classification decision steps, readily understood and repeatable. The approach allows mixing unsupervised, supervised and stratification decision rules such that requirements for training data were minimized.

The general advantages of this kind of approaches are: i) those linked to the use of non-parametric algorithms (Tso and Mather 2009), for example less restrictions with the size of the training sample; ii) the use of key input variables combined with key algorithms defined following specific characteristics of individual habitats; iii) The use of a type of geospatial input data (nominal, ordinal, interval or ratio data like forest inventory maps, biophysical and derived maps) that are difficult to incorporate into a statistical classifier.

5.4 New model based on the identification of ecological-units and on the selection of habitat-key variables (Advances from Martinez et al., 2010)

The methodology proposed in this model is summarized in Figure 2. The approach includes two main steps: the adaptation of an international classification scheme and the generation of *ecological unit* maps. The concept of *ecological unit* goes farther than land cover notion: through ecological expert knowledge it is possible identify in the study area ecological units directly related to Annex I habitats (Table5). Each ecological unit is linked to a distinctive set of characteristic habitats through ecological expert knowledge. Consequently, the system allows identifying and assessing the habitats listed in Annex I of Habitats Directive 92/43/EC through the identification of land covers by remote sensing.

Ecologically significant units of analysis were defined (Table 5), based on the *EUNIS* pan-European classification proposed by the EEA and the CORINE Biotopes System of Classification (www.eea.europa.eu, last accessed, May 2011). By using these systems, the approach can be applicable to other European regions and it will produce cross-comparable results.

The generation of *ecological unit* mapping is based on the selection of key input variables (spectral, derived and ancillary variables) as a function of the main characteristics of the target habitats and on the use of a standard maximum likelihood classification (MLC) algorithm (Swain and Davis 1978). The target habitats are those defined in the Annex I of the Habitats Directive for the Atlantic Biogeographical Region. Because of the *not direct* correspondence between spectral classes and habitat types, we propose the combination of ecological expert knowledge to find the relationship between both of them (step 1) along with the selection of suitable input variables (step 2) in order to achieve the best possible classification of habitats.

The study area was the Biosphere Reserve of *Terras do Miño* in the Northwest of Iberian Peninsula (Figure 3).The classification process was undertaken by the use of multi-temporal Landsat ETM+ images and ancillary data.

Input variables were rectified to the Universal Transverse Mercator Projection (UTM 29T) using the European Datum 1950 (ED50) and resampled to 30m grid size. Training samples were taken by fieldwork. They were located with a global positioning system (GPS) differential receiver. Training sites were selected as to be large and representative enough to characterize each target class and provide efficient and unbiased estimators using stratified sampling. At least 50 additional points per ecological unit were surveyed on the field for results assessment. These data were not used in the training process.

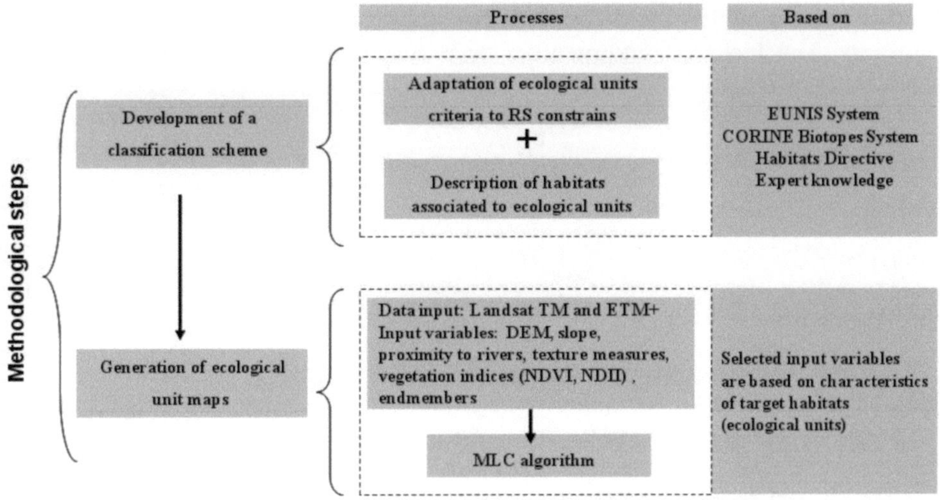

Fig. 2. Flowchart of the methodological steps for the model applied to Biosphere Reserve *Terras do Miño*

Early spring (March 26, 2002), late spring (May 26, 2001) and summer (August 17, 2002) images were selected for this model to account for the seasonal trends in vegetation communities. Images were geometrically registered using ground control points (GCP), first order transformations and nearest neighbor interpolation. The August 2002 image was geo-referenced to 1/25,000 digital maps, produced by the National Geographical Institute of Spain and used as a reference for geometrical correction of the other images. Atmospheric correction was based on the dark-object technique proposed by Chavez(Chavez 1996). Correction for effects of ground slope and topographic orientation was computed using the Lambertian cosine method initially proposed by Teillet et al. (Teillet et al. 1982) and later modified by Civco (Civco 1989). To model illumination conditions a Digital Elevation Model (DEM) was generated using contour lines from 1/5000 digital cartography. The thermal band was not included in the classification processes.

It was hypothesized that derived and ancillary variables would provide critical information for landscape classification and enable the identification of complex habitats. For example, topographic features and vicinity to fluvial corridors have an important influence on the distribution of natural and semi-natural habitats; therefore, the discrimination of this type of habitats should be favored by those variables.

The input dataset for classification processes included satellite derived variables (reflectance, vegetation indices, texture measures and spectral mixture analysis) along with continuous ancillary data. Reflectance bands were included using principal components transform (Mather 2004) in order to reduce the dimensionality of the dataset and optimize the number of training samples. Vegetation indices were: NDVI (Normalized Difference Vegetation Index) (Rouse et al. 1974) and NDII (Normalized Difference Infrared Index; using Landsat TM Band 5) (Hunt and Rock 1989). Texture measures (homogeneity using Band 3 of each Landsat ETM+ image) were calculated using the co-ocurrence matrix as designed by Haralick et al. (Haralick et al. 1973). The co-ocurrence matrix was computed from a window of 3x3 pixels, which was considered an optimum size for measuring neighbor conditions. Linear spectral mixture analysis (SMA) (Mather 2004) generated endmember spectra, which are defined as the proportion of each pixel covered by a basic spectral class. The included endmembers were water, soil, green vegetation (GV), and

ECOLOGICAL UNIT		NATURA 2000 Code (Main Habitat)
Natural –Seminatural Landscape		
Standing Water		3110/3120/3130
Running Water		3140/3150/3160
W	Water courses	3260/3270
Inland no-wooded wetlands		
WH	Bogs (Raised and blanket bogs) and Atlantic wet heaths	7130/7110*/7120 7230, 4020*
HM	Tall and mid-herb humid meadows	6430,6410
Inland wooded wetlands		
RF	Alluvial and riparian forests	91D0*/91E0*/91F0
Other natural and seminatural forests		
DF	Deciduous oak forests	9230
Rocky habitats and other heaths		
DH	Siliceous rocky habitats and dry heaths	4030/8220
Anthropic Landscape		
Forest plantations		
P	*Pine sp.* groves	
E	*Eucalyptus sp.* plantations	
Transformed rural landscape		
TF	Rural system mainly made up of pastures	
TR	Rural system mainly made up of corn and pasture in rotations	
BL	Bare land	
Traditional rural landscape		
CG	Traditional rural mosaic with fenced fields, dominated by crops and grasslands	6410/6510
WG	Traditional rural mosaic with fenced fields, dominated by wet grasslands	6410/6510
Man-made landscape		
Ur	Urban areas (villages, towns)	
ME	Mining exploitations	
I	Communication infrastructures	
B	Buildings for agricultural, forestry and industrial use	

Table 5. Ecological units directly related to CD 92/43/EEC habitats in the proposed model

non_photosynthetic vegetation (NPV). SMA models the reflectance of each pixel as a linear combination of reflectance of those four components. It was assumed constrains of no negative values and that the four components explained the whole variation of reflectance, although the model was allowed to produce a residual image.
Slope gradient was calculated from the DEM. Proximity to rivers was calculated from the 1/5000 river map using raster processing.
Results were assessed by cross-tabulation with a sample of pixels included in test plots. Global, user and producer accuracies were evaluated using an error matrix (Congalton and Green 2009). Additionally the Kappa analysis (Congalton and Green 2009) was used to evaluate the accuracy of the results: we used KHAT statistic to measure how well the remotely sensed classification agrees with the reference data, the Z statistic to determine the significance of a matrix error and the Z pairwise comparison to decide if two KHAT values are significantly different.

CODE	VARIABLES	GA (%)	KHAT
MLC_1	PCs (may+march+august)	75,56	0,733
*MLC_2	PCs (may+march+august) + NDVI (my,ag) + $NDII_5$ (my,ag) + prox. streams + SLOPE+ Homogeneity- B_3 (my, mz, ag)+ WATER (endmember-mz)+ FMo 3x3	82,75	0,811

Legend: [GA] Gloabl accuracy (%). [MLC]: Maximun likelihood classification, [*] without *Bare Land* class, [PCs]: Principal Components,
[FMo]: modal filter

Table 6. Global accuracies for parametric multi-temporal processes (MLC algorithm)

CODE	GA (%)	KHAT	Z	PAIR-WISE Z SCORES**
MLC_1	75,56	0,733	50,126*	
MLC_2	82.75	0,811	62,089*	3,496569*

Legend: [MLC]: Maximun likelihood classification [*] Significant at the 95% confidence level. [**] Comparison with MLC_1

Table 7. Kappa analysis for parametric multi-temporal processes (MLC algorithm)

The results of this model showed 82.75% global accuracy after the application of a modal filter. The best result provided a Kappa value (Congalton and Green 2009) of 0.811 with a Z value indicating very good agreement between classification results and the reference data.
Tables 6 and 7 shows the accuracy assessment for two processes: i) one of them based on the principal components of the three images (MLC_1); ii) the second one also includes the group of ancillary and derived variables (MLC_2). Some variables like slope, distance to rivers, NDVI, NDII and homogeneity showed its valuable potential (Table 9). The combination of all of them in the best MLC trial produced a significant increase in global accuracy along with an increase on user and producer accuracies for the most part of the classes of habitats (Table 8). MLC_2 showed user and producer accuracies above 70% and 80% in the most part of habitats. Only WH and forests showed producer or user accuracies less than 60%.

ECOLOGICAL UNITS	MLC$_1$		MLC$_2$	
	user	producer	user	producer
WG	60,41	87,00	74,22	95,00
TR	97,19	100,00	96,30	100,00
CG	84,62	74,00	85,85	87,50
HM	43,94	72,5	73,81	77,50
DF	59,74	44,23	76,39	52,88
RF	52,88	55,0	59,84	73,00
WH	53,33	26,67	73,91	56,67
P	97,92	90,38	96,23	98,08
E	94,74	90,0	90,91	100,00
TF	84,26	91,0	91,01	81,00
TI	100,00	91,67	100,00	91,67
W	100,00	100,00	100,00	100,00
ME	100,00	100,00	100,00	83,33
Ur+I+B	83,33	100,00	86,96	100,00
BL	66,67	30,0	----	----
DH	82,28	81,25	83,56	76,25

Legend: [MLC]: Maximun likelihood classification

Table 8. User and producer accuracies for parametric multi-temporal processes (MLC algorithm)

Variables	Global accuracy	Improvement in multi-temporal***	Kappa analysis		
			KHAT	Z	Pair-wise Z scores***
DEM + Slope	76,05	0,49	0,738	50,792*	0,256709
Prox. to streams	76,34	0,78	0,742	51,213*	0,412576
NDVI and NDII**	77,91	2,35	0,758	53,603*	1,248175
Homogeneity**	77,22	1,66	0,751	52,467*	0,867394
Endmember Water	75,75	0,19	0,735	50,442*	0,102807

Legend: [*] Significant at the 95% confidence level. [**] Three images. [***] In relation to the trial with principal components of the three images

Table 9. Improvement in global accuracy for multi-temporal analyses after adding to the classification the layers showed in the table.

The contribution of topographical variables and vegetation indices to the habitat mapping accuracy is appropriate for the analyses; the combination of vegetation indices was relevant in the analyses, with improvements in global accuracy (to a maximum of 2.35% of accuracy increase when both NDVI and NDII were combined in the process) (Table 9). The topographical variables (Slope and MDE) improved also the global accuracy of multi-temporal classifications, although to a lesser extent than variables like homogeneity. Texture measures and SMA components did not provide significant improvements in global accuracy in parametric methods. However they showed to be suitable for improving the discrimination of some particular classes like HM, WH or Ur. Therefore, they could be considered as interesting and helpful variables for nonparametric methods in order to get good discrimination of some classes.

The special interest of this approach comes from the use of an international classification system (EUNIS) that will allow cross-comparable spatial and temporal assessments and make the methodology extrapolated to other regions. The definition of ecological units goes farther than the simple *land cover* idea and it allows the definition of a direct relation win AnnexI habitats and consequently with species. Finally the use of a standard maximum likelihood algorithm based on the selection of key input variables makes possible the accurate identification of many ecological units.

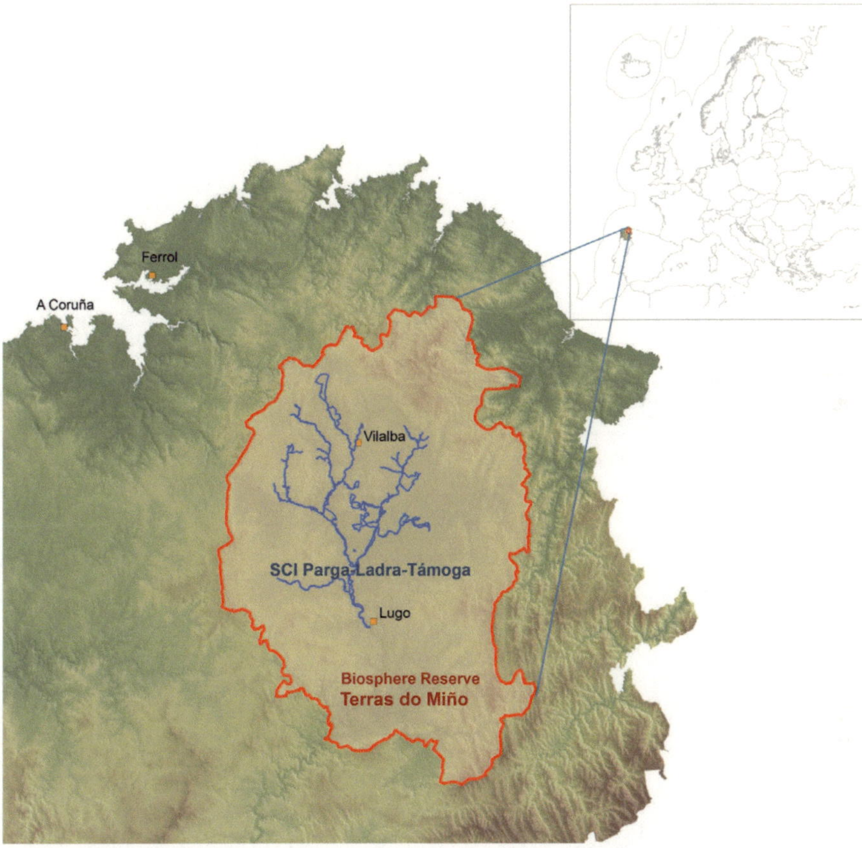

Fig. 3. Localization of the Biosphere Reserve *Terras do Miño* and de SCI *Parga-Ladra-Támoga* in Galicia (NW Spain)

The output classification of this model and the CLC map (5th level) were spatially compared with a habitat map of the Site of Community Importance (SCI) Parga-Ladra-Támoga in the Northwest of Iberian Peninsula (which belongs the Biosphere Reserve of *Terras do Miño).* This map was elaborated by photo interpretation through aerial photography with different scales ranging from 1/20000 until 1/2000 (Ramil et al. 2005). It also based on expertise fieldwork and its minimum mapping unit was 0,5ha. The map was the reference to evaluate this site as a candidate to belong to *Natura 2000* ecological network.
Again, spatial inconsistencies between both sources (CLC and the model applied to Biosfere Reserve *Terras do Miño*) were evaluated using this map by the analysis of coincidences in the cells of two different grids (UTM based): 1 and 10 km^2.

HAB	10 KM GRID (UTM) CLC	10 KM GRID (UTM) MODEL	1 KM GRID (UTM) CLC	1 KM GRID (UTM) MODEL
3110	0,00	77,78	0,00	35,14
3120	0,00	100,00	0,00	100,00
3130	0,00	83,33	0,00	70,00
3140	0,00	100,00	0,00	100,00
3150	0,00	100,00	0,00	80,00
3160	0,00	100,00	0,00	50,00
3260	6,67	73,33	1,40	5,83
3270	6,67	66,67	2,41	9,64
4020*	90,91	100,00	18,63	88,24
4030	91,67	100,00	26,16	78,48
6410	38,46	100,00	1,78	100,00
6430	-	100,00	-	91,29
6510	46,15	100,00	2,08	100,00
7110*	0,00	100,00	0,00	89,47
7230	0,00	100,00	0,00	100,00
8220	0,00	100,00	0,00	100,00
91D0*	-	100,00	-	88,89
91E0*	0,00	100,00	0,00	78,88
91F0	0,00	100,00	0,00	96,97
9230	100,00	100,00	30,77	99,04

Red	**Total gap**: less than 10% of coincidences
Orange	**Very high gap**: coincidences between 10-30 %
Yellow	**High gap**: coincidences between 30-50%
Green	**Moderate gap**: coincidences between 50-90%
Blue	**No gap**: coincidences upper 90%

Table 10. Spatial correlations between CD 92/43/EEC habitats cartography, Corine Land Cover cartography (5th level) and the MODEL *Terras do Miño* in the SCI Parga-Ladra-Tamoga (NW Spain)

At 10km CLC shows a total gap in the most part of the habitats (Table 10). Only heaths and the woodlands with *Quercus spp.* (9230) have good correspondence. At 1 km CLC shows total or very high gap in any case. On the other hand, the model of *Terras do Miño* shows good results at 10km. At 1km, the most part of the habitats present good correspondence or moderate gap. Only two habitats present total gap which corresponds to water courses (3260 and 3270) which can be assigned to the constraints of the spatial resolution of the images.

Results show that, although both sources (CORINE and the model) are based on LANDSAT images with 30m of spatial resolution, the inclusion of decisive variables in the classification processes along with the identification of ecological units was crucial. And it is again proved that it is uncertain to use CLC as a proxy of habitat maps.

6. Conclusions

To meet the requirements of European policies such as *Natura 2000* Network and the 2020 EU Biodiversity Strategy the development of more cost and time effective monitoring strategies are mandatory. Remote sensing (RS) techniques contribute significantly to biodiversity monitoring and several approaches have been proposed to get on-going requirement for spatially explicit data on the ecological units, and the value and threats against natural and semi-natural habitats (Bock et al. 2005; Weiers et al. 2004), but no definite nor any that has been standardized across Europe.

The major obstacles to get standardized scientific monitoring methodologies for habitat monitoring form a complex patchwork. The immense versatility of RS, the full range of RS techniques and products, has led to numerous potential approaches but all of them are dependent of many factors: i) firstly the large variability in the quality of input variables, their semantic, thematic and geometrical accuracy; many approaches have assumed the suitability and representativity of the selected geospatial data; ii) secondly, the possible variability of the spectral, spatial and temporal resolutions; iii) finally, the availability of suitable RS and ancillary data.

There is no a simple relationship between habitats and biophysical parameters like land covers (Groom et al. 2006). *Habitat classes* are not the same that *land cover classes* and the inconsistencies and gaps when a land cover map, as CORINE Land Cover, is used as a surrogate of a habitat map are significant and it should be evaluated in each case. It is necessary to develop *ad hoc* criteria to get the objective of identifying and monitoring habitats from remote sensing. It should be found the optimal way (cost effective and in an acceptable time, and with an optimal level of accuracy) to get from one unit of land cover (which can definitely be detected directly by remote sensing) to a unit of habitat (which may be, at least not in a direct way).

At the European Community level the appropriate criteria for getting that relation should be achieved through EUNIS system (Martínez et al. 2010; Moss and Davies 2002) since it is a common denominator that is compatible with the requirements of Annex I of the Habitat Directive. It will support the standardization because it makes possible cross-comparable data: at spatial and temporal levels.

In regard to habitat identification through RS recent researches have suggested different relevant considerations and requirements: study areas specific approaches; ecological expert knowledge implemented as decision rules; the implementation/inclusion of key input variables selected following specific characteristics of individual habitats; the integration of ancillary data into the classification processes, related to shape, texture, context; the use of non-parametric algorithms implemented through binary classifications or decision trees that allow to include nominal, derived and ancillary geospatial data and also are advantageous with scarce training samples; (Bock et al. 2005; Boyd et al. 2006; Foody et al. 2007; Franklin et al. 2001; Kerr and Ostrovsky 2003; Martínez et al. 2010; Mücher et al. 2009).

On the other hand, insufficient integration at different scales is one of the constraints of the current biodiversity monitoring programmes (Pereira and Cooper 2006) and it is also urgent to advance in this issue. Remote sensing analyses of ecological phenomena at global scale are too general to meet regional and local monitoring requirements. Medium and high spatial resolution remotely sensed data, like Landsat TM and ETM+ sensors, have been widely used in ecological investigations and applications, because their suitability at regional and landscape scales. But there is a mismatch between broad-scale remote sensing and local scale field ecological data (Kerr and Ostrovsky 2003): the synoptic view of the remote sensing should be enhanced with *in situ* data and regional assessments should combine high spatial resolution satellite RS with on-the ground monitoring, and aerial photography or very high spatial resolution satellite RS in studying some habitats which are best monitoring at small scales (Hansen et al. 2004; Pereira and Cooper 2006).

To conclude, the upcoming standardized methodology should incorporate these recommendations. For habitat mapping through RS, expert knowledge and field measurements should be combined with key input variables and optimal algorithms related to each individual target habitat, implemented in a decision structure like a tree. At European level the new methodology should be based on the EUNIS system that meets the objectives and requirements of the Habitat Directive, the Convention of Biological Diversity and the new 2020 biodiversity targets. It should also look at the new possibilities of medium and high resolution satellite images.

7. Appendix 1

CLC 311 - Forest and semi natural areas / Forests / Broad-leaved forest

HC 9120 - Atlantic acidophilous beech forests with *Ilex* and sometimes also *Taxus* in the shrublayer (*Quercion robori-petraeae* or *Ilici-Fagenion*)

EUNIS CORRELATION: G1.62, G1.6

HC 9180* - *Tilio-Acerion* forests of slopes, screes and ravines

EUNIS CORRELATION: G1.A, G1.A4

HC 91E0* - Alluvial forests with *Alnus glutinosa* and *Fraxinus excelsior* (*Alno-Padion, Alnion incanae, Salicion albae*)

EUNIS CORRELATION: G1.1, G1.2

HC 91F0 - Riparian mixed forests of *Quercus robur, Ulmus laevis* and *Ulmus minor, Fraxinus excelsior* or *Fraxinus angustifolia,* along the great rivers (*Ulmenion minoris*)

EUNIS CORRELATION: G1.2, G1.22, G1.223

HC 9230 - Galicio-Portuguese oak woods with *Quercus robur* and *Quercus pyrenaica*

EUNIS CORRELATION: G1.7, G1.7B

HC 9330 - *Quercus suber* forests

EUNIS CORRELATION: G2.1, G2.11

HC 9340 - *Quercus ilex* and *Quercus rotundifolia* forests

EUNIS CORRELATION: G2.1, G2.12

HC 9380 - Forests of *Ilex aquifolium*

EUNIS CORRELATION: G2.6

CLC 312 - Forest and semi natural areas / Forests / Coniferous forest

HC 9410 - Acidophilous *Picea* forests of the montane to alpine levels (*Vaccinio-Piceetea*)

EUNIS CORRELATION: G3.1

CLC 321 - Forest and semi natural areas / Scrub and/or herbaceous vegetation associations / Natural grasslands

HC 6170 - Alpine and subalpine calcareous grasslands

EUNIS CORRELATION: E4.4,

HC 6410 - *Molinia* meadows on calcareous, peaty or clayey-silt-laden soils (*Molinion caeruleae*)

EUNIS CORRELATION: E3.5, E3.51

HC 6510 - Lowland hay meadows (*Alopecurus pratensis, Sanguisorba officinalis*)

EUNIS CORRELATION: E2.2

CLC 322 - Forest and semi natural areas / Scrub and/or herbaceous vegetation associations / Moors and heathland

HC 4020* - Temperate Atlantic wet heaths with *Erica ciliaris* and *Erica tetralix*

EUNIS CORRELATION: F4.1

HC 4030 - European dry heaths

EUNIS CORRELATION: F4.2

HC 5130 - *Juniperus communis* formations on heaths or calcareous grasslands

EUNIS CORRELATION: F3.1, F3.16

CLC 323 - Forest and semi natural areas / Scrub and/or herbaceous vegetation associations / Sclerophyllous vegetation

HC 5130 - *Juniperus communis* formations on heaths or calcareous grasslands

EUNIS CORRELATION: F3.1, F3.16

CLC 331 - Forest and semi natural areas / Open spaces with little or no vegetation / Beaches, dunes, sands

HC 2120 - Shifting dunes along the shoreline with *Ammophila arenaria* ('white dunes')

EUNIS CORRELATION: B1.3, B1.32

HC 2130* - Fixed coastal dunes with herbaceous vegetation ('grey dunes')

EUNIS CORRELATION: B1.4

CLC 332 - Forest and semi natural areas / Open spaces with little or no vegetation / Bare rocks

HC 8130 - Western Mediterranean and thermophilous scree

EUNIS CORRELATION: H2.5, H2.5

HC 8220 - Siliceous rocky slopes with chasmophytic vegetation

EUNIS CORRELATION: H3.1

CLC 411 - Wetlands / Inland wetlands / Inland marshes

HC 7230 - Alkaline fens

EUNIS CORRELATION: D4.1

CLC 412 - Wetlands / Inland wetlands / Peat bogs

HC 7110* - Active raised bogs

EUNIS CORRELATION: C1.4, D1.1, G5.6

HC 7120 - Degraded raised bogs still capable of natural regeneration

	EUNIS CORRELATION: D1.1, D1.121
	HC 7130 - Blanket bogs (* if active bog)
	EUNIS CORRELATION: D1.2
CLC 421 - Wetlands / Maritime wetlands / Salt marshes	
	HC 1310 - *Salicornia* and other annuals colonizing mud and sand
	EUNIS CORRELATION: A2.5
CLC 423 - Wetlands / Maritime wetlands / Intertidal flats	
	HC 1140 - Mudflats and sandflats not covered by seawater at low tide
	EUNIS CORRELATION: A2.1, A2.4, A2.6
CLC 511 - Water bodies / Inland waters / Water courses	
	HC 3260 - Water courses of plain to montane levels with the *Ranunculion fluitantis* and *Callitricho-Batrachion* vegetation
	EUNIS CORRELATION: C2.1, C2.1B, C2.2, C2.28, C2.3, C2.34
	HC 3270 - Rivers with muddy banks with *Chenopodion rubri pp* and *Bidention pp* vegetation
	EUNIS CORRELATION: C3.5, C3.53
CLC 512 - Wetlands / Inland waters / Water bodies	
	HC 3150 - Natural eutrophic lakes with *Magnopotamion* or *Hydrocharition* -type vegetation
	EUNIS CORRELATION: C1.3, C1.33
CLC 521 - Wetlands / Marine waters / Coastal lagoons	
	HC 1150* - Coastal lagoons
	EUNIS CORRELATION: A1.3, A2.2, A2.3, A2.4, A2.5, A3.3, A5.1, A5.2, A5.3, A5.4, A5.5, A5.6, A7.1, A7.2, A7.3, A7.4, A7.5, A7.8, C1.5, C3.4
CLC 522 - Wetlands / Marine waters / Estuaries	
	HC 1130 - Estuaries
	EUNIS CORRELATION: A1.2, A1.3, A1.4, A2.1, A2.2, A2.3, A2.4, A2.5, A2.6, A2.7, A3.2, A3.3, A3.7, A4.2, A4.3, A5.1, A5.2, A5.3, A5.4, A5.5, A5.6, A7.1, A7.3, A7.4, A7.5, A7.8 ,X01

Table 11. Correspondences between Corine Land Cover classification (3rd level) and habitats of Community interest, and correspondences between habitats of Community interest (HD) with EUNIS classification (only overlap, same and narrow relation) (Source: www.eunis.eea.europa.eu, last accessed May 2011)

CODE	DENOMINATION
31110	Forest and semi natural areas / Forests / Broad-leaved forest / Evergreen broad-leaved woodlands
31120	Forest and semi natural areas / Forests / Broad-leaved forest / Deciduous and marcescent forest
31150	Forest and semi natural areas / Forests / Broad-leaved forest / River forest
31210	Forest and semi natural areas / Forests / Coniferous forest / Needle coniferous forests
32111	Forest and semi natural areas / Scrub and/or herbaceous vegetation associations / Natural grasslands / High-productive alpine grasslands / High-productive alpine grasslands of temperate-oceanic climate areas

32112	Forest and semi natural areas / Scrub and/or herbaceous vegetation associations / Natural grasslands / High-productive alpine grasslands / Mediterranean high-productive grasslands
32121	Forest and semi natural areas / Scrub and/or herbaceous vegetation associations / Natural grasslands / Other grasslands / Other grasslands of mild-oceanic climate
32122	Forest and semi natural areas / Scrub and/or herbaceous vegetation associations / Natural grasslands / Other grasslands / Other mediterranean grasslands
32312	Forest and semi natural areas / Scrub and/or herbaceous vegetation associations / Sclerophyllous vegetation / Mediterranean sclerophyllous bushes and scrubs / Not very dense Mediterranean sclerophyllous bushes and scrubs
33110	Forest and semi natural areas / Open spaces with little or no vegetation / Beaches, dunes, sands / Beaches and dunes
33210	Forest and semi natural areas / Open spaces with little or no vegetation / Bare rocks / Steep bare rock areas (cliffs, etc.)
33220	Forest and semi natural areas / Open spaces with little or no vegetation / Bare rocks / Rocky outcrops and screes
41100	Wetlands / Inland wetlands / Inland marshes
41200	Wetlands / Inland wetlands / Peat bogs
42100	Wetlands / Maritime wetlands / Salt marshes
42300	Wetlands / Maritime wetlands / Intertidal flats
51110	Water bodies / Inland waters / Water courses / River and natural water courses
51210	Water bodies / Inland waters / Water bodies / Lakes and lagoons
52100	Water bodies / Marine waters / Coastal lagoons /
52200	Water bodies / Marine waters / Estuaries

Table 12. Corine Land cover European Nomeclature

CODE	DENOMINATION
1130	Estuaries
1140	Mudflats and sandflats not covered by seawater at low tide
1150*	Coastal lagoons
1310	*Salicornia* and other annuals colonizing mud and sand
2120	Shifting dunes along the shoreline with *Ammophila arenaria* ('white dunes')
2130*	Fixed coastal dunes with herbaceous vegetation ('grey dunes')
3110	Oligotrophic waters containing very few minerals of sandy plains (*Littorelletalia uniflorae*)
3120	Oligotrophic waters containing very few minerals generally on sandy soils of the West Mediterranean, with *Isoetes spp.*
3130	Oligotrophic to mesotrophic standing waters with vegetation of the *Littorelletea uniflorae* and/or of the *Isoëto-Nanojuncetea*
3140	Hard oligo-mesotrophic waters with benthic vegetation of *Chara spp.*
3150	Natural eutrophic lakes with *Magnopotamion* or *Hydrocharition* -type vegetation
3160	Natural dystrophic lakes and ponds
3260	Water courses of plain to montane levels with the *Ranunculion fluitantis* and *Callitricho-Batrachion* vegetation
3270	Rivers with muddy banks with *Chenopodion rubri pp* and *Bidention pp* vegetation
4020*	Temperate Atlantic wet heaths with *Erica ciliaris* and *Erica tetralix*

4030	European dry heaths
5130	*Juniperus communis* formations on heaths or calcareous grasslands
6170	Alpine and subalpine calcareous grasslands
6410	*Molinia* meadows on calcareous, peaty or clayey-silt-laden soils (*Molinion caeruleae*)
6430	Hydrophilous tall herb fringe communities of plains and of the montane to alpine levels
6510	Lowland hay meadows (*Alopecurus pratensis, Sanguisorba officinalis*)
7110*	Active raised bogs
7120	Degraded raised bogs still capable of natural regeneration
7130	Blanket bogs (* if active bog)
7230	Alkaline fens
8130	Western Mediterranean and thermophilous scree
8220	Siliceous rocky slopes with chasmophytic vegetation
9120	Atlantic acidophilous beech forests with *Ilex* and sometimes also *Taxus* in the shrublayer (*Quercion robori-petraeae* or *Ilici-Fagenion*)
9180*	*Tilio-Acerion* forests of slopes, screes and ravines
91D0*	Bog woodland
91E0*	Alluvial forests with *Alnus glutinosa* and *Fraxinus excelsior* (*Alno-Padion, Alnion incanae, Salicion albae*)
91F0	Riparian mixed forests of *Quercus robur, Ulmus laevis* and *Ulmus minor, Fraxinus excelsior* or *Fraxinus angustifolia,* along the great rivers (*Ulmenion minoris*)
9230	Galicio-Portuguese oak woods with *Quercus robur* and *Quercus pyrenaica*
9330	*Quercus suber* forests
9340	*Quercus ilex* and *Quercus rotundifolia* forests
9380	Forests of *Ilex aquifolium*
9410	Acidophilous *Picea* forests of the montane to alpine levels (*Vaccinio-Piceetea*)
[*], Priority natural habitat types as Habitats of the Council Directive 92/43/EEC (Annex 1)	

Table 13. Habitats of Community interest (Council Directive 92/43/EEC) denomination

8. Acknowledgments

The authors thank the Research Group GI-1934 TB. The Autonomous Region of Galicia (Spain) has financed this study through the Project 10MDS276025PR of the Research Program *PGIDT-INCITE-Xunta de Galicia, Consellería de Economía e Industria.*

9. References

Balmford, A., Bennun, L., et al. (2005). The Convention on Biological Diversity's 2010 Target. *Science,* 307, pp. (212-213),ISSN 1095-9203

Blondel, J. (1979). *Biogéographie et Écologie,*Masson,ISBN 2225639213,Paris.

Bock, M., Panteleimon, X., et al. (2005). Object-oriented methods for habitat mapping at multiple scales. Case studies from Northern Germany and Wye Downs, UK. *Journal for Nature Conservation,* 13, pp. (75-89),ISSN 1617-1381

Boyd, D. S., Sánchez-Hernández, C., et al. (2006). Mapping a specific class for priority habitats monitoring from satellite sensor data. *International Journal of Remote Sensing,* 27, pp. (2631-2644),ISSN 1366-5901

Civco, D. L. (1989). Topographic Normalization of Landsat TM digital imagery. *Photogrammetric Engineering and Remote Sensing,* 55,9, pp. (1303-1309),ISSN 0099-1112

Cohen, W. B. and Goward, S. N. (2004). Landsat's role in ecological applications of remote sensing. *BioScience,* 54, pp. (535-545),ISSN 0006-3568

Congalton, R. G. and Green, K. (2009). *Assessing the accuracy of remotely sensed data. Principles and practices,*Taylor & Francis Group,ISBN 978-1-4200-5512-2,Boca Raton.

Chavez, P. S. (1996). Image-based atmospheric corrections revisited and improved. *Photogrammetric Engineering and Remote Sensing,* 62,9, pp. (1025-1036),ISSN 0099-1112

Chhabra, A., Geist, H. J., et al. (2006). Multiple impacts of land-use/cover change,In: *Land-use and land-cover change. Local processes and global impacts*, Lambin, E. F. and Geist, H. J., pp.(71-116), Springer-Verlag, ISBN 3-540-32201-9, Berlin

Davies, C. E., Moss, D., et al. (2004). *EUNIS habitat classification revised 2004,* European Environment Agency, European Topic Centre on Nature Protection and Biodiversity: pp. (310),

DeFries, R. S. (2008). Terrestrial Vegetation in the Coupled Human-Earth System: Contributions of Remote Sensing. *Annual Review of Environment and Resources,* 33, pp. (369-390),ISSN 1543-5938

Devillers, P., Devillers-Terschuren, J., et al. (1991). *CORINE biotopes manual. Vol.2. Habitats of the European Community,*Office for Official Publications of the European Communities,Luxembourg.

Devillers, P., Devillers-Terschuren, J., et al. (1992). *Habitats of the European Community, Central Europe, Northern Europe. A preliminary List,*IRSNB,Brussels.

Díaz Varela, R. A., Ramil Rego, P., et al. (2008). Automatic habitat classification methods based on satellite images: A practical assessment in the NW Iberia coastal mountains. *Environmental Monitoring and Assessment,* 144, pp. (229-250),ISSN 1573-2959

Dirzo, R. and Raven, P. H. (2003). Global state of biodiversity and loss. *Annual Review of Environmental and Resources,* 28, pp. (137-167),ISSN 1543-5938

Duro, D. C., Coops, N. C., et al. (2007). Development of a large area biodiversity monitoring system driven by remote sensing. *Progress in Physical Geography,* 31,3, pp. (235-260),ISSN 1477-0296

EuropeanCommission (2006). *Assessment, monitoring and reporting under article 17 of Habitats Directive. Explanatory Notes & Guidelines. Final Draft 5*: pp. (64)

EuropeanCommission (2007). *The InterpretationManual of EuropeanUnionHabitats - EUR27,*DG Environment,Brussels.

Eurostat (2001). *Manual of concepts on land cover and land use,*Office for Publications of the European Communities,ISBN 92-894-0432-9,Luxembourg.

Foody, G. M. (2008). GIS: biodiversity applications. *Progress in Physical Geography,* 32,2, pp. (223-235),ISSN 1477-0296

Foody, G. M., Boyd, D. S., et al. (2007). Mapping a specific class with an ensemble of classifiers. *International Journal of Remote Sensing,* 28, pp. (1733-1746),ISSN 1366-5901

Franklin, S. E., Peddle, D. R., et al. (2002). Evidential reasoning with Landsat TM, DEM and GIS data for landcover classification in support of grizzly bear habitat mapping. *International Journal of Remote Sensing*, 23,21, pp. (4633-4652),ISSN 1366-5901

Franklin, S. E., Stenhouse, G. B., et al. (2001). An Integrated Decision Tree Approach (IDTA) to mapping landcover using satellite remote sensing in support of grizzly bear habitat analysis in the Alberta Yellowhead Ecosystem. *Canadian Journal of Remote Sensing*, 27, pp. (579-592),ISSN 1712-7971

Glenn, E. M. and Ripple, W. J. (2004). On using digital maps to assess wildlife habitat. *Wildlife Society Bulletin*, 32, pp. (852-860),ISSN 0091-7648

Groom, G., Mücher, C. A., et al. (2006). Remote sensing in landscape ecology: experiences and perspectives in a European context. *Landscape Ecology*, 21, pp. (391-408),ISSN 1572-9761

Hansen, A. J., DeFries, R. S., et al. (2004). Land use change and biodiversity: a synthesis of rates and consequences during the period of satellite imagery,In: *Land Change Science: observing, monitoring and understanding trajectories of change on the Earth's surface*, Gutman, G., Janetos, A. C.et al, pp.(277-300), Springer Verlag, ISBN 978-1-4020-2561-7, New York

Haralick, R. M., Shanmugan, K., et al. (1973). Textural features for image classification. *IEEE Transactions on Systems, Man and Cybernetics*, 3, pp. (610-621),ISSN 0018-9472

Hellawell, J. M. (1991). Development of a rationale for monitoring,In: *Monitoring for conservation and ecology*, Goldsmith, F. B., pp.(1-14), Chapman and Hall, ISBN 0412356007, London

Hunt, E. R. and Rock, B. N. (1989). Detection of changes in leaf water content using near and middle-infrared reflectances. *Remote Sensing of Environment*, 30, pp. (43-54),ISSN 0034-4257

Izco Sevillano, J. and Ramil Rego, P. (2001). *Análisis y Valoración de la Sierra de O Xistral: un modelo de aplicación de la Directiva Hábitat en Galicia*,Consellería de Medio Ambiente. Xunta de Galicia,ISBN 84-453-3158-2,Santiago de Compostela.

Jackson, D. L. and McLeod, C. R., Eds. (2000). *Handbook on the UK status of EC Habitats Directive interest features: provisional data on the UK distribution and extent of Annex I habitats and the UK distribution and population size of Annex II species*, NHBS,

Jennings, M. D. (2000). Gap analysis: concepts, methods, and recent results. *Landscape Ecology*, 15, pp. (5-20),ISSN 1572-9761

Kerr, J. T. and Ostrovsky, M. (2003). From space to species: ecological applications for remote sensing. *TRENDS in Ecology and Evolution*, 18,6, pp. (299-305),ISSN 0169-5347

Lefsky, M. A., Cohen, W. B., et al. (2002). Lidar Remote Sensing for Ecosystem Studies. *BioScience*, 52,1, pp. (19-30),ISSN 0006-3568

Lengyel, S., Déri, E., et al. (2008). Habitat monitoring in Europe: a description of current practices. *Biodiversity and Conservation*, 17,14, pp. (3327-3339),ISSN 1572-9710

Martínez, S., Ramil, P., et al. (2010). Monitoring loss of biodiversity in cultural landscapes. New methodology based on satellite data. *Landscape and Urban Planning*, 94, pp. (127-140),ISSN 0169-2046

Mather, P. M. (2004). *Computer processing of remotely-sensed images. An introduction*,John Wiley & Sons, Ltd,ISBN 0-470-84918-5,West Sussex.

McDermid, G. J., Hall, R. J., et al. (2009). Remote sensing and forest inventory for wildlife habitat assessment. *Forest Ecology and Management,* 257, pp. (2262-2269),ISSN 0378-1127

Moss, D. and Davies, C. E. (2002). *Cross-references between the EUNIS habitat classification and the nomenclature of CORINE Land Cover*, European Environment Agency and Centre for Ecology & Hydrology pp. (21),

Mücher, C. A., Hennekens, S. M., et al. (2004). *Mapping European habitats to support the design and implementation of a pan-European Ecological Network. The PEENHAB project*,Alterra,ISBN 1566-7197,Wageningen, The Netherlands.

Mücher, C. A., Hennekens, S. M., et al. (2009). Modelling the spatial distribution of Natura 2000 habitats across Europe. *Landscape and Urban Planning,* 92, pp. (148-159),ISBN 0169-2046

Pereira, H. M. and Cooper, H. D. (2006). Towards the global monitoring of biodiversity change. *TRENDS in Ecology and Evolution,* 21,3, pp. (123-129),ISBN 0169-5347

Ramil Rego, P., Rodríguez, M. A., et al. (2005). La expresión territorial de la biodiversidad. Paisajes y hábitats. *Recursos Rurais,* 2, pp. (109-128),1698-5427

Ramil Rego, P., Rodríguez Guitián, M. A., et al. (2008). *Os Hábitats de Interese Comunitario en Galicia. Descrición e Valoración Territorial.*,Universidade de Santiago de Compostela - IBADER,ISSN 1988-8341,Lugo.

Rouse, J. W., Haas, R. W., et al. (1974). Monitoring vegetation systems in the Great Plains with ERTS. *Third earth resources techonology satellite-I symposium, Volume 1: technical presentation,* ISBN CI-08791, Washington D.C., NASA SP-351.

Scott, J. M., Davis, F., et al. (1993). *Gap Analysis: a geographical approach to protection of biological diversity*,Wildlife Society,ISBN 0084-0173.

Soule, M. E. and Terborgh, J. (1999). *Continental conservation: scientific foundations of reserve networks*,Washington, D.C.

Strand, H., Höft, R., et al., Eds. (2007). *Sourcebook on remote sensing and biodiversity indicators,* Secretariat of the Convention on Biological Diversity, ISBN 92-9225-072-8, Montreal

Swain, P. H. and Davis, S. M. (1978). *Remote sensing: the quantitative approach*,McGraw-Hill,ISBN 9780070625761 New York.

Teillet, P. M., Guindon, B., et al. (1982). On the slope-aspect correction of multispectral scanner data. *Canadian Journal of Remote Sensing,* 8,2, pp. (84-106),ISSN 1712-7971

Thogmartin, W. E., Gallant, A. L., et al. (2004). A cautionary tale regarding use of the National Land Cover Dataset 1992. *Wildlife Society Bulletin,* 32,3, pp. (970-978),ISSN 0091-7648

Tso, B. and Mather, P. M. (2009). *Classification methods for remotely sensed data*,Taylor & Francis Group,ISBN 978-1-4200-9072-7,Boca Raton.

Turner, W., Spector, S., et al. (2003). Remote sensing for biodiversity science and conservation. *TRENDS in Ecology and Evolution,* 18,6, pp. (306-314),ISSN 0169-5347

Weiers, S., Bock, M., et al. (2004). Mapping and indicator approaches for the assessment of habitats at different scales using remote sensing and GIS methods. *Landscape and Urban Planning,* 67,1-4, pp. (43-65),ISSN 0169-2046

Zhao, K., Popescu, S., et al. (2011). Characterizing forest canopy structure with lidar composite metrics and machine learning. *Remote Sensing of Environment,* 115, pp. (1978-1996),ISSN 0034-4257

3

Biodiversity Stability of Shallow Marine Benthos in Strait of Georgia, British Columbia, Canada Through Climate Regimes, Overfishing and Ocean Acidification

Jeffrey B. Marliave, Charles J. Gibbs,
Donna M. Gibbs, Andrew O. Lamb and Skip J.F. Young
Vancouver Aquarium (JM, DG, SY) and Pacific Marine Life Surveys Inc. (CG, DG, AL)
Canada

1. Introduction

The highest human population density in British Columbia, Canada is situated around the shores of the Strait of Georgia, where current government policy is focusing early efforts toward achieving ecosystem-based management of marine resources. Climate regime shifts are acknowledged to have affected commercial fishery production in southern British Columbia (McFarlane et al., 2000), and overfishing is well documented in the Strait of Georgia region for a variety of important species, to the extent that Rockfish Conservation Areas have been created (Marliave & Challenger, 2009). As CO_2 levels rise in the atmosphere, the oceans become progressively more acidic. While ocean acidification is predicted to be a great threat to marine ecosystems, little is known about its ecosystem impacts. Few taxpayer-funded studies have committed to long-term monitoring of full ecosystem biodiversity. This document presents results of over forty years of private taxonomic monitoring of shallow seafloors in the region centering on the Strait of Georgia. Also presented are records of ambient ocean acidity levels (pH), documented continuously by the Vancouver Aquarium through the same time period. Biodiversity data are summarized in ways that enable visualization of possible relationships to climate regimes and ocean acidification. This work does not attempt statistical analyses, in the hope that the data trends can be incorporated into future models.

Biodiversity survey data can reveal fundamental differences in community function, as with the disparate trophic complexity and rockfish nursery capacity of glass sponge gardens versus reefs (Marliave et al., 2009). Trophic cascades can be elucidated when coupling biodiversity surveys with transect abundance surveys (Frid & Marliave, 2010). It has been suggested that biodiversity provides more accurate definition of climate regime shifts than does physical oceanographic data (Hare & Mantua, 2000) and the abundance, survival and spawning distribution of commercial fish species have been linked to decadal-scale changes in ocean and climate conditions (McFarlane et al., 2000).

Ocean acidification can detrimentally impact anti-predator behaviors of fish (Dix et al., 2010). Ocean acidification is most intensive in the geographic area of the NE Pacific Ocean centering on the present area of study, surrounding the Strait of Georgia (Byrne et al., 2010). Ecosystem impacts of ocean acidification trends have not, however, been segregated from climate impacts such as El Niño winters or climate regime shifts. Indeed, it is difficult to segregate shorter term El Niño and La Niña years from climate regimes (Hare & Mantua, 2000), but there is consensus that the regime shift of 1976/1977 was major, followed by a prominent shift in 2000/2001 (Tsonis et al., 2007). McFarlane et al. (2000) provide evidence of another possible regime shift in 1989.

This data presentation summarizes results of 44 years of biodiversity monitoring in the Strait of Georgia region of southern British Columbia, in comparison with monitoring results for surrounding inland sea and outer coast regions at the same latitude and to the immediate north and south (Figure 1). The data treatment accommodates a continual increase in the knowledge base for identification of benthic nearshore marine life. A principal focus of this analysis is the possible climate shifts that have been proposed as regimes for the NE Pacific Ocean. The contention that biodiversity can serve to define climate regime shifts is implicitly tested in this study for the shallow seabeds of coastal NE Pacific regions. As well, perhaps the first long-term documentation of ocean acidification in this region is presented to permit comparison with any possible trends in biodiversity.

2. Methods

Biodiversity monitoring with SCUBA diving centered in the Strait of Georgia region has been conducted by Pacific Marine Life Surveys, Inc. (PMLS) from 1967 to the present, with over 4,500 dives entered into a database from which different data summaries can be extracted. Programming details are explained below for this PMLS database. A total of 1,185 taxa have been documented, but analyses of different climate regime periods are limited to the 328 more prominent species that were identified during the first regime period of 1967-1977.

The area covered by PMLS surveys is depicted in Figure 1. The Strait of Georgia is central to this region, with two other inland seas, Puget Sound and Johnstone Strait, to the south and north. Offshore of the Strait of Georgia is the west coast of Vancouver Island, with northern British Columbia and Alaska to the north, and the outer coast of Washington to the south.

This monitoring by PMLS was not derived from a traditional research program involving designated and pre-determined sampling sites visited at regular intervals. All species documented were observed underwater, during the actual dive profiles. The results are derived from a long-term monitoring effort involving sites selected for their accessibility and convenience for the three participating divers recording data. This effort was largely based on recreational SCUBA trips and relying on shore access or boat availability (charter or private).

An important confounding factor has been the increase over time of published taxonomic identifications, as well as the successive publication of increasingly useful taxonomic keys and identification guidebooks. Figure 2 shows the species accumulation at one dive site in the first five dives of each climate regime period. A wide field of specialist taxon experts gradually associated with PMLS as well, so that the PMLS team continuously increased the species list. Early focus was on fishes and larger crustacean, mollusc and echinoderm invertebrates. Sponges in particular required time to develop expertise with, and various species remain unidentified. It should also be noted that seaweeds were not a focus for identifications during the first 15 years of the survey. For these reasons, comparative

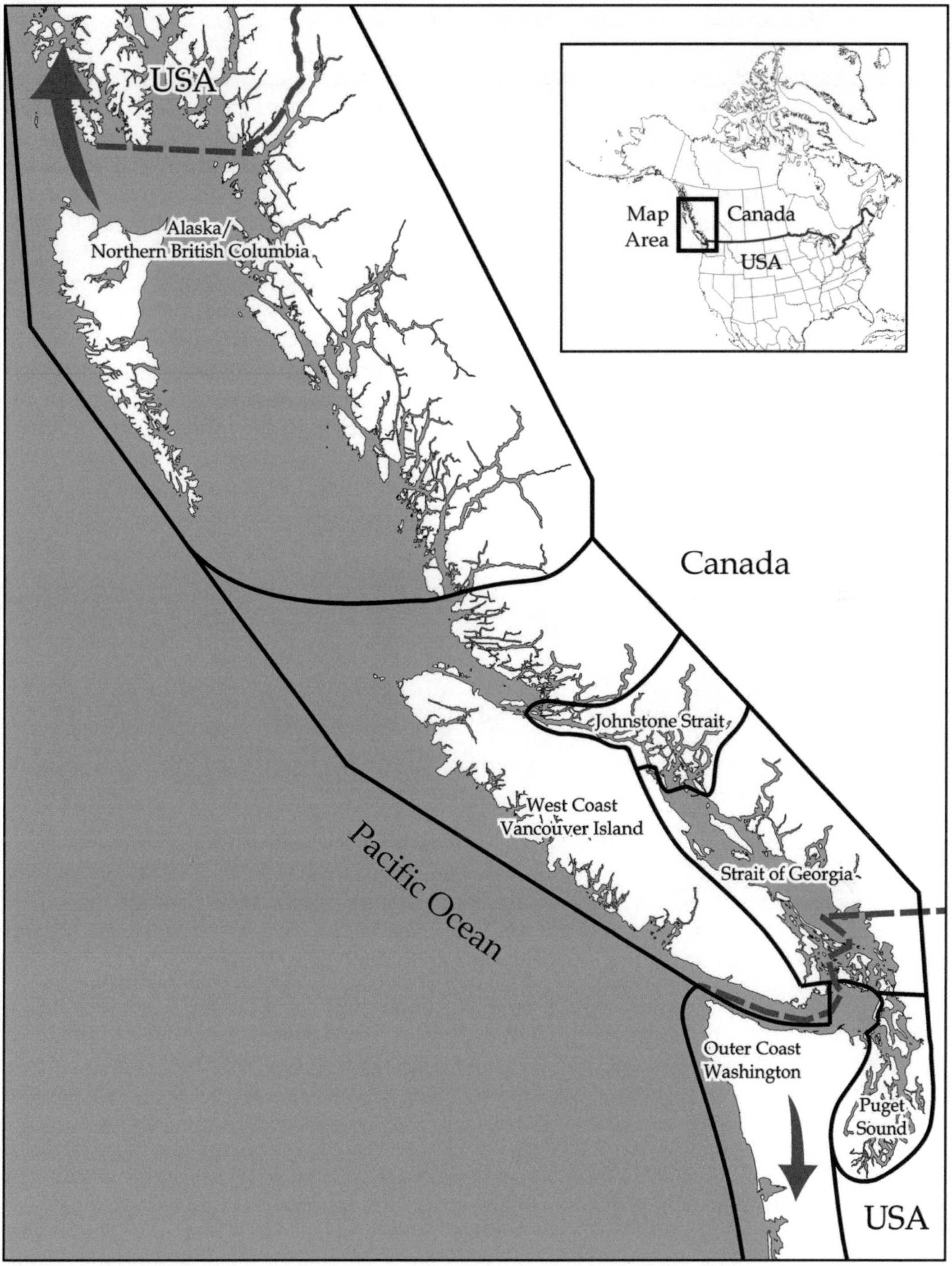

Fig. 1. Map of the coast of the eastern North Pacific Ocean, centering on the Strait of Georgia, British Columbia, with the six regions for which taxonomic data collations were organized for shallow marine benthos species.

analysis of different time periods (i.e. climate regimes) must be limited to the earliest species list of 328 species.

Another variable is the number of dives in a given region. It typically takes about 6-10 dives for the taxon list at a site to start to plateau, but that is assuming comparable depth profiles and time of day (Figure 2). The site with the greatest number of dives, Whytecliff Park, in West Vancouver, BC, involved the entire duration of the survey, considerable deep diving and many night dives as well, all factors that need to be assessed with the results, as the taxon list has not reached a plateau at that site. To demonstrate this asymptotic level of diversity for a site, dives were arranged in descending order of number of species identified on a dive, then the total cumulative species number for that site graphed next to the number of species for that dive (Figure 3). Some dives were oriented to other tasks so that only unusual species were recorded.

Lookout Point is adjacent to Whytecliff, but Lookout primarily involved shallow daytime dives, so reached a biodiversity plateau in a more typical number of dives than for Whytecliff, which received the highest number of dives and achieved the highest biodiversity list.

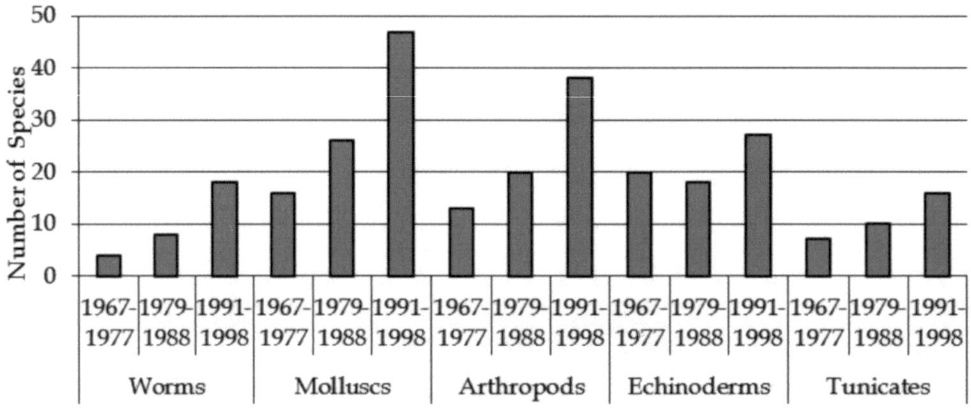

Fig. 2. Species were successively accumulated in these higher taxa at one dive site, Whytecliff (near Vancouver, in the Strait of Georgia), in the top five spp.-count dives of each climate regime period, illustrating artefacts from continually increasing taxonomic expertise. This graph is based on the ultimate total list of 1,185 species, versus the 382 for 1967-1977.

In recording relative abundance, a quotient is used. For each species recorded on each dive, the quotient is developed as follows: 0 = none sighted; 1 = few sighted (<10); 2 = some sighted (<25); 3 = many sighted (<50); 4 = very many sighted (<100); 5 = abundant sighted (<1,000); 6 = very abundant sighted (thousands). For each species, the values for all dives are averaged and then scaled so that the tabulated relative abundance rating is a number from 0 to 100 rather than 0 to 6. Species abundance ratings are calculated by averaging abundance scores for all dives, then dividing by 6 (highest score) and multiplying by 100.

All reports are driven directly from dive log data, which are stored as a set of CSV files, one per dive. Each log contains general information such as date, time, location, depth, diver name, and overall comments. In addition, an entry is made for each species observed on the dive; each entry consists of at least the species name, and may also include abundance estimates or comments (e.g. age or behaviour of specimens). Although this structure may

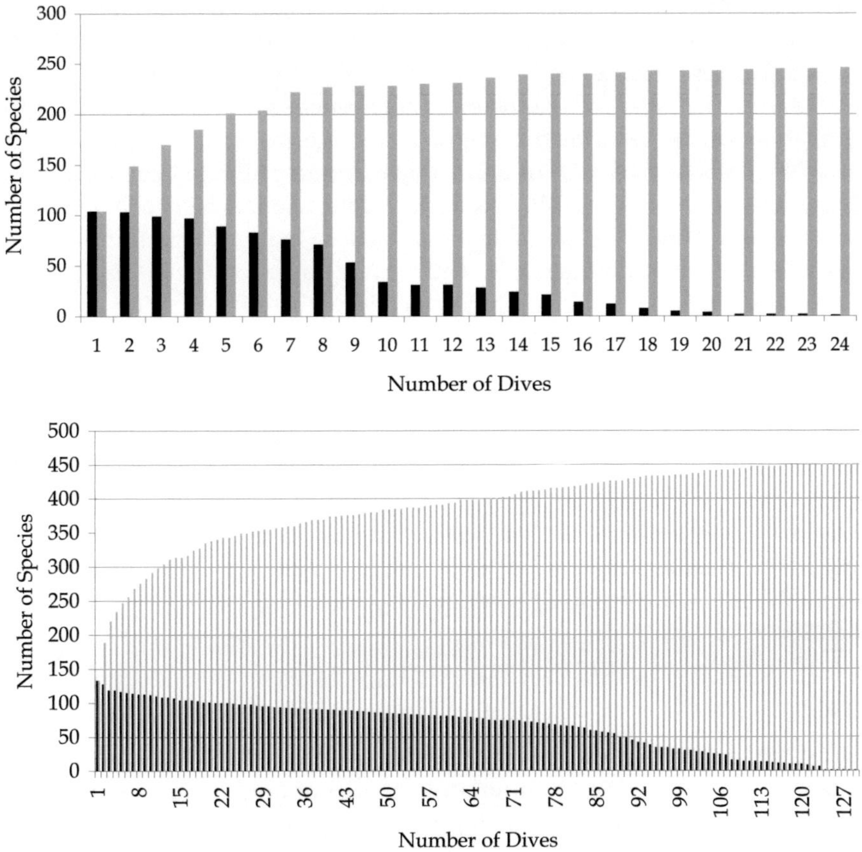

Fig. 3. Addition of cumulative species to site species lists through continued diving. Results are ordered according to number of species recorded, with cumulative total depicted in gray bars adjacent to bars for total species per dive. Note that only shallow (<20m) day dives were conducted at Lookout Point (above), within 0.5 km of Whytecliff (below), where day, night, shallow and deep (<40m) dives took place over the full time period of these surveys.

seem primitive in comparison with relational databases, it has the advantage of simplicity. Individual dive logs can be examined (and, if necessary, damaged portions repaired) with simple tools such as text editors and standard command-line utilities. Any updates to the dive logs will be reflected immediately in subsequent reports. The structure also makes it easy to write custom programs to analyze the data and generate results in any desired form. A species list table file provides detailed general information for each species (e.g. author and date). Another table file, keyed by location name, allows locations to be grouped into regions.

The search program is run once per region. All species' data are written to an intermediate file; each species is looked up in the species list, and its corresponding phylum is appended. The order of listing species is according to phylogenetic relationships within a phylum. The merge program counts all species by phylum to generate the final result. Note: all seaweed and flowering plant divisions (phyla) are grouped together for most compilations.

Seawater acidity was measured at the Vancouver Aquarium by colorimetric (titration) methods from 1954-1979, by adding a selective reagent to a sample of water so that a color was produced, the intensity of which was proportional to the concentration of Hydronium ions (H+) in the water, then matched to calibrated color standards. From 1954-1967, Winkler titration methods were employed on weekly grab samples taken by the veterinary department. Starting in 1967, the engineering department's seawater monitoring included pH determinations using colorimetric methods for the swimming pool industry. These methods employed a selective reagent and comparison to calibrated color standards. From 1980-1994, portable conventional pH electrodes were used, which allowed temperature compensation. From 1994-2009, portable field instruments with platinum free-diffusion junctions provided faster and more stable readings. A limitation to portable pH probes was that with time, the junction would become clogged with silver chloride or contaminants, causing large variation in the reference potential; clogged or fouled junctions could cause drift along with inaccurate, noisy, erratic and sluggish pH measurements. Some adjustments to data records during this period, for outlier data points related to fouled probes, was required in preparing data presentations. Therefore, annual minimum and maximum pH levels were tabulated for graphing only when at least five measures at that level were recorded on different dates in a year. Starting in 2010 professional lab bench instrumentation was introduced, with a free-flowing liquid-to-liquid junction that provides stable, drift-free measures from an easily cleaned junction that never clogs (a double junction design). Results are reported here for the period of the biodiversity survey for data from 1968-2010.

3. Results

From 1967 to 2010, when shallow dives <20m are undertaken in daylight, the biodiversity list for a site reached an asymptotic cumulative species number within 7-9 dives (Figure 2), whereas a higher overall biodiversity listing is obtained within 9-25 dives if a larger set of dives at both shallow and deep (<40m) depths was conducted during both daylight and night, as at Whytecliff Park. Different benthic habitat types at different locations had divergent biodiversities. It was necessary to restrict temporal analysis of trends for climate regimes to the species list that was generated during the earliest climate regime of 1967-1976, with a total of 328 species (versus 1,185 for the most recent period).

The species occurring at high relative abundance (rating of 6 or more in at least one region) for the overall study region are listed for different major regions in Table 1, based on the original 328 species identified during the earliest climate regime (1967-1976). Puget Sound had the greatest absence of species (26 species), and both Puget Sound and Johnstone Strait had the highest numbers of species (23 and 22, respectively) occurring at trace abundance. The only species absent from the Strait of Georgia was *Astraea gibberosa*, an exposed coast snail particularly associated with the kelp *Macrocystis integrifolia*.

In contrast to the Strait of Georgia, Puget Sound was lacking 23 species that occurred in all other regions. Puget Sound is a broad fjord with much less hard substrate compared to other regions in this study. In addition, it possesses few high current passages. These two factors combine to provide less habitat for organisms requiring rock and tidal current. Some of these missing species include the zoanthid *Epizoanthis scotinus* that occurs abundantly elsewhere and the hydroid *Garveia annulata* that also occurs abundantly everywhere else (least so in Strait of Georgia). Similarly, the chiton *Katharina tunicata* is absent from Puget Sound and only at trace abundance in Strait of Georgia, but this species is at high abundance in nearby Johnstone Strait, and also occurs at comparably high levels in outer coastal areas.

The red urchin *Strongylocentrotus franciscanus* is very abundant everywhere except Puget Sound, where it occurs at trace levels.

The Strait of Georgia and Alaska/north BC had the lowest numbers of species occurring at the highest levels of abundance. Although Puget Sound had the lowest biodiversity, it was the only region with high abundance of the anemone *Anthopleura artemesia*, a species that attaches to rock surrounded by sand or shell hash. The rockfish *Sebastes auriculatus* is very common in Puget Sound but usually rare in southern Strait of Georgia, where it is known for only a few areas. Similarly, the sculpin *Artedius fenestralis* is only abundant in Puget Sound. The sculpin *Chitonotus pugetensis* is primarily nocturnal, so its recorded abundance is affected by access to night diving, which took place mostly in Howe Sound (Strait of Georgia) and Puget Sound. The sole *Pleuronichthys coenosus* was also at high abundance in Puget Sound. Embiotocid perches were notably more abundant in Puget Sound (especially *Rhacochilus vacca* and *Embiotoca lateralis*) and only in trace numbers in Alaska/north BC. The northern range limits for NE Pacific embiotocid perches are in northern BC and southeastern Alaska. Strait of Georgia is closest to Puget Sound in overall embiotocid abundance. The sculpin *Enophrys bison* is much more abundant in Puget Sound and the outer coast of Washington than elsewhere and only occurs at trace abundance in Alaska and northern BC. Puget Sound differs considerably in biodiversity from the Strait of Georgia, as does Johnstone Strait at the northern end of the Strait of Georgia.

Some species like the anemone *Metridium farcimen*, the tubeworm *Serpula columbiana*, the shrimp *Pandalus danae*, the sea star *Pycnopodia helianthoides*, the sea cucumber *Parastichopus californianus*, the tunicate *Boltenia villosa* and the greenling *Hexagrammos decagrammus* are abundant in all the regions monitored in this project (Table 1). Most of the common species tend to be more abundant in one or more regions than in others. The only species uniformly occurring at limited abundance levels in all regions are several nudibranchs (*Doris montereyensis*, *Diaulula sandiegensis* and *Flabellina triophina*), the octopus *Enteroctopus dofleini*, the sea star *Pteraster tesselatus* and the tunicate *Aplidium solidum*.

A north to south trend can be detected from species absence where the outer coast of Washington and Puget Sound are both lacking species that occur everywhere else, including in the Strait of Georgia. These species more abundant in the north include the soft coral *Gersemia rubiformis* (prefers high current), the hydrocoral *Stylaster norvigicus*, the hydroid *Ectopleura marina*, the bryozoan *Phidolopora pacifica*, the sea anemone *Urticina lofotensis*, the snail *Astraea gibberosa*, the nudibranch *Tochuina tetraquerta*, the sea star *Stylasterias forreri*, the basket star *Gorgonocephalus eucnemis*, the feather star *Florometra serratissima* and the rockfish *Sebastes nebulosus*. As mentioned, the sculpin *Enphrys bison* is a southern species, as is the gunnel *Apodichthys flavidus*. The tunicate *Styela montereyensis* (a species ranging S to Mexico), is abundant on all outer coasts, but at trace levels in all inland seas.

Some abundant species peak at extremely high abundance in one area or another. The shrimp *Pandalus danae* is abundant everywhere, as mentioned, but considerably higher in abundance in Puget Sound than anywhere else. Other species are extremely abundant in only one region, absent in one other region, and moderately abundant elsewhere, as for the anemone *Cribrinopsis fernaldi*, very abundant in Johnstone Strait, absent in Puget Sound, and frequent in other regions. *Gersemia rubiformis* is extremely abundant in Johnstone Strait, at a trace in Strait of Georgia, absent from Puget Sound, and moderately abundant in outer coastal regions. Another cnidarian, *Garveia annulata*, is also abundant in Johnstone Strait and absent from Puget Sound

	ALNC	WCVI	OCW	JSTR	SoG	PS
Green algae (Chlorophyta)						
Ulva intestinalis	1	*	3	1	2	8
Ulva spp. / *Ulva lactuca*	8	7	19	8	13	25
Codium setchellii	5	6	3	20	2	-
Brown algae (Ochrophyta)						
Pterygophora californica	6	14	17	16	4	3
Red algae (Rhodophyta)						
Misc. branching red seaweeds	3	8	11	11	5	7
Sponges (Porifera)						
Rhabdocalyptus dawsoni	1	*	-	1	7	-
Aphrocallistes vastus	2	*	-	2	6	-
Cliona californiana	14	18	12	17	16	2
Myxilla lacunosa	2	9	4	18	2	*
Ophlitaspongia pennata	7	10	6	13	12	1
Misc. demo sponges	7	6	6	6	4	2
(Cnidaria)						
Metridium senile	9	23	20	27	9	28
Metridium farcimen	24	18	18	15	23	15
Cribrinopsis fernaldi	8	6	5	27	8	-
Urticina crassicornis	9	11	27	26	9	17
Urticina lofotensis	15	16	1	1	*	*
Urticina piscivora	11	14	6	*	*	-
Stomphia didemon	1	2	*	*	8	7
Anthopleura artemisia	1	1	*	*	*	7
Epiactis prolifera	*	6	7	19	1	4
Pachycerianthus fimbriatus	11	14	*	3	17	3
Epizoanthus scotinus	8	15	10	22	9	-
Gersemia rubiformis	14	13	2	29	*	-
Ptilosarcus gurneyi	8	11	2	1	10	11
Stylantheca spp.	8	6	9	21	6	*
Stylaster norvigicus	2	1	2	21	*	-
Aglaophenia spp.	4	7	8	11	3	1
Abietinaria spp.	12	14	19	21	11	3
Plumularia setacea	11	12	4	19	3	1
Obelia spp.	9	6	5	7	6	2
Garveia annulata	15	10	9	23	3	-
Ectopleura marina	4	6	2	20	4	*
Tubularia indivisa	2	*	*	9	*	-

	ALNC	WCVI	OCW	JSTR	SoG	PS
Segmented worms (Annelida)						
Protula pacifica	5	6	1	5	6	*
Serpula columbiana /vermicularis	31	29	21	30	18	26
Pileolaria spp. (spirorbids)	2	4	5	6	2	*
Dodecaceria fewkesi	6	21	12	30	11	2
Eudistylia vancouveri	4	11	6	8	2	19
Schizobranchia insignis	2	1	1	2	1	9
Bispira sp. (*Sabella crassicornis*)	7	5	7	4	3	9
Spiochaeopterus costarum	2	4	2	*	4	11
Bryozoans (Bryozoa)						
Membranipora serrilamella	14	9	18	11	8	5
Schizoporella unicornis	2	5	9	5	8	9
Bugula californica	5	7	6	*	2	3
Crisia spp.	5	11	11	14	7	1
Phidolopora pacifica	5	11	3	*	2	-
Heteropora pacifica	15	14	18	7	6	-
Diaperoecia californica	15	15	14	1	7	-
Hippodiplosia insculpta	7	7	4	*	1	*
Brachiopods (Brachiopoda)						
Terebratalia transversa	6	5	15	17	8	3
Molluscs (Mollusca)						
Tonicella lineata	16	21	20	31	15	3
Mopalia spp.	1	7	13	16	7	13
Katharina tunicata	1	3	9	8	*	-
Cryptochiton stelleri	12	13	20	24	7	2
Mytilus trossulus	8	3	3	3	8	15
Chlamys spp.	6	11	16	16	10	10
Crassadoma gigantea	12	22	13	21	10	3
Pododesmus macrochisma	8	6	10	7	13	17
Clinocardium nuttalli	2	1	2	*	1	6
Panopea abrupta	4	3	4	-	1	7
Entodesma navicula	2	5	6	13	3	*
Acmaea mitra	15	15	19	20	5	*
Diodora aspera	11	14	11	23	4	1
Haliotis kamtschatkana	6	11	6	9	1	-
Nucella lamellosa	3	4	12	6	9	10
Ceratostoma foliatum	11	17	15	12	11	2
Calliostoma ligatum	21	24	22	25	9	3
Calliostoma annulatum	8	6	5	*	1	-
Astraea gibberosa	9	13	*	*	-	-
Euspira lewisii	2	4	3	*	2	11
Fusitriton oregonensis	9	8	16	10	2	2
Peltodoris nobilis	4	12	5	3	6	4

	ALNC	WCVI	OCW	JSTR	SoG	PS
Molluscs, continued						
Doris montereyensis	2	6	5	3	4	8
Diaulula sandiegensis	2	6	5	8	3	6
Cadlina luteomarginata	5	8	6	5	12	1
Triopha catalinae	5	5	8	8	1	3
Tochuina tetraquetra	2	6	*	10	*	-
Dirona albolineata	4	7	11	*	6	11
Hermissenda crassicornis	18	14	6	9	5	6
Flabellina triophina	5	3	1	6	5	3
Enteroctopus dofleini	5	5	6	3	3	8
Arthropods (Arthropoda)						
Caprella spp.	6	2	1	2	1	1
Pandalus danae	10	10	19	14	16	28
Cancer oregonensis	3	7	16	14	5	6
Cancer productus	3	6	10	4	8	22
Cancer magister	1	1	2	*	4	14
Telmessus cheiragonus	*	*	6	*	*	1
Pugettia producta	*	2	5	*	2	19
Pugettia gracilis	4	6	9	13	2	6
Scyra acutifrons	4	17	16	15	8	19
Oregonia gracilis	8	6	12	5	6	8
Phyllolithodes papillosus	2	3	8	3	1	-
Lopholithodes mandtii	5	7	3	7	2	-
Pagurus beringanus	8	10	17	21	10	17
Elassochirus gilli	6	1	3	4	1	-
Pagurus armatus	2	1	3	*	3	17
Balanus glandula	6	6	8	2	17	29
Balanus nubilus	16	23	29	30	15	13
Echinoderms (Echinodermata)						
Pisaster ochraceus	8	9	4	1	14	10
Pisaster brevispinus	4	7	1	1	14	9
Evasterias troschelii	10	11	15	10	15	23
Orthasterias koehleri	17	20	10	9	14	*
Stylasterias forreri	7	8	*	1	3	-
Dermasterias imbricata	11	19	10	9	17	8
Asterina miniata	1	6	*	-	*	-
Mediaster aequalis	7	13	3	5	12	7
Pteraster tesselatus	4	6	3	3	9	3
Henricia leviuscula	10	5	4	4	4	*
Henricia aspera	1	13	15	18	11	12
Leptasterias spp. complex	1	3	8	*	*	-
Pycnopodia helianthoides	24	20	22	15	26	26

	ALNC	WCVI	OCW	JSTR	SoG	PS
Echinoderms, continued						
Crossaster papposus	7	6	*	2	7	5
Solaster dawsoni	8	9	7	10	11	9
Solaster stimpsoni	5	13	15	15	8	14
Ophiopholis aculeate	24	14	9	22	10	*
Ophiura lutkeni	6	1	-	-	11	*
Gorgonocephalus eucnemis	3	5	1	21	1	-
Florometra serratissima	*	1	*	1	11	-
Strongylocentrotus franciscanus	25	25	15	32	18	*
Strongylocentrotus droebachiensis	8	8	17	26	17	7
Parastichopus californicus	16	23	18	14	24	21
Cucumaria miniata	13	21	20	9	12	7
Eupentacta quinquesemita	8	16	15	16	9	11
Psolus chitonoides	12	18	18	26	14	3
Tunicates (Urochordata)						
Corella willmeriana	8	8	2	5	14	5
Ascidia paratropa	5	6	3	4	5	1
Cnemidocarpa finmarkiensis	9	15	10	6	14	*
Halocynthia aurantium	8	4	*	6	8	*
Halocynthia igaboja	9	13	4	20	9	*
Pyura haustor	8	12	14	8	8	13
Styela montereyensis	4	10	5	*	*	*
Boltenia villosa	11	12	11	8	13	13
Chelyosoma productum	8	8	8	*	5	3
Metandrocarpa taylori	17	13	21	12	8	*
Distaplia occidentalis	14	16	8	12	4	3
Aplidium solidum	4	7	4	9	2	2
Vertebrates (Chordata)						
Aulorhynchus flavidus	1	2	10	2	4	11
Ammodytes hexapterus	1	*	8	-	*	1
Rhinogobiops nicholsii	4	18	7	4	19	12
Ronquilus jordani	6	3	*	5	5	4
Chirolophis nugator	2	2	7	3	2	4
Apodichthys flavidus	-	*	4	*	*	6
Pholis laeta	2	*	5	1	*	12
Embiotoca lateralis	*	10	17	2	16	26
Rhacochilus vacca	*	6	6	1	9	20
Cymatogaster aggregata	*	4	5	1	10	17
Brachyistius frenatus	*	3	3	*	6	7
Sebastes caurinus	9	19	18	13	24	20
Sebastes maliger	12	18	9	17	19	13
Sebastes nebulosus	12	14	2	2	*	-
Sebastes auriculatus	*	*	-	-	1	13

	ALNC	WCVI	OCW	JSTR	SoG	PS
Vertebrates, continued						
Sebastes melanops	19	25	14	13	2	4
Sebastes flavidus	13	22	5	20	1	*
Sebastes emphaeus	6	15	14	12	7	3
Hexagrammos decagrammus	19	26	26	28	19	10
Ophiodon elongatus	7	14	14	6	15	9
Oxylebius pictus	4	10	9	3	8	20
Artedius harringtoni	8	17	20	27	11	23
Artedius fenestralis	*	*	1	*	1	12
Jordania zonope	9	15	18	13	13	2
Scorpaenichthys marmoratus	1	3	3	4	3	15
Hemilepidotus hemilepidotus	6	7	5	15	3	13
Leptocottus armatus	*	*	*	*	1	7
Enophrys bison	*	1	11	3	2	17
Myoxocephalus polyacanthocephalus	2	*	2	1	*	7
Chitonotus pugetensis	1	*	-	*	2	8
Nautichthys oculofasciatus	1	1	1	2	2	6
Citharichthys stigmaeus	*	2	5	1	5	9
Pleuronichthys coenosus	*	*	1	-	2	10

Table 1. Average abundance rating of most frequently observed species with a rating 6 or more in at least one region (asterisk = trace, dash = absent) for ALNC (Alaska and north coast British Columbia), WCVI (west coast Vancouver Island), OCW (outer coast Washington), JSTR (Johnstone Strait), SoG (Strait of Georgia) and PS (Puget Sound). See region locations on map in Figure 1. Within a higher taxon, species are listed according to phylogenetic relationships.

When biodiversity of these regions is considered for more abundant species (rating of 6 or more) in terms of two prominent climate regime shifts (1977, 2000) for the original 328 species from the first regime (Figure 4), it appears that biodiversity increased in Puget Sound and Johnstone Strait during the 1977-2000 regime, but that is likely an artefact of greater numbers of dives in that period. Biodiversity remained stable in Strait of Georgia and west coast Vancouver Island. There were too few dives in the other regions (Alaska/northern BC, outer coast Washington) to permit comparisons. It should be noted that Johnstone Strait had only a single dive in the first period. Another program run collated all species of abundance rating of 2 or more and showed the same very stable pattern of biodiversity as for the abundance rating of 6 or more depicted in Figure 4, indicating that, not considering the lowest trace abundances, species biodiversity is quite stable for animal phyla in the Strait of Georgia and nearby regions.

If climate regimes are considered to have shifted in 1977, 1989 and 2000, then it appears that biodiversity still remained relatively stable in Strait of Georgia and west coast Vancouver Island through at least the last three of four regimes (Figure 5), even when including species at all abundance levels. The biodiversity in the first regime for every area involved a lesser expertise on the part of observers as well as the lowest level of effort in every region except Puget Sound, where the fewest dives were conducted during the last regime. The single dive for the first regime period in Johnstone Strait necessarily limited the number of species recorded there. Nonetheless, the evident drop in biodiversity during the last, 2001-2010, regime, occurred in every region including Strait of Georgia, where the highest level of effort (and arguably the greatest level of expertise) was during that last regime. Thus, when all species including trace levels of occurrence are included for the list of the original 328 species (from the first regime), it appears that the regime shift of 2000 did lead to reduced biodiversity, but probably only for more rare species (considering the constant biodiversity stability for more abundant species depicted in Figure 3).

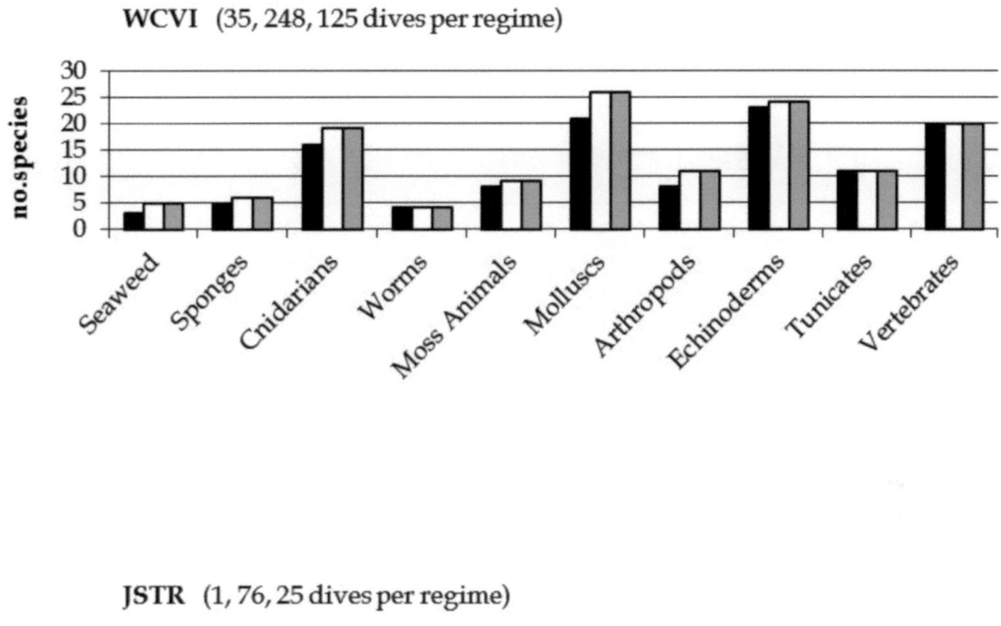

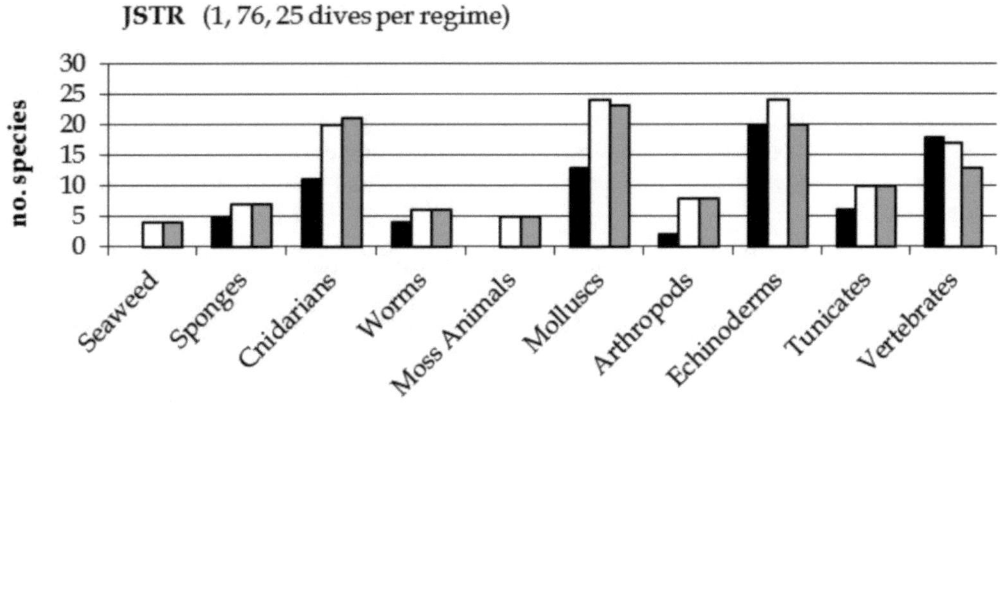

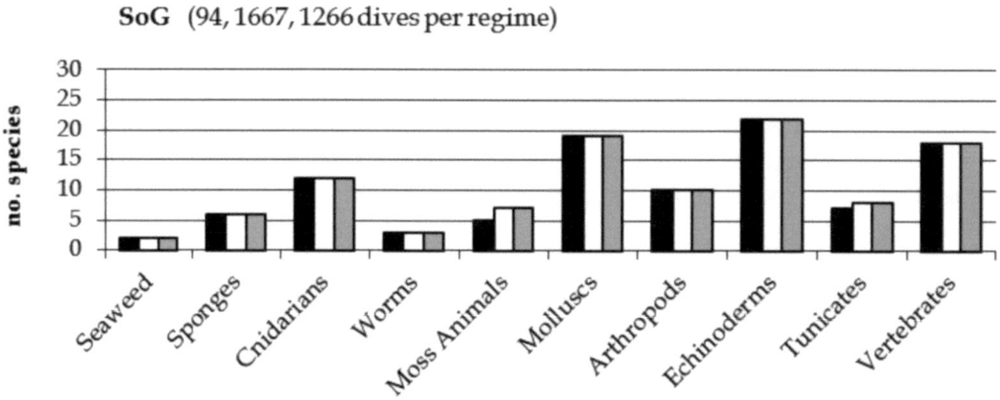

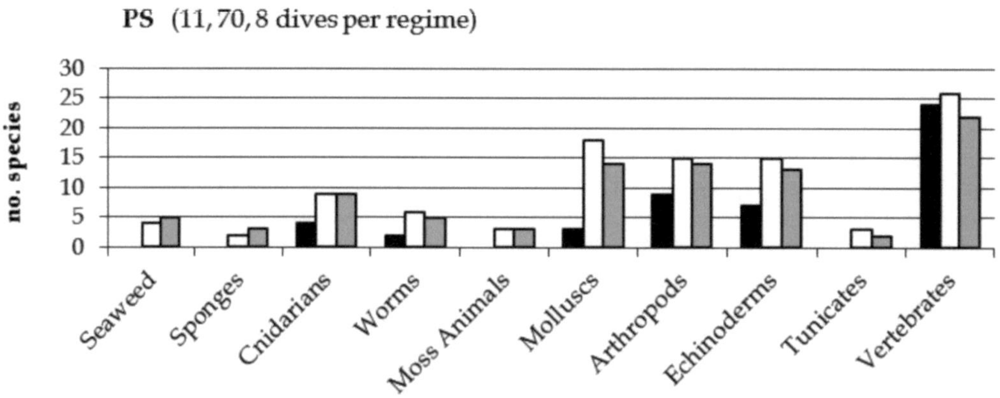

Fig. 4. Biodiversity in four Pacific coast regions according to higher taxon groupings, for three climate regimes: 1967-1976 (black bars), 1977-2000 (white bars), and 2001-2010 (gray bars). Number of species is on the vertical axis. Data are for species with relative abundance rating of 6 or more. Note that Johnstone Strait had only one dive for the first regime. WCVI = west coast Vancouver Island, JSTR = Johnstone Strait, SoG = Strait of Georgia and PS = Puget Sound.

Because seaweeds were not emphasized in the first climate regime, they ranked low diversity in that regime. Life forms with variable morphometry, like sponges and moss animals (bryozoans) were also poorly identified during the first regime, with gradual increases in diversity documented through the second, to the third regime (Figure 5, which includes all abundance ratings including trace occurrence). A less gradual increase in identification capability was evident for molluscs, for which a more marked drop in biodiversity occurred in the last, fourth regime. Similarly, although to a lesser extent, echinoderms and fishes peaked in diversity during the third regime.

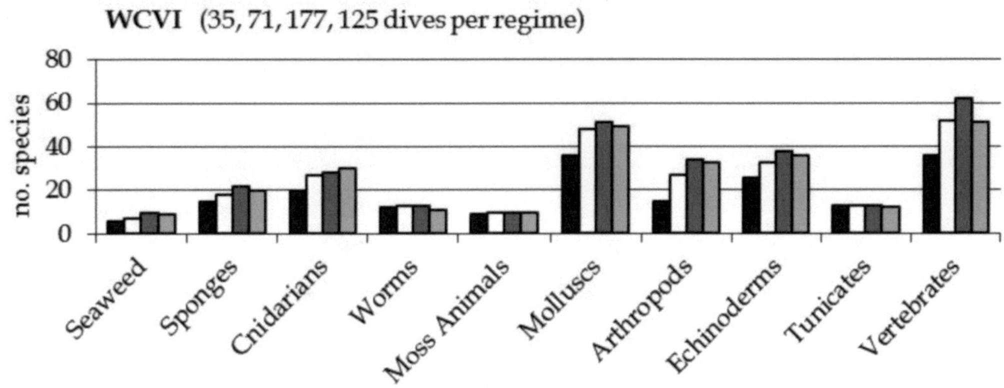

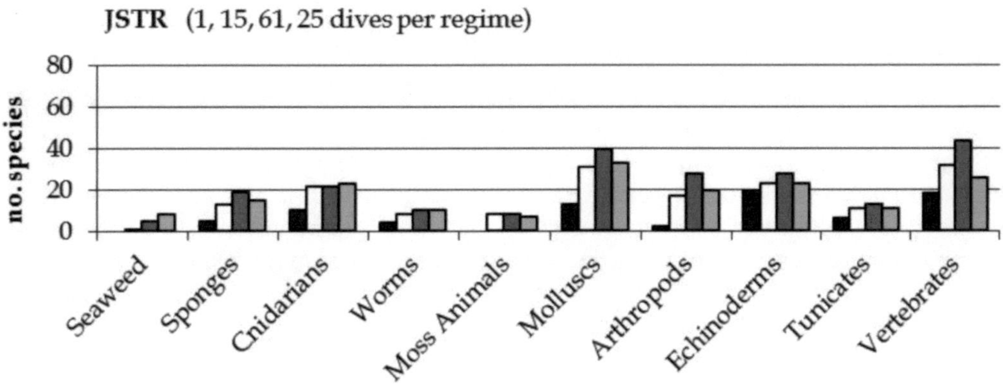

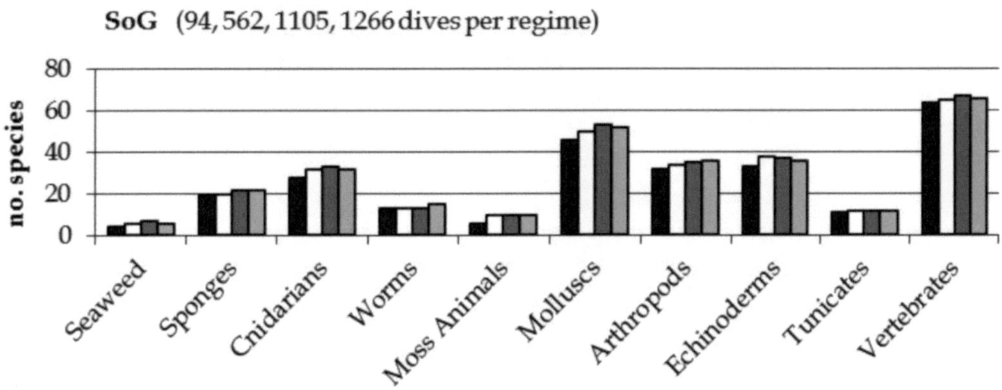

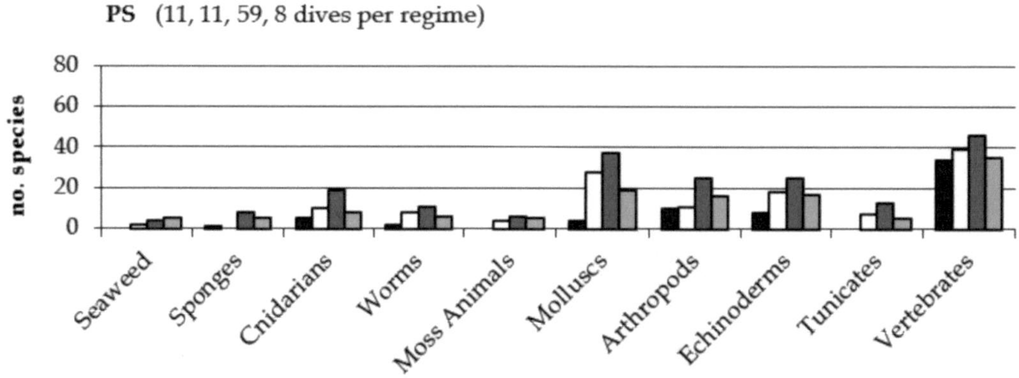

Fig. 5. Biodiversity in Pacific coast regions according to higher taxon groupings, for four climate regimes: 1967-1976, 1977-1988, 1989-2000, 2001-2010 (bars arranged left to right by time period). Numbers of dives for each period, in order, are in parentheses following each area name. Number of species is on the vertical axis. Data are for species at all relative abundance levels, including trace occurrence. WCVI = west coast Vancouver Island, JSTR = Johnstone Strait, SoG = Strait of Georgia and PS = Puget Sound.

Examining only one specific location, Whytecliff (in Strait of Georgia) for equal numbers of dives in each of four climate regimes (Figure 6) indicates that the third regime from 1989-2000 had a greater biodiversity than either the preceding or subsequent regimes. These data demonstrate that there does appear to have been a regime shift in 1989, and that biodiversity increased at this particular location during that third regime, then decreased somewhat during the fourth regime. The compilation of data for all sites in the Strait of Georgia (Figure 5) shows a slight tendency for the same trends, but not as distinctly as at a single site, probably owing to confounding effects of pooling biodiversity data from many locations

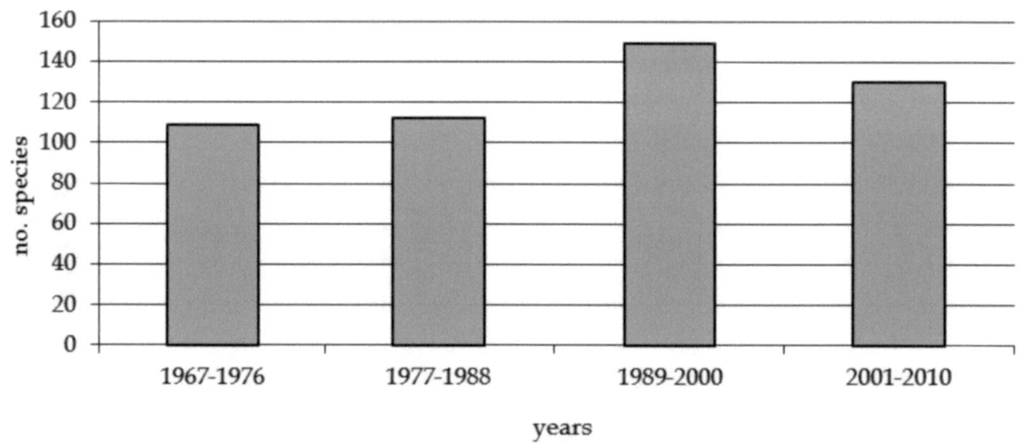

Fig. 6. Biodiversity based on the original 328 species during four regimes for just five dives (the total number for 1977-1988) at Whytecliff, in the Strait of Georgia.

	ALNC		WCVI		OCW		JSTR		SoG		PS	
	89-	01-	89-	01-	89-	01-	89-	01-	89-	01-	89-	01-
Flowering plants (Anthophyta)												
Zostera marina	-	2	4	5	8	9	*	5	8	2	22	3
Phyllospadix spp.	-	*	1	2	6	-	-	-	*	*	-	-
Green algae (Chlorophyta)												
Prasiola meridionalis	-	1	*	-	-	3	-	-	*	*	*	7
Cladophora spp.	-	2	-	1	*	-	-	-	*	*	2	1
Ulva intestinalis	-	1	1	*	4	6	-	4	3	2	9	7
Ulva spp./*U. lactuca*	4	8	8	10	21	43	6	14	16	15	29	28
Acrosiphonia coalita	-	2	-	*	-	-	-	-	*	*	-	-
Bryopsis spp./*B. corticulans*	-	*	-	-	*	4	-	-	*	*	*	-
Codium setchellii	-	5	7	8	4	-	17	36	3	2	-	-
Kornmania leptoderma	-	*	-	*	*	3	-	-	*	*	-	-
Brown algae (Ochrophyta)												
Ectocarpus complex	-	-	*	*	1	-	-	-	*	*	2	-
Fucus gardneri	-	10	5	7	3	18	-	6	12	12	3	7
Leathesia difformis	-	-	*	1	*	4	-	-	*	*	*	-
Hedophyllum sessile	-	-	*	3	1	-	1	4	*	-	-	-
Egregia menziesii	-	2	3	5	4	1	-	-	*	-	-	-
Alaria nana	-	3	*	2	*	-	-	1	*	*	-	-
Alaria marginata	-	13	1	9	6	12	-	21	1	5	*	3
Costaria costata	4	13	6	10	10	15	2	31	4	2	2	-
Cymathere triplicata	-	8	1	1	7	21	-	1	*	*	-	-
Laminaria saccharina	8	13	8	11	19	31	8	10	15	13	16	15
Laminaria setchellii	-	20	7	18	3	-	-	13	*	*	-	-
Pleurophycus gardneri	-	5	-	4	2	-	-	9	-	*	-	-
Lessoniopsis littoralis	-	5	*	4	4	-	-	-	2	-	*	-
Sargassum muticum	4	*	2	1	2	1	*	*	13	14	13	12
Desmarestia aculeata complex	-	17	3	9	6	16	1	24	1	1	2	-
Desmarestia ligulata/*D. munda*	4	19	9	18	17	9	5	29	2	2	7	13
Pterygophora californica	16	5	14	23	20	10	17	20	5	5	4	6
Eisenia arborea	-	*	3	9	-	-	-	1	*	-	-	-
Macrocystis integrifolia	8	4	3	9	4	1	-	1	-	-	-	-
Nereocystis luetkeana	12	24	18	20	29	13	22	55	10	8	11	4
Dictyota binghamiae	-	7	1	2	-	-	-	-	*	*	-	-
Agarum clathratum	-	-	-	1	-	-	-	5	-	*	-	-
Agarum fimbriatum	14	19	12	18	12	1	10	25	21	26	2	6
Red algae (Rhodophyta)												
Bangia spp.	-	-	*	-	*	-	-	-	*	*	2	-
Porphyra spp.	-	9	2	3	5	3	-	4	3	2	1	-
Endocladia muricata	-	2	*	1	-	-	-	*	*	*	-	-
Hildenbrandia spp.	4	11	17	21	4	10	7	34	10	23	*	3
Mastocarpus papillatus	-	-	-	1	-	1	-	*	*	*	*	7
Microcladia borealis	-	-	-	1	-	3	-	2	*	1	*	3
Halosaccion glandiforme	-	2	*	3	-	-	-	1	*	*	-	-
filamentous red algae	-	6	2	9	1	7	-	14	*	10	*	12
Prionitis lyallii	-	-	*	*	*	-	-	-	*	*	-	3
Clathromorphum etc.	16	49	33	55	23	13	26	54	21	23	8	15

	ALNC		WCVI		OCW		JSTR		SoG		PS	
	89-	01-	89-	01-	89-	01-	89-	01-	89-	01-	89-	01-
Melobesia / Mesophyllum	-	2	*	2	*	-	-	1	-	-	-	-
Bossiella / Calliarthron	8	29	23	43	12	6	23	35	4	1	*	-
Corallina vancouveriensis	-	4	2	3	*	-	-	6	*	*	-	-
Bossiella spp., *Calliarthron* spp.	-	15	7	15	1	3	-	20	*	*	-	-
Palmaria sp.	-	-	-	*	-	-	-	2	-	2	-	3
Ceramium pacificum	-	-	*	2	-	3	-	-	*	*	*	-
Callophyllis spp.	-	11	2	10	5	16	-	17	3	8	2	10
Chondracanthus exasperatus	-	1	2	7	5	19	1	*	4	6	11	24
Mazzaella splendens	4	*	2	4	7	7	3	1	3	2	1	10
Cryptopleura spp.	-	5	-	6	-	17	-	5	*	6	*	6
Hymenena spp.	-	2	-	*	-	-	-	*	*	1	-	3
Erythrophyllum delesseroides	-	2	*	1	-	-	-	1	*	*	-	-
Gracilaria/Gracilariopsis	-	-	*	*	-	3	-	1	*	1	-	-
Odonthalia floccosa	-	1	-	*	-	-	-	2	*	*	-	-
Odonthalia washingtoniensis	-	1	-	*	*	3	-	-	*	-	-	-
Laurencia spectabilis	-	1	-	1	*	7	-	-	*	*	*	3
Sarcodiotheca gaudichaudii	-	-	1	2	4	19	1	1	2	2	8	19
Rhodymenia californica	-	*	-	2	-	-	-	2	*	*	-	-
Schizymenia pacifica	-	-	-	1	-	-	-	4	-	1	-	-
Smithora naiadum	-	1	1	1	5	-	-	-	*	*	2	3
Polyneura latissima	-	1	-	*	-	4	-	-	*	*	*	-
Sparlingia pertusa	-	3	1	1	1	6	*	3	3	1	3	3
Constantinea subulifera	-	2	*	1	-	-	*	1	*	*	-	-
Constantinea simplex	-	1	*	2	*	-	*	1	2	1	-	-
Delesseria decipiens	-	2	1	1	-	-	-	9	*	1	1	-
Membranoptera platyphylla	-	1	*	1	-	4	-	2	*	1	1	-
Bonnemaisonia nootkana	-	2	*	1	*	-	-	1	*	*	-	-
Fauchea laciniata	-	15	10	13	12	1	5	18	1	1	2	-
Botryocladia pseudodichotoma	-	*	1	*	-	-	*	-	2	2	*	-
Opuntiella californica	-	5	5	7	4	-	4	6	2	2	-	6

Table 2. Average abundance rating of seaweed species with an abundance rating 2 or more in at least one region (asterisk = trace, dash = absent) for ALNC (Alaska and north coast British Columbia), WCVI (west coast Vancouver Island), OCW (outer coast Washington), JSTR (Johnstone Strait), SoG (Strait of Georgia) and PS (Puget Sound). See region locations on map in Figure 1. Abundance rating is listed for each of the last two climate regimes (89- = 1989-2000, 01- = 2001-2010) for each region. Within a higher taxon, species are listed according to phylogenetic relationships.

with different habitat attributes that tend toward different community species compositions at those various sites. Whytecliff actually had the most sampling of any site, but only five dives during the second regime period of 1977-1988. The five dives with the highest species counts were used for the other regimes.

For seaweeds, all species were being identified during the latest two climate regimes. In Alaska/northern BC, outer coast Washington and Puget Sound, however, too few dives were conducted in either the earlier or later regime, so that graphs summarizing biodiversity for those areas would not show valid trends. That is, for Alaska / northern BC, only 5 dives were conducted in 1989-2000, versus 103 dives in 2001-2010; on the outer coast of Washington, 75 dives took place in 1989-2000 versus only 5 in 2001-2010, and in Puget Sound, 59 dives were in 1989-2000, but only 8 dives in 2001-2010; thus these three regions are not presented graphically. The seaweed abundance ratings are listed for all marine plants occurring at abundance ratings of 2 or more in Table 2 and the relative biodiversity of different plant groups is depicted, including all abundance ratings, for Strait of Georgia, west coast Vancouver Island and Johnstone Strait in Figure 7. It can be seen from Table 2 that if the areas with disproportionate dive focus were graphed, there would be artefact appearance of biodiversity shifts, as with a decrease in all marine plant biodiversity in Puget Sound (resulting from only 8 dives there in 2001-2010).

Considerable confidence can be placed in identifications of the brown algae *Fucus gardneri, Hedophyllum sessile, Egregia menziesii, Alaria nana, Alaria marginata, Costaria costata, Cymathere triplicata, Laminaria saccharina, Laminaria setchellii, Pleurophycus gardneri, Lessionopsis littoralis, Sargassum muticum, Desmarestia lingulata/munda, Pterygophora californica, Eisenia arborea, Nereocystis luetkeana, Dictyota binghamae, Agarum clathratum* and *Agarum fimbriatum*. For red algae, some have been observed for many years as they were easily recognized, including *Porphyra* spp., *Hildenbrandia* spp., *Mastocarpus papillatus, Halosaccion glandiforme, Prionitis lyallii, Clathromorpha* etc. (encrusting corallines), *Callophyllis* spp., *Chondracanthus exasperatus, Mazzaella splendens, Sarcodiotheca gaudichaudii, Smithora naiadum, Sparlingia pertusa, Bonnemaisonia nootkana, Fauchea laciniata, Botryocladia pseudodichotoma* and *Opuntiella californica*. A considerable number of red algae, however, are not as readily identified by SCUBA divers in the field, particularly the branching and bladed forms. For that reason, diversity shifts in red algae, as depicted in Figure 7, are not as likely to represent genuine changes as are shifts depicted for the diversity of brown algae.

The data for the later two climate regimes in Figure 7 illustrate apparent increases in seaweed biodiversity for red algae in both the west coast of Vancouver Island and in Johnstone Strait, as well as an increase in brown algae diversity in Johnstone Strait during the latest regime. Considering that there were 61 dives during the 1989-2000 regime in Johnstone Strait, compared to just 25 dives there from 2001-2010, it seems that there may have been a genuine, significant increase in seaweed biodiversity in that region in particular. Note as well that the indication from limited diving in the more southerly regions is for decreasing, not increasing seaweed biodiversity over that time period, although those trends are not as likely to be valid.

The seaweed biodiversity in the Strait of Georgia remained very stable through the two most recent climate regimes (Figure 7), considering the increasing expertise in identification of red algae. Note from Table 2 that the red algae *Palmaria* sp., for example, was identified in the Strait of Georgia only during the last regime, but also in three other regions during the last regime, but nowhere during the previous regime. That species of intertidal dulse is very shallow and is difficult to identify, so probably does not represent a new appearance but rather a newly established identification capacity. For that reason, more confidence can be placed in apparent changes in brown algae biodiversity than in the reds, but considerable increase in seaweed biodiversity is apparent. This is in contrast to the overall appearance of loss of biodiversity in the last regime for the original list of 328 species, which included very few seaweeds.

Many dives (including the Whytecliff site) have been conducted near Vancouver, BC in Howe Sound, an area of fjord geography which experienced heavy sport fishing pressure

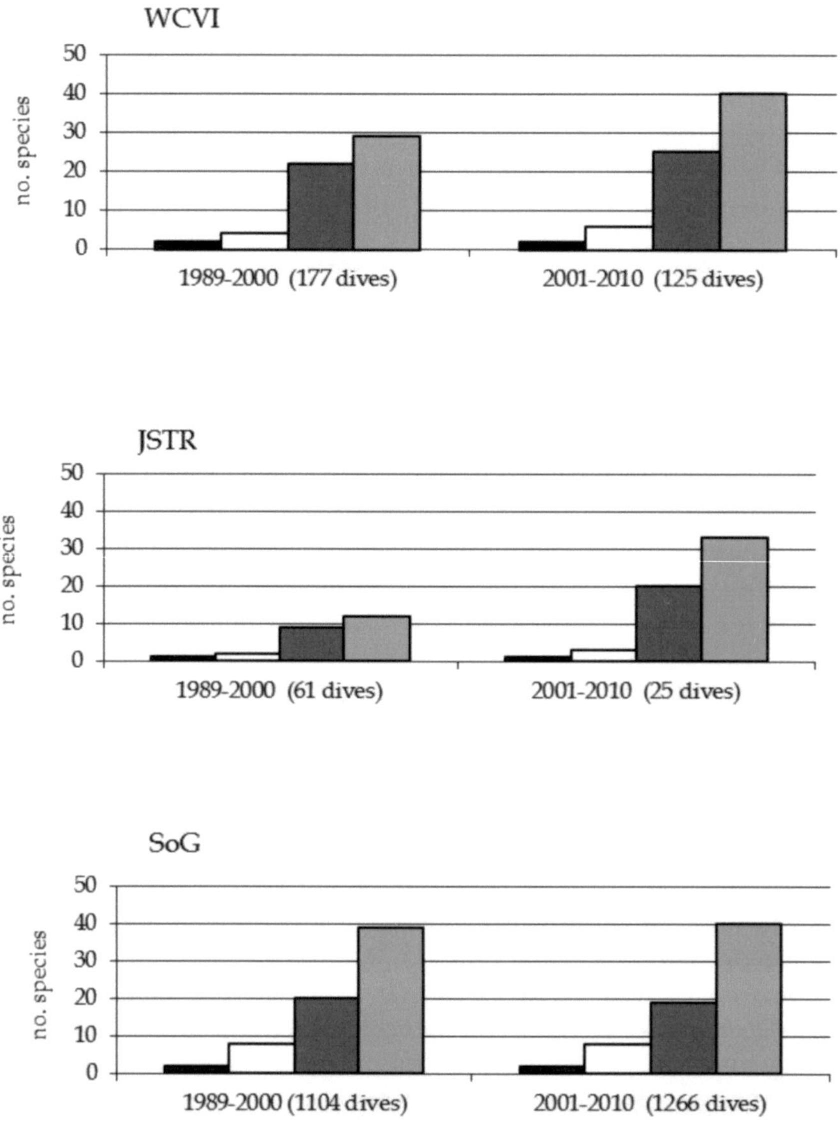

Fig. 7. Biodiversity of seaweeds is listed for the last two climate regimes (1989-2000 and 2001-2010). All abundance ratings are included. Considerable effort took place in these three regions. WCVI = west coast Vancouver Island, JSTR = Johnstone Strait and SoG = Strait of Georgia. Numbers of dives are listed in parentheses for regime periods. Black bars = flowering plants, white = green algae, dark gray = brown algae, light gray = red algae (see Table 2 for species with abundance rating of 2 or more).

before and during the early years of this survey. Howe Sound is now considered more overfished than the remainder of the Strait of Georgia (Marliave & Challenger, 2009). Howe Sound is contiguous with Vancouver Harbor, where the Vancouver Aquarium has maintained ocean acidity records through the period of this survey. The Vancouver Aquarium seawater records reveal that ocean acidification has steadily occurred through this period (Figure 8), with the range of pH in Vancouver Harbor shifting from typically pH 7.8-8.1 during the early period from 1954-1974, then increasingly varying to lower pH levels until the recent period when the range has often been from pH 7.3-7.9, sometimes varying to greater extremes. The most extreme high pH, in 1987, was during May-June, and the most extreme low pH, in 2001, was in July-August.

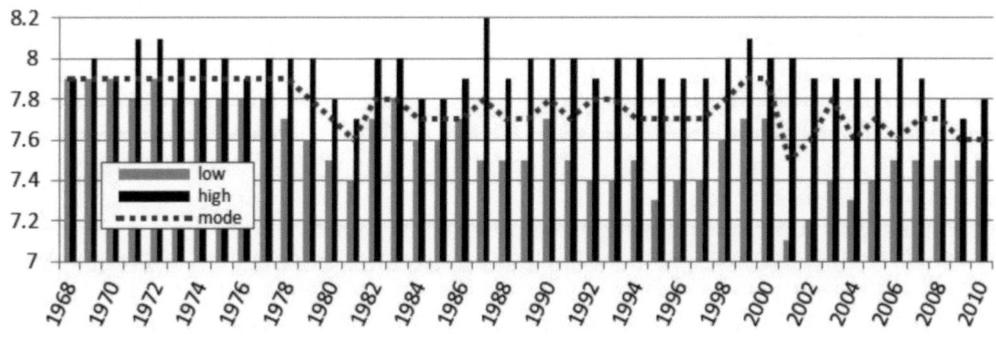

Fig. 8. Modal pH and extreme range (minimum 5 measures for low or high value) for Vancouver Harbor from 1968-2010.

This continual trend toward decrease in pH contrasts to the four reversing climate regime periods during the overall 1967-2010 time period for taxon records. The overall biodiversity remained stable in the Strait of Georgia in the presence of this declining pH level. The last three climate regimes, from 1978-2010, have included a steady decline in ocean pH in the middle latitudes of the Strait of Georgia, as seen in the data for Vancouver Harbor (Figure 8), yet the biodiversity has remained very stable, in fact more stable than for some of the adjacent areas, as for seaweeds in other regions.

4. Discussion

Overall, biodiversity was quite stable in the two most heavily investigated regions, west coast Vancouver Island and Strait of Georgia, for the study period of 1967-2010. The long duration of this biodiversity monitoring has involved an expanding network of experts and the discovery and description of new species so that the biodiversity list has continually expanded. For that reason, most of the results presented necessarily dealt with a curtailed species list of 328 species identified during the first climate regime period. This biodiversity stability has occurred through successive climate regimes and despite the continuous reduction in seawater pH, through global warming and through the continued state of stock depletion of fished groundfish and other fish species during that period. It is beyond the scope of this chapter to review the full literature on fisheries sustainability in the Strait of Georgia, but the reader may consult the review by Levy et al. (1996) to see a summary of historic declines in shellfish and finfish stocks in that region. There appear to have been no concomitant declines in overall biodiversity as one or another species has fluctuated in abundance. Consistent differences, however, persisted between adjacent regions, compared to the Strait of Georgia.

The region designations presented here are not based on any existing literature or governmental statistical areas for fisheries surveys. In fact, it may seem exceptional that smaller areas that seem to be inland seas are lumped together with outer coast areas such as the west coast of Vancouver Island (here including Queen Charlotte Strait inside the north end of Vancouver Island) and the extreme eastern end of the Strait of Juan de Fuca within the region of the outer coast of Washington. In the data compilation of Lamb et al. (2011) the occurrence of the exposed coast indicator species *Pyllospadix scouleri* and *Strongylocentrotus purpuratus* (Lamb & Hanby, 2005) on the southern coast of San Juan and Lopez Islands and the west coast of Whidbey Island was a rationale for designating these areas part of the fully wave-exposed outer coast rather than the protected inland seas of either Puget Sound or the Strait of Georgia, where all remaining portions of the San Juan Islands were included. The different, yet stable, biodiversities of the communities in these broad regions attest to the validity of the present region designations, and contrast rather markedly with some fisheries statistical areas. It might be well to conduct analyses of fisheries data in accordance with the present regional boundaries (Figure 1).

The literature is equivocal on whether a climate regime shift occurred in 1989, so results have been presented for both three and four climate regimes during the 1967-2010 period of this present study. Note that the first regime (here, 1967-1976) is actually considered to have persisted for a much longer time, from 1947-1976 (McFarlane et al., 2000). Whether the middle study period of 1977-2000 is considered to consist of just one climate regime or of two regimes (1977-1988, 1989-2000), the biodiversity appears (Figures 4, 5) to have been higher during that period than during either the first or last regimes. As well, differences between the second and third of the four regimes occurred (Figure 5), so that 1989 does appear to mark a climate regime shift, as proposed by McFarlane et al. (2000), from the standpoint of biodiversity, supporting the contention of Hare & Mantua (2000) that biodiversity accurately defines regime shifts. This is despite the evidence for increased observer expertise and improved capacity generally for taxonomic identifications based on field observation. Thus, there does appear to have been a slight overall reduction in biodiversity in recent years, in terms of the basic list of 328 species (mostly animals), in contrast to possible increases in seaweed biodiversity (based on the full species list).

The collation of species for one site at Whytecliff in four different climate regimes enabled a view of trends without concern over effects of habitat differences between sites within a region. The detailed data compilations for different areas within the Strait of Georgia region (Lamb et al., 2011) revealed that broad differences in biodiversity exist between different areas, apparently related to presence of high-energy tidal passes in areas like the southern Gulf Islands (Active Pass, Gabriola Pass, Porlier Pass) and Burrard Inlet (First Narrows, Second Narrows), where seaweed biodiversity is elevated in comparison to areas in middle latitudes of the Strait of Georgia. That study also showed considerable stability through time for those smaller areas in terms of their biodiversity differences from adjacent areas of different habitat characteristics. As well, that paper discusses the effects of sea urchins in creating "urchin barrens" where seaweeds are essentially absent in a confined area for some period of years. The effect of sea otters (*Enhydra lutris*) on reducing sea urchin densities and enhancing abundance and diversity of seaweeds is well demonstrated (Estes & Duggins, 1995) and is considered to relate to ecological stable states in terms of community persistence, resilience and stability. Such studies are typically conscribed to encompass specific study sites, whereas the present data compilation pools data from many sites for many years. It is likely that persistence of communities fluctuates with episodes of herbivore densities, as with urchin barrens, and that the present large-scale compilation masks shorter-term, more localized fluctuations between stable states.

The assessment of the 1977, 1989 and 2001 climate regime shifts may be confounded by the effects of overfishing and changing seawater acidity, but there is no basis for integrating those different effects. Nonetheless, this data presentation demonstrates that long-term records of pH do exist and that acidification trends can be related to trends in taxonomic diversity. Many fishing effects took place before this entire survey period, however (Levy et al. 1996). There has been no decline in biodiversity correlated to the trend of ocean acidification, with the possible exception of the disappearance of bull kelp (*Nereocystis luetkeana*) from mid-latitudes of the Strait of Georgia, especially the Sunshine Coast, where decline of bull kelp occurred during the time of dropping pH (Lamb et al., 2011). It should be noted, however, that the disappearance of bull kelp in middle latitudes of the Strait of Georgia has coincided with the establishment of the highest densities of human population in that part of the Strait of Georgia and where currents are generally least powerful (Lamb et al., 2011). Finally, the Strait of Georgia has warmed by one degree Centigrade over this same period from 1970 (Beamish et al., 2010). The greatest challenge in ecosystem-based management is to determine causal relationships where multiple correlations appear to be evident. Climate regime shifts were selected as the parameter on which to base comparisons since regimes have supposedly reversed two times (possibly three) during this overall period, but the biodiversity of this region has emerged as quite stable despite all of these possible influences.

There is only a very limited signal of climate regime shifts, particularly as reflected in changes in seaweed biodiversity after the regime shift of the year 2000. Seaweed identification was not well established by the present team during the first two regimes encompassed in the period from 1967-2010, so there is no long-term baseline for assessing seaweed biodiversity through time. Diversity of various seaweeds, however, showed signs of increasing at more northern latitudes and decreasing at the more southern latitudes in Washington state (outer coast Washington and Puget Sound), but the sampling effort may well have led to spurious appearance of lower seaweed occurrence in these southern areas. The most conservative data for the original list of 328 species shows stability. Note that few seaweeds occur in that list, so the apparent increase in seaweed diversity in recent years may relate to increased expertise in field identification. It can be discerned from Figures 4 and 5 that there was not only limited focus on seaweed identification in the early years, but that it took some time for groups of more plastic morphology like sponges and bryozoans to be registered in all areas. The greatest effort took place in the Strait of Georgia, which shows the greatest stability of biodiversity, probably to a considerable degree because of the level of effort. This was also a region where there was a tendency for continuing accumulation of additional species with repeated monitoring (Figure 3).

It is important to have continuity for single surveys such as the present in order to be able to evaluate relational databases that may in the future be derived from disparate studies that do not have equivalent metadata. The PMLS database has been used to assess bull kelp abundance (Lamb et al., 2011) and it has been found that biodiversity does not change when bull kelp disappears from a location. That publication by Lamb et al. (2011) also includes a complete species list with abundance data for the greater Strait of Georgia region, and serves as an adjunct to this present publication for reference purposes.

5. Conclusion

The biodiversity of the shallow seabeds of the Strait of Georgia and surrounding regions including Johnstone Strait, Puget Sound and the west coast of Vancouver Island show considerable stability through time. The most obvious biodiversity shifts appear to be in seaweeds in the most recent climate regime, but advances in taxonomic identification suggest that more monitoring effort is required. There are stable and reliable differences, however, in the biodiversity of shallow benthos in the different regions. The Strait of Georgia has a very high biodiversity in comparison to the adjacent inland seas of Johnstone Strait (to the north) and Puget Sound (to the south), but those other two inland seas have subsets of their biodiversity at uniquely high levels of abundance. The present data compilation, together with the more exhaustive records for Strait of Georgia in Lamb et al. (2011) will provide a baseline for comparison to future trends. In all, the news is encouraging that marine biodiversity can demonstrate such resilience in the face of documented fisheries exploitation, climate change and ocean acidification. The present study also serves to demonstrate that it is important to evaluate existing data archives for continuous records of such important aspects of marine biology as species identification and abundance estimates, as well as records of physical seawater quality, such as seawater pH. Models for ecosystem-based management need to minimize the assumptions made and incorporate as much quantified information as possible in order to ensure the greatest possible precision and accuracy of predictions. This work presents descriptive data without any attempt at statistical analyses, in the hope that the obvious data trends can be incorporated into future models. The current academic trend toward producing expertise in quantitative scientific methods needs to be balanced with training of taxonomists for the front line, in order that existing marine biodiversity can be fully monitored now and into the future. The work presented here would not have been produced with any continuity in any existing academic or government biological monitoring programs in this part of the world, and perhaps not anywhere.

6. Acknowledgments

The authors (CG, DG, AL) gratefully acknowledge years of expert taxonomic verifications by Roland Anderson, Bill Austin, David Behrens, Sheila Byers, Chris Cameron, Roger Clark, Jim Cosgrove, David Denning, Daphne Fautin, Jayson Gillespie, Daniel Gotshall, Rick Harbo, Leslie Harris, Mike Hawkes, Lea-Anne Henry, Fumio Iwata, Greg Jensen, Tom Laidig, Charles Lambert, Gretchen Lambert, Philip Lambert, Robert Lea, Neil McDaniel, Valerie McDonald, Catherine Mecklenburg, Sandra Millen, Claudia Mills, Chris Pharo,

Perry Poon, Eugene Ruff, Paul Scott, Kelly Sendall, Bob Van Syoc, Gary Williams, Bruce Wing and Bruce Whitaker. Also, from among many dive partners, thanks are specially owed to Bernie Hanby, Steve Martell, Paul Malcolm, Conor McCracken and Alejandro Frid for assistance in the field. Kris Moulton created the map of study regions. Portions of the diving for this work were supported by donations from members of the Howe Sound Research and Conservation Group and the Vancouver Aquarium Board of Directors, as well as by a grant from the SeaDoc Society through the Wildlife Health Center (School of Veterinary Medicine, University of California, Davis).

7. References

Beamish RJ, Sweeting RM, Lange KL, Noakes DJ, Preikshot D & Neville CM. (2010) Early marine survival of coho salmon in the Strait of Georgia declines to very low levels. *Marine and Coastal Fisheries: Dynamics, Management, and Ecosystem Science* Vol. 2, No. 1, pp.424-439, ISSN 1942-5120

Byrne RH, Mecking S, Feely RA & Liu X. (2010) Direct observations of basin-wide acidification of the North Pacific ocean. *Geophysical Research Letters* Vol. 37, No. 2, L02601, doi: 10.1029/2009GL040999, ISSN 0094-8276

Dixson DL, Munday PL & Jones GP. (2010) Ocean acidification disrupts the innate ability of fish to detect predator olfactory cues. *Ecology Letters* Vol. 13, No. 1, pp.68-75, ISSN 1461-0248

Estes JA & Duggins DO. (1995) Sea otters and kelp forests in Alaska: generality and variation in a community ecological paradigm. *Ecological Monographs,* Vol. 65, No. 1, pp. 75-100, ISSN 0012-9615

Frid A & Marliave J. (2010) Predatory fishes affect trophic cascades and apparent competition in temperate reefs. *Biology Letters* Vol. 6, No. 4, pp. 533-536, ISSN 1744-957X

Hare SR & Mantua NJ. (2000) Empirical evidence for North Pacific regime shifts in 1977 and 1989. *Progress in Oceanography* Vol. 47, Nos. 2-4, pp. 103-145, ISSN 0097-6611

Lamb A & Edgell P. (2010) *Coastal Fishes of the Pacific Northwest* (2nd edition), Harbor Publishing, ISBN 978-1-55017-471-7, Madeira Park BC

Lamb A, Gibbs D & Gibbs C. (2011) *Strait of Georgia Biodiversity in Relation to Bull Kelp Abundance.* Pacific Fisheries Resource Conservation Council, ISBN 1-897110-70-6, Vancouver, BC

Lamb A & Hanby BP. (2005) *Marine Life of the Pacific Northwest,* Harbor Publishing, ISBN 1-55017-361-8, Madeira Park BC

Levy DA, Young LU & Dwernychuk LW (Eds.). (1996) *Strait of Georgia Fisheries Sustainability Review.* West Coast Reproduction, ISBN 0-9680214-0-9, Vancouver BC

Marliave J & Challenger W. (2009) Monitoring and evaluating rockfish conservation areas in British Columbia. *Canadian Journal of Fisheries and Aquatic Sciences,* Vol. 66, No. 6, pp. 885-1006, ISSN 1205-7533

Marliave JB, Conway KW, Gibbs DM, Lamb A & Gibbs C. (2009) Biodiversity and rockfish recruitment in sponge gardens and bioherms of southern British Columbia, Canada. *Marine Biology* Vol. 156, No. 11, pp. 2247-2254, ISSN 0025-3162

McFarlane GA, King JR & Beamish RJ. (2000) Have there been recent changes in climate? Ask the fish. *Progress in Oceanography,* Vol. 47, No. 2, pp. 147-169, ISSN 0097-6611

Tsonis AA, Swanson K & Kravtsov S. (2007) A new dynamical mechanism for major climate shifts. *Geophysical Research Letters,* Vol. 34, No. 13, L13705, doi: 10.1029/2007GL030288, ISSN 0094-8276

4

Coral Reef Biodiversity in the Face of Climatic Changes

Stéphane La Barre
Université Pierre et Marie Curie-Paris 6, UMR 7139 Végétaux marins et Biomolécules, Station Biologique F-29682, Roscoff
CNRS, UMR 7139 Végétaux marins et Biomolécules, Station Biologique F-29682, Roscoff
France

1. Introduction

Loss of marine biodiversity seems inevitable in the 21st century. In benthic marine systems, survivors will have to acclimatize to seawater constantly increasing in temperature and evolving chemically, while also needing to out-compete new opportunistic neighbors and possibly facing increased predation pressure. Last but not the least, with a metabolism already pushed to its limits, survivors will have to fight against emerging diseases.

Recent studies have shown that thermal stresses on coral reef scleractinians (Vega Thurber et al., 2009), sponges (Webster et al., 2008) and coralline algae (Webster et al., 2010a) induce changes from a balanced, functional associated microflora to a pathogen-dominated one, causing disease before the physiological tolerance limits of the hosts are reached. Changes in water chemistry (loss of bioavailable calcium carbonate due to acidification) and other stress factors (temperature, salinity, oxygen, sedimentation, etc., due to the greenhouse effect or to its climatic consequences) will affect biodiversity and community structure, and in the long term induce disaggregation of limestone scaffolds.

The first section of this chapter is devoted to a presentation of the mechanisms involved in the climate-driven loss of coral reef biodiversity predicted within the next few decades in response to increasing anthropogenic pressure. Most of the arguments developed in the following sections reflect recent work published on reef-building corals, sponges and algae and their associated macro- and micro- biota, as major reef "engineers" (Wild et al., 2011).

Biodiversity and *chemodiversity* have always been linked in the history of our planet. Both have undergone explosively creative periods, and at other times suffered dramatic losses or even extinctions followed by the emergence of better-adapted forms of life. Coral reefs have existed for many millions of years and are no exception to this. The final part of this chapter is a reflection on how a few generations of humans have been able to overexploit the planet's biodiversity for their own immediate benefit, and harm it by producing and disseminating freak molecules and genomes for which the ocean is the final depository. The threat to coral reefs comes more from effluents of highly industrialized nations than from the daily activities of low-revenue populations living on site (Donner & Potere, 2007). We now need to apply our creativity or *"intello-diversity"* to preserving existing natural equilibriums to make the planet safe for future generations.

2. Global chemistry and coral reef biostructures

2.1 Healthy coral reefs are efficient ecosystems

Coral reefs are marine structures made of biogenic calcium carbonate mineralized from dissolved bicarbonate and calcium ions. Scleractinian corals are the largest contributors to the formation of huge and solid limestone scaffolds that are home to countless communities of marine benthos and fish. Often equated to primary tropical rainforests in terms of overall biodiversity, coral reefs concentrate an estimated quarter of all identified marine species in only about 0.1% of the total oceanic surface. Restricted to pan-tropical and subtropical zones where shallow waters are warm enough throughout the year to allow photosynthesis-assisted biomineralization to occur, reef-forming scleractinian corals thrive in clear nutrient-limited waters. Optimal efficiency in nutrient cycling is achieved when overall equilibrium between all trophic levels, i.e. from higher predators to microbes, is reached. The higher the biodiversity, the more efficient the carbon and nitrogen fluxes, and the lower the probability of epidemics of bacterial pathogens or of predators upsetting the equilibrium between compartments. Azam & Malfatti (2007) stress on the importance of studying the nano and micro scale microbial structuring of marine ecosystems in order to better understand food webs and biogeoclimatic cycling. Coral reefs are of various types in relation to their location relative to land (fringing reefs, lagoon reefs, barrier reefs) or to geology (remote atoll reefs). Coral reef biodiversity hotspots occur where optimal equilibrium benefits from land-associated ecosystems and the fluxes they generate. More generally, the association of coral reefs with coastal ecosystems is both beneficial and also a source of problems when the above equilibrium is broken (Dinsdale et al., 2008). Like forests, reefs represent a substantial source of revenue to humans in goods and services (Moberg & Folke, 1999), estimated at a global 29.8 billion dollars per year in 2003 (Cesar et al., 2003), but this revenue is already being heavily compromised due to overexploitation (Cesar, 2002; Lough, 2008). The preservation of coral reef biodiversity is becoming a central ecological concern, and an economic issue to hundreds of millions of islanders and coast-dwellers living off the associated resources.

2.2 Climate change and human interference affect natural equilibria

Production of carbon dioxide and other greenhouse effect promoters, like methane and nitrogen oxide, together with water vapor, contribute to shielding nocturnal infra-red re-emissions that are necessary for the cooling of surface seawater and land. This phenomenon, known as the greenhouse effect or global warming, is generated by natural phenomena and by human activities. Natural production of greenhouse gases (by volcanism in particular) can usually be buffered by carbon and nitrogen fixing organisms, mainly bacteria, phytoplankton and forest trees, albeit at the expense of temporary and localized loss of biodiversity, as witnessed by coral skeleton biomarkers. Atmospheric enrichment of greenhouse gases by agricultural and industrial practices is accelerating at an alarming rate and its effects are now perceivable and measurable, but there is currently no means of accurately predicting how much reef biodiversity will be lost in 20, 50 or 100 years from now. In all of the proposed scenarios, estimates of recovery of existing taxa vs. replacement by more resistant or more adaptable taxa are highly speculative.

2.2.1 Greenhouse effects on coral reef biota

The shielding effect of greenhouse gases leads to three categories of damage to coral reefs:

- overheating of air and seawater that causes bleaching of corals, sponges and photosymbiotic invertebrates, and mortality in case of lasting episodes. In addition, destruction of the high altitude ozone layer by fluorinated volatiles is likely to reduce the shielding of harmful short-wave radiations that are genotoxic to exposed biota,
- gradual acidification of seawater causes gradual decalcification in biomineralizers, like scleractinian corals, coralline algae, foraminiferans, and calcifying sponges,
- increased evaporation and condensation of water are already causing more severe and more frequent hurricanes that may destroy entire portions of reefs, both mechanically by wave action and by osmotic damage to polyps by abundant rainfall,
- The combination of low tides, warmer waters and increased UV irradiance is likely to increase the strength and the persistence of the bleaching phenomenon, hence diminishing the chances of recovery of coral colonies.

2.2.2 Man has colonized most reef environments, denaturing them in the process

Human influence on coral reefs is enormous, multifaceted and expanding at a fast rate. Apart from the generation of gases producing the greenhouse effect, "contact" influences result from (i) natural landscape remodeling, (ii) industrial dumping, and (iii) household pollution. All have direct and readily observable effects on marine biota, with alien molecules killing sensitive species and microbial pathogens plaguing entire populations to extinction.

2.3 Bleaching of shallow water photosymbiotic systems

Bleaching has been defined as the loss of integrity of the photosymbiont - host relationship (hermatypic corals and some sponges) or the loss of photosynthetic pigments by the photosymbiont (zooxanthellae). Following the first large-scale bleaching events, Brown (1997) classified the causes of bleaching in corals as (i) elevated/decreased seawater temperature, (ii) solar irradiation, (iii) reduced salinity, and (iv) microbial infection. Long-term bleaching leading to mortality of entire expanses of shallow-water reefs was clearly identified as pathological in contrast to short-term episodes of occasional bleaching that allow corals to renew their resident zooxanthellae with better adapted *clades* (Suggett & Smith, 2011), some of which, e.g. *Symbiodinium* clade D may be regarded as indicators of habitat degradation more than agents of adaptation to warming (Stat & Gates, 2011). Research over the last decade has benefited from two major analytical developments: (i) functional genomics and transcriptomics that allow exploration of stress responses at cellular and whole-organism levels (Reitzel et al., 2008), and (ii) microbial metagenomics that allow culture-independent comparative analyses of bacterial and viral (Vega Thurber et al., 2008) profiles of impacted vs. healthy organisms (Vega Thurber et al., 2009), based on robust database on the former (e.g. Wegley et al., 2007). Various scenarios have been proposed to account for coral bleaching, leading to debate as to the respective importance of causative factors of mortality of corals (Bourne et al., 2009; Leggatt et al., 2007; Rosenberg et al., 2007; Rosenberg et al. 2007b), while the functional importance of bacteria in essential coral life processes is emerging from multiple examples (Mouchka et al., 2010) that also reveal their evolutionary significance (Fraune et al., 2010).

2.4 Decalcification of reef-structuring biomineralizers

Decalcification is the decrease or loss of the ability of marine invertebrates and of calcifying algae and plankton to perform accretion of calcium into adapted and functional skeletal structures, due to increasing seawater protonation. Biomineralization is a finely-tuned process requiring proper equilibrium between external (seawater) and internal (body fluid) chemistries. Individual susceptibilities to seawater acidification vary between organisms.

2.4.1 Coral reefs as contributors or sinks to atmospheric CO_2

There has been controversy as to whether the global impact on atmospheric CO_2 by coral reefs is positive or negative. Coral respiration produces carbon dioxide and so does calcification on a mole-to-mole basis (Gattuso et al., 1995), a fact that tends to place corals as net contributors to atmospheric carbon dioxide. On the other hand, communities dominated by coralline algae in temperate seas may act as carbon sinks (Bensoussan & Gattuso, 2007), an important consideration in estimating carbon dioxide fluxes in reef systems with high algal biomass. Furthermore, factors such as the influence of land runoff on inshore reefs may explain that in some areas, coral communities act as carbon dioxide sinks rather than sources (Chisholm & Barnes, 1998). The debate on carbon cycling in coastal marine environments is now taking a new dimension with consideration of functional interactions of marine microorganisms with their hosts. Useful information about calcification and ocean acidification is found in pages of the website of the European EPOCA project (http://epoca-project.eu/) and of the Woods Hole Oceanographic Institute website (http://www.whoi.edu /OCB-OA/FAQs/).

2.4.2 Key parameters in marine biomineralization

Biomineralization is widespread in marine eukaryotes and in all marine ecosystems. Carbonates, phosphates, oxides, silicates, etc. are produced by marine organisms to form tissue supporting, defensive or protective structures such as shells, spicules, skeletons, tests, or teeth in invertebrates (e.g. Bentov et al., 2009). Calcification in the form of carbonates is the most widespread form of biomineralization, and calcite (coralline red algae and foraminifers) and aragonite (corals and green calcifying algae) represent the most important contributors to the hard substrata and lagoon sand of coral reefs. Biomineralization results from a finely controlled interfacial chemistry between the organisms and seawater. Basically, the carbonate system in seawater is defined by four master variables, total or dissolved inorganic carbon (DIC), total alkalinity (TA), the partial pressure of CO_2 in water (pCO_2w), and pH (Blackford, 2010). Knowledge of any two of these along with basic physical properties is sufficient to derive the other two and the carbonate saturation state omega (Ω), bicarbonate ion concentration ($[HCO_3^-]$) and carbonate ion concentration ($[CO_3^{2-}]$). Marine calcifiers do not all respond in the same way to Ω thresholds, and acidification will most strongly affect species and their larvae that are the most susceptible to lowering carbonate saturation state.

2.4.3 Biomineralization and short and long term consequences on coral reefs

Biomineralization of $CaCO_3$ (calcification) in the oceans, undertaken by planktonic eukaryotes in the photic zone of open oceans and numerous plants, invertebrates and protists in coastal zones, is one of the major processes that control the global carbon cycle.

The atmospheric partial pressure of carbon dioxide (pCO_2) will almost certainly be double that of pre-industrial levels by the year 2100 and will be considerably higher than at any time during the past few million years. The oceans are a principal sink for anthropogenic CO_2 with an estimated 30% increase in surface water protonation since the early 1900s and with a projected drop in seawater pH of up to 0.5 units by 2100 (Hall-Spencer et al., 2008), i.e. down to 7.6 to 7.8 in 100 years time (Clark et al., 2009).

Open ocean life forms such as nanoplankton represent a considerable biomass of calcifying organisms with very short generation times, and are thus particularly susceptible to acidification and are potential bio-indicators of the progress of decalcification. Coccolith fossil records indicate that responses to past volcanism-related seawater acidification included malformations and dwarfism, and that carbonate recovery is a very long process (Erba et al., 2010).

Bioavailability of calcium and bicarbonate ions in seawater is therefore central to the question of calcification by corals, coralline algae and other biomineralizers, and to the success of their larvae or spores in metamorphosing into viable adults. A progressive lowering of the pH of seawater (acidification) occurring as a result of greenhouse effects would tend to keep divalent cations, e.g. calcium and magnesium, in soluble form in sea water, thus requiring an ever increasing metabolic effort on behalf of the organisms or of their larvae or spores to achieve an acceptable level of calcification for their needs.

Pluteus larvae of sea-urchins from various latitudes subjected to different pH levels (6.0 to ambient) suffer from reduced size and survival time, some also showing degradation of fine skeletal structures (Clark et al., 2009). Knowing the importance of urchins (e.g. *Diadema*) in regulating algal proliferation on coral reefs, imbalances between trophic fluxes are to be expected if key grazers are eradicated. An interesting question is that of the effects acidification will have on settling larvae of calcifying benthic organisms that often rely on a combination or a sequence of physical, chemical and contact cues to initiate their establishment and induce cementation. Barnacle larval proteome responds to ambient acidification by producing unique protein signatures, an expression considered of adaptive value by Wong et al (2011, in press). Little is known about the conditions in which calcification inducing genes are activated in settling coral planulae. The physiological mechanisms of the responses of adult coral colonies to the direct effects of CO_2 or to changes in HCO_3^- concentration have been investigated (Marubini et al., 2008), the net effect being a decrease in the calcification of coral skeletons. The development and survival of mollusks has been considerably affected since preindustrial CO_2 concentrations of 250 ppm, and the survival of commercially important species will be compromised in year 2100 with CO_2 levels expected to reach 750 ppm (Talmage & Glober, 2010).

Loss of benthic diversity will occur as soon as biomineralizing organisms (essentially corals and coralline algae, foraminiferans and other benthos to a lesser extent) will no longer be able to adequately turn dissolved cations into insoluble cement, meaning their larvae may no longer be able to settle, metamorphose and build a strong mineral matrix upon a steadily cemented holdfast. Solid substratum being one of most important resources in shallow marine habitats (Jackson & Buss, 1975; Connell, 1978), biodiversity on coral reefs will inevitably be affected by the non-replacement of disaggregating limestone structures, dwarfism or crooked shells. Gradual seawater acidification may lead to a new mass extinction of marine species, this time as a result of human interference rather than of cataclysmic events at planetary scale (Veron, 2008). There is an urgent need for studying the effects of climatic changes on sensitive species that may act as bioindicators (Sammarco & Strychar, 2009), and to have coordinated environmental policy-making and management (Sammarco et al., 2007).

2.5 Human exploitation of coral reefs

When the soil-fixing coastal vegetation is destroyed, the physico-chemistry of lagoon waters is upset with negative impacts on all stages of the adult and plankton instars of marine biota. When mangroves are removed, the nutrient cycling efficiency needed to maintain biodiversity along a seaward gradient is overproductive for heterotrophic microorganisms and underproductive for eukaryotic consumers. Biodiversity loss in coastal waters is the result of ever-increasing activity of sea farming, fishing and tourism in tropical countries. Poor household waste management and the use of fertilizers cause oxygen depletion of seawater with suffocation of reef invertebrates and fish. In the longer term, nutrient enrichment, by favoring algal growth, will cause a strong imbalance between trophic compartments of coral reefs, to the detriment of coral survival.

3. Macrophytic algae

3.1 Macrophytic algae are well-adapted to coral reefs

Macrophytic algae are major benthic contributors to the living intertidal and subtidal biomasses of most temperate and subpolar coastal regions. In contrast to kelp forests dominating nutrient-rich colder waters, the presence of macrophytic algae appears secondary to that of invertebrates in coral reefs that naturally develop in oligotrophic conditions. However the three macroalgal lineages (red, brown and green) have many representatives that are well adapted to the numerous communities found in all reef types and zones, as witnessed by their extremely varied growth forms. In addition, the rhodophyte (red) and the chlorophyte (green) algae have calcifying forms that actively contribute to substrate-forming calcification (by calcareous green algae) and to the cementation of loose aggregates (by encrusting coralline red algae). If stony corals are usually regarded as the framework architects, algae certainly play a role in the building and consolidation of reef assemblages, along with other biomineralizers such as sponges, tube worms and foraminifera.

3.1.1 Calcareous chlorophytes and coralline rhodophytes are essential components of reef ecosystems

Calcification, mostly in the form of calcium carbonate, has arisen independently in the three algal lineages (though to a much lesser extent in the brown algae), reflecting different growth strategies to those of non-calcifying forms. On coral reefs, heavily calcifying algae can adopt two forms: (i) geniculate, i.e. made up of calcified segments separated by flexible joints called genicula, and (ii) crustose , i.e. encrusting forms that lack genicula and grow as thin encrusting patches on hard substratum. The articulated chlorophyte *Halimeda* provides a major calcium carbonate contribution to the substratum of reef flats, i.e. the sand, and to sea grass beds which host many juvenile forms of fish and invertebrates. Crustose coralline red algae are arguably the most abundant organism (plant or animal) to occupy hard substrata within the world's marine photic zone. Unattached ball-like rhodoliths and maerl are responsible for reef formations called algal ridges in tropical wave-exposed environments (Steneck & Martone, 2007), while encrusting forms are responsible for the cementation of rubble and mineral debris into larger structures onto which other benthos can attach. Coralline algae provide food to sea urchins, parrot fish, limpets and chitons, and together with scleractinian corals and sponges they provide the framework for the development of complex communities of invertebrates. Furthermore, crustose coralline algae (CCA) are known to play a crucial role in the settlement and metamorphosis of larvae of sea urchins (Huggett et al., 2006), starfish (Johnson & Sutton, 1994), mollusks (Williams et al., 2008), corals (Negri et al., 2001; Webster et al., 2004), and possibly sponges (Carballo & Avilla, 2004). Several CCA metabolites have been identified as potential inducers, e.g. 11-deoxyfistularin-3, a bromotyrosine derivative which stimulates settlement of coral planulae in the presence of algal carotenoids (Kitamura et al., 2007), δ-aminovaleric acid and other salts which induce competent abalone larvae (Stewart et al., 2008), and dibromomethane which induces sea-urchins and the invasive slipper limpet *Crepidula fornicata* (Taris et al., 2010). Bacterial consortia which form specific biofilm-like CCA-associated assemblages, and/or their products have also been shown to act as inducers (Johnson et al., 1991) in the settlement and metamorphosis of larvae (reviewed by Hadfield, 2011). Coralline algae do not grow in sediment-rich coastal waters. Organisms with larvae that specialize in settling on CCA are naturally excluded from zones that are unsuitable for the growth of their host (Fabricius & De'ath 2001). This is a possible explanation for the lower species abundance and community composition of CCAs in coastal environments receiving sediments from rivers or from heavy rainfall washing unstabilized top soil (e.g. mining sites), in addition to insufficient bioavailability of calcium for calcification. Increasing metabolic cost associated with global climate change may offset the advantages of calcification, and recent increases in disease may indicate that oceans are becoming a more stressful environment for calcified algae. Dessication, temperature and light are capable of inducing 50% pigment loss within 24 minutes of emersion in the model species *Calliarthron tuberculosum*, and prediction-oriented models based on the combined effects of these parameters are being developed to evaluate the effects of progressive climatic changes (Martone et al., 2010). Taking *Neogoniolithon fosliei* (a primary reef-builder) as a case study, Webster et al. (2010a) demonstrated a strong correlation between elevated sea water temperature (32°C) and (i) de-pigmentation, (ii) a large shift in the structure of microbial communities associated with CCAs, (iii) development of chlorophytic endophytes, and (iv) a dramatic decrease in the ability to induce metamorphosis of coral planulae, i.e. loss of bacterial surface flora. Physiological experiments have recently shown that temperature stress induced bleaching of the coralline alga *Corallina officinalis* was the result of pigment loss following an oxidative burst, i.e. an increase in production of H_2O_2 and other reactive oxygen species and decrease in quenching capacity of the haloperoxidase system (Latham, 2008).

3.2 The phycosphere and its associated microbiome

3.2.1 Algae are phylogenetically ancient and have coevolved in association with microbes

Macrophytic algae belong to three distinct lineages that diverged early in the history of eukaryotic evolution. Red and green algae (including plants) arose from a common ancestor (Keeling, 2010) through primary endosymbiosis, i.e. engulfing of a cyanobacterium by an aerobic eukaryote, the former becoming the original plastid. Fossils as old as 1 to 1.2 billion years (mid-Proterozoic) indicate the existence of filamentous forms of both red and green algae at the root of present day plants (Javaux et al., 2004). Brown algae (stramenopiles) evolved much later as a consequence of secondary endosymbiosis (engulfing of a unicellular red alga by an aerobic eukaryote). Analysis of the genome of the filamentous brown alga *Ectocarpus siliculosus* (Cock et al, 2010) has provided evidence of the independent evolution of multicellularity within the stramenopile lineage which also includes diatoms. All three lineages have since developed into thousands of different forms with sizes ranging from microscopic (e.g. endophytes) to gigantic (e.g. brown kelp *Macrocystis*) and adaptation to most known aquatic ecosystems. Their capacity to survive eons of biotic and abiotic stresses and their adaptability to colonize extreme environments has emerged through finely tuned interactions and co-evolution with microbial organisms (La Barre & Haras, 2007).

3.2.2 Macrophytic algae control surface microbial colonization and biofilm formation

A recent paper by Goecke et al. (2010) reviews the various ways in which extant marine macroalgae and bacteria interact positively and negatively via chemical mediators. Adult macroalgae are strongly susceptible to surface colonization, as they provide potential surfaces for settlement of epibionts. As well as increasing the weight of the thallus and making it mechanically fragile, epibiosis significantly reduces the surface area available for photosynthesis. On inert substrata, macrofouling is generally facilitated by the presence of bacterial biofilms, as demonstrated by numerous studies (Fusetani, 2011). Marine algae produce photosynthates that serve as prime carbon sources for bacteria, as witnessed by the surprisingly large microbial biodiversity found on their apparently "clean" surfaces, but the surfactant nature of these carbohydrates strongly discourages adhesion. Biofilm formation is also discouraged by molecules that (1) interfere with bacterial quorum-sensing activation, e.g. hypobromous acid in the brown kelp *Laminaria digitata* (Borchardt et al., 2001) or halogenated furanones produced by the rhodophyte *Delisea pulchra*, and (2) act as decoy analogs of bacterial acyl-homoserine lactones which induce quorum sensing (Manefield et al., 2002). Surface compounds produced by the phaeophyte *Fucus vesiculosus* are capable of modulating both epibiotic development and bacterial biofilm production (Lachnit et al., 2010). Flow-cell experiments (this author, unpubl. results) have shown that the thickness and 3-D architecture of monospecific biofilms of model bacteria isolated from inert substrata and from *Laminaria digitata* blades are differentially affected by exposure to exudates and surface compounds extracted from this kelp. Antifouling programs are now integrating biofilm denaturation studies into their strategies, an ecologically sound alternative to the use of toxic ingredients.

3.2.3 Microbial pathogenic infection results in coordinated immune responses from infected algae

Bacteria interact with algae mostly in a non-invasive manner until older tissues can no longer resist saprophytic degradation by bacteria or bacterial consortia that have the appropriate lytic enzymes. However, fortuitous bacterial intrusion following mechanical damage or exposure to specific signal substances (elicitors) may trigger an oxidative burst followed by intracellular responses comparable to classical inflammation, as part of an innate immune mechanism (Weinberger, 2007). In brown algae, microbial infection is strongly inhibited by the emission of halogenated compounds that may have bactericidal or bacteriostatic activities in the immediate hydrosphere of the thallus, due to an efficient apoplastic enzymatic machinery (Butler & Carter-Franklin, 2004) that can be activated in a non-systemic fashion. Indeed, red and brown algae are reported to use both animal-like (eicosanoid) and higher plant-like (octadecanoid) oxylipins in the regulation of defense metabolism for protection against pathogens and grazers or in response to elicitors of defense responses (Cosse et al., 2009). Tetracyclic brominated diterpenes active against multi-resistant strains of the nosocomial bacteria *Staphylococcus aureus* have recently been isolated from the red alga *Sphaerococcus* (Smyrniotopoulos et al 2010), suggesting efficient antibiotic responses against potentially infective strains. Kornprobst (2010a) provides a comprehensive review of bioactive metabolites of various chemical classes from red, brown and green algae. Compounds with have antibacterial activities are now inspiring ecological alternatives to classical antifouling paints and coatings based on genotoxic heavy metal formulations.

Certain parasitic endophytes (fungi and minute filamentous algae) are capable of penetrating the outer cuticle of red and brown algae and causing modifications of the growth of the host (e.g. Gauna et al., 2009). Some bacteria have the enzymatic machinery capable of breaking down cell wall polymers into absorbable energy sources, as well as being resistant to the defense compounds elicited following their intrusion. Their biotechnological potential for the production of low MW bioactive components from bulk tissue is being investigated (e.g. Kim et al., 2009; Colin et al., 2006). Bacterial epidemics are occasionally reported, leading to the complete destruction of local natural populations (e.g. *Laminaria hyperborea* in western Brittany), or of seaweed farms of brown (e.g. *Laminaria japonica*, Wang et al., 2008), or of tropical red algae (e.g. carrageenophytes, Largo et al., 1999).

3.2.4 Bacteria are also essential to the development, fitness and defense of algae

The life cycle of Ulvales (Chlorophyta) is clearly dependent upon the presence or the association of adequate bacterial strains. Production by a bacterium of thallusin, a N-containing carboxylated diterpene, induces the morphogenesis of foliose thalli of the green ulvophyte *Monostroma oxyspermum* from filamentous cell aggregates (Nishizawa et al., 2007), and more generally from *Enteromorpha*-like filamentous forms into *Ulva*-like foliose forms (Matsuo et al., 2005). Swimming *Ulva* tetrazoospores are strongly influenced by specific quorum-sensing signals liberated from bacterial biofilms (review by Joint et al., 2007) resulting in settlement and metamorphosis in their vicinity. *Ulva australis* favors colonization of its surfaces by *Roseobacter gallaeciensis* and by *Pseudoalteromonas tunicata* which coexist as segregated microcolonies, and modulate the settlement of other competing strains essentially by (i) production of the antibacterial protein AlpP by *P. tunicata* and (ii) biofilm invasion and dispersion by *R. gallaeciensis* (Rao et al., 2006). Molecular investigations on *Ulva australis* showed that alphaproteobacteria (dominated by the Roseobacter clade) and *Bacteroidetes* are a regular and stable component of this alga's functional microbiome (Tujula et al., 2010). The *Roseobacter* clade is now regarded as an essential functional component of the microbiome of phytoplankton (both intra- and extracellularly), assisting in both primary and secondary roles (Geng & Belas, 2010), as well as seemingly in a number of macrophyte associations of symbiotic nature. Brown algae shelter bacterial communities that may vary according to their location on the thallus, partly due to differences in surface chemistries (q.v. review by Goecke et al., 2010), and to the mode of colonization, i.e. by planktonic bacteria on distal surfaces and by surface contamination of the holdfast by epibionts and complex microbial biofilms. Among the numerous and unidentified non-cultivatable strains, it can be assumed that some of have neutral or beneficial interactions with their host.

3.3 Algae and their chemical language

3.3.1 Thalli control their immediate hydrosphere…and the atmosphere

Lam & Harder (2007) have described how macroalgae can selectively control their immediate surroundings by diffusing waterborne chemicals that prevent settlement by potentially fouling microbiota and epibiota. The emission of cocktails of volatile halogenated C1-C3 compounds has been extensively reported in red (e.g. Kladi et al., 2004) and in brown (e.g. La Barre et al., 2010) algae, with impacts beyond the proximal underwater effects on atmospheric chemistry. Green algae are responsible for massive releases of dimethylsulfoproprionate (DMSP) and diffusion of their breakdown products (acrylate and dimethylsulfide) into the atmosphere, which may create conditions favorable for the dispersal of their gametes (Welsh et al., 1999). This is reminiscent of cloud-forming emissions of iodinated compounds above kelp beds (Ball et al., 2010). Thus benthic algae and phytoplankton actively participate in the cycling of iodine, bromine, chlorine and sulfur at the water-air interface, while responding to requirements at both cellular and population levels. Biomass breakdown following massive blooms of DMSP algae and Prymnesophytes may cause severe anoxia and hydrogen sulphide intoxications to local fauna and seashell farming, pointing out the necessity for proper control of nutrient enrichment of coastal waters.

3.3.2 Algal metabolites alter the fitness and growth of their benthic neighbors

In addition to volatile compounds, macroalgae owe their competitiveness to the production of whole arrays of metabolites that are bioactive (i) by contact interactions with adjacent alien tissues, using lipid-soluble compounds (Rasher & Hay, 2010a), (ii) by diffusing water-borne chemicals, or (iii) by altering the functional microbial flora of their invertebrate neighbors thereby encouraging the development of pathogenic strains. Damage to the scleractinian cover on the outer reef slopes in various localities of the Central and South Western Pacific by blooms of *Asparagopsis taxiformis* may be the result of one or more of these modes of action. Toxic volatile or diffusible halogenated compounds, like haloforms, methanes, ketones, acetates and acrylates, were described for *A. taxiformis* and its sibling species *A. armata* (Mc Connell & Fenical, 1977; Woolard et al., 1979; Kladi et al., 2004). Thick mats can overrun live scleractinian colonies and diffuse a range of volatile halocarbons that are considered toxic in addition to having various antimicrobial activities (Genovese et al., 2009). Hypoxia and tissue disruption of polyps at coral-algal tuft and coral-macroalgae interfaces led Barott et al. (2009) to the conclusion that erect (i.e. non crustose) algae were a constant cause of stress to adjacent coral colonies in contact or close vicinity in pairwise experimental associations.

3.4 Algae as a crucial ecological link between corals and microbes

3.4.1 Control of biomass of algae by grazers is essential to coral reef diversity

Algae are regarded as superior space competitors on hard substrata. However, predation is an important pressure on non-calcifying algae (Hay, 1997). The epilithic algal community is grazed by herbivorous fish, echinoderms (sea urchins), mollusks, crustaceans and worms, themselves serving as food to carnivores in a bottom-up succession of predators. Thus, small turf-like species, sporelings of macrophytes, and the unicellular forms that are associated with surface slime on sand and rubble, e.g. protein-generating cyanobacteria, represent an essential primary trophic component of the reef, generating more than half of the edible biomass of the whole food chain (Hay, 1997). A number of larger algae produce chemicals (terpenes, polyphenols and halogenated compounds) that have an inhibitory effect on grazers (see recent review by Paul et al., 2011). Fish being the largest consumers of algae and being selective in their food source, the biodiversity of seaweeds on the reef is a reflection of both the diversity of grazing modes (Burkepile & Hay, 2010) and of the chemodiversity of the defence compounds they produce. To complete the picture, aggressive fish such as the common Pomacentrid damselfish tend to fend intruders off their territory, including foraging herbivores, thus promoting spatial and taxonomic diversity in the distribution of reef algae (Brawley & Adey, 1977).

3.4.2 Overfishing is linked to algal and microbial development and to coral decline

The Philippines and Indonesia include the western half of the Coral Triangle of marine biodiversity, which extends eastwards to New Guinea and the Solomon islands. One quarter or more of the human populations of these islands live in coastal areas and derive their revenues from coral reef production, on a non-sustainable basis.

Overfishing, besides causing mechanical damage to coral biota (wading, use of explosives and use of dragnets) modulates the predation pressure on benthic algae in many reefs worldwide, with consequences for corals and their associated biodiversity. Near extinction of reef sharks and other carnivorous fish and of grazing herbivores has a dramatically positive incidence on algal biomass and average size in overexploited reefs worldwide (Hay, 1997 , Sotka & Hay, 2009). The resulting increase in the production of photosynthates feeds a bacterial population that is potentially pathogenic to corals. Disrupting the coral–microbe relationship by organic carbon loading (dissolved organic carbon (DOC), i.e. mainly carbohydrates) can directly cause coral mortality by over-stimulating growth of coral mucus-associated microbes (Kuntz et al., 2005). An analysis of the gradual decline of Jamaican coral reefs by the marine microbiologist Forest Rohwer (2010) led him to define this self-feeding loop as the DDAM model (DOC>disease>algae>microbes), which can only be broken by top down herbivory that reduces the algal biomass to levels compatible with the development of a functional scleractinian microbiome. Future management policies should ensure that key algae consumers be identified and protected in reef areas susceptible to recurrent blooming of coral damaging species (Rasher & Hay, 2010b).

3.4.3 Farming of macroalgae in tropical regions also suffers from climatic changes

Farming of macrophytic algae for the food and the pharmacological industries provides an alternative activity to fishing and tourism for entire communities in tropical islands (Indonesia, Philippines, Madagascar and the Caribbean islands). However, algal monoculture can be risky. The carrageenophytes *Eucheuma* and *Kappaphycus*, that have the capacity to adjust to hyper and hypo salinity changes encountered in shallow tropical waters (q.v. Teo et al., 2009), are prone to bacterial plague disease ("ice-ice"). Such epidemics have caused occasional eradication of entire populations (Largo et al., 1999), with total loss of the livelihood of farmers. Seasonal infestations by filamentous endophytes of Malaysian *Kappaphycus/Eucheuma* farms which are attributed to seasonal changes (Vairappan, 2006) may be an aggravate bacterial infections.

3.5 Exotic algae may adapt to new environments as climate changes

3.5.1 Invasive species

The large-scale introduction of non-indigenous species and homogenization of the world's biota has long been considered among the greatest threats to species diversity (Carlton & Geller, 1993). Before global warming became the central issue in the coral reef biodiversity literature, the necessity for proper management of reef resources had already been highlighted (Maragos et al., 1996), including the issue of species of commercial interest imported into developing countries due to the low cost of local labor. Introductions may result in competition with native biota, eventually affecting human livelihood. Whether deliberately imported (introduced) or accidentally established (invasive), alien species, like endangered species, are now identified on periodically upgraded lists. For example, the International Coral Reef Initiative (ICRI) devotes a whole section of its website (http://www.icriforum.org/) to this topic. Alien seaweed, fish, mollusks and even corals are listed as invasive species competing with native species for resources and being potential causes of new diseases of resident scleractinian corals.

The primary cause of invasion of new territories by alien algae is not always easy to determine, and a combination of factors usually contribute to the success of their expansion in new territory. Regular de-ballasting of huge amounts of seawater by container ships is often held responsible for the spread of exotic algae and microalgae along major maritime trade routes, eventually leading to endemic settlement points in regions where acclimation is possible (Carlton, 1996). On the other hand, massive invasions along the Mediterranean shores of the toxic Australian green alga *Caulerpa taxifolia* (and later of *Caulerpa racemosa*) are examples of accidental contaminations starting with a few individuals (Klein & Verlaque, 2008). The displacement by these two *Caulerpa* species of resident halophyte meadows (e.g. *Posidonia*) which are home to juveniles of a number of important demersal fish is considered a threat to local biodiversity. Rhodophytes (red algae) include a number of "cosmopolitan" species that have become established in tropical environments. An example is the territorial expansion of *Asparagopsis taxiformis* in the South Pacific that is currently considered a serious threat to corals in New Caledonia and in French Polynesia, like its sibling species *A. armata* in other parts of the world.

3.5.2 Cataclysmic events may favor supremacy of algal communities

In tropical seas, hurricanes are becoming more frequent and more severe with global warming, resulting in near-total destruction of the coral cover in the most exposed localities, with negative impacts on larval recruitment (Crabbe et al., 2008) and hence on biodiversity. Initially, only the most resilient and hardy scleractinian species are likely to reestablish and create the replacement coral cover. Also, colonies killed by sudden and massive rainfall in shallow lagoons, or broken into fragments by brutal wave action, will provide clean substrate for fast-growing opportunists, initially mostly algal species, usually via some mediation by microbial biofilms. Once established, and provided adequate nutrients are available, algae tend to replace lost coral cover. A "side effect" of cyclone damage is the temperature-associated blooming (Chateau-Degat et al., 2005) of the neurotoxic dinoflagellate *Gambierdiscus toxicus* growing on algal turf that colonizes newly available substrate. With the observed trend of the increase in frequency and in severity of tropical cyclones associated with global warming, important edible fish species may increasingly become unfit for human consumption in impacted areas, a tragic situation in remote islands with low revenue populations with little choice for food substitutes.

3.6 Nutrient enrichment favors algal development to the detriment of corals

Non-calcifying green algae and in particular the ulvaceans, a ubiquitous group of non-calcifying chlorophytic algae, represent the most visible sign of pollution due to organic enrichment. They are distributed worldwide, from cold temperate waters to warm tropical latitudes, and from fresh or brackish to saline coastal environments, as they are both opportunistic and resilient to environmental stresses. Their taxonomy may prove difficult (Loughlane et al., 2008) with occasional reassessments made necessary owing to their phenotypic flexibility (e.g. Kang & Lee, 2002). In temperate regions, green algae tend to be seasonal, their development dependant not only on the amount of sunlight, but also on nutrient availability and on microbial consortia associated to particular developmental stages. Spectacular biomass explosions attributed to nutrient enrichment generated by agricultural and farming practices and long daylight exposure occur during spring and summer. Though considered efficient at absorbing excess nutrient enrichment, if not collected they can themselves become a source of chemical pollution (hydrogen sulfide in particular) through massive bacterial degradation of the decaying biomass. In tropical environments, developing economies associated with fishing and farming of marine resources generate massive effluxes of untreated urban sewerage leading to localized blooming of chlorophytic algae. In the vicinity of reefs, bacterial enrichment by resident algae may be detrimental to coral polyps by promoting the growth of pathogenic strains at the expense of the regular associated bacterial microflora (Rohwer, 2010).

4. Sponges

4.1 Sponges and reef structuration

In association with spongin (proteinaceous) fibres, marine sponges typically biomineralize non-aragonitic calcium and magnesium carbonates (subphylum Calcispongia) or silica (subphylum Silicispongia). One of the most ancient eumetazoan lineages, sponges have adopted different modes of biomineralization and reef-forming capacities through geological time, as seawater chemistry and the bioavailability of dissolved salt species has changed periodically due to tectonic events (Stanley & Hardie, 1998). Most reef-forming sponges have disappeared since the Phanerozoic, and present-day sponges are mostly siliceous Demosponges which are ubiquitous worldwide in their distribution (see Hooper & Van Soest, 2002).

4.2 Why are sponges so unique?

The position of sponges at the root of metazoan evolution is still a hotly debated topic (see eg. Maldonaldo, 2004). Choanoflagellates, with which sponge choanocytes and metazoan lineages share genomic similarities (King et al., 2008), have emerged as model organisms for studies of early metazoan evolution. For example, the choanoflagellate genome carries the markers of three types of molecules that cells use to achieve phospho-tyrosine signaling proteins, involved in important processes (cell-cell communication, immune system responses, hormonal stimulation, etc.) in metazoans. These molecules are tyrosine kinases (TyrK), protein tyrosine phosphatases (PTP) and Src Homolgy 2 (SH2) molecules that operate as a tandem system to achieve signal recognition (Manning et al., 2008). Recent studies have also demonstrated the role of associated bacteria in the colony-forming behavior of otherwise free-living choanoflagellates (e.g. a glycosphingolipid produced by the bacterium *Algoriphagus* that affects the choanoflagellate *Salpingoeca rosetta* (Alegado et al., 2010). Other studies mention the role bacteria may have had in the transfer of genes between unicellular eukaryotes. Nedelcu et al. (2008) provide an example of bacterially mediated lateral transfer of four stress-related genes of algal origin to a choanoflagellate host, supposedly providing the recipient cell the capacity to adapt to stress under environmental changes.

Going one evolutionary step further, Srivstava et al. (2010) have shown that the genome of the demosponge *Amphimedon queenslandica* contains the set of genes that correlates with critical aspects of more evolved metazoans (body plan, cell cycle control and growth, development somatic and germ-cell differentiation, cell adhesion innate immunity and allorecognition). Adaptive responses to environmental changes are known to be more rapid in microbial symbionts than in sponge cells, and bacteria may act as environmental sentinels to marine holobionts as they are particularly sensitive to environmental stressors such as heavy metal pollution (Webster et al., 2001), elevated seawater temperature, sedimentation and disease (Webster et al., 2011), and tolerance to eutrophication (Turque et al., 2010). Garderes et al. (2011) showed that specific bacterial quorum sensing signals can be recognized by sponge cells, triggering phagocytosis, a response proposed as part of a symbiont population-regulating mechanism.

4.3 The sponge holobiont

4.3.1 Sponges shelter complex consortia of microorganisms, akin to some biofilms

Modern reef-dwelling sponges are studied as important contributors to the constructive and bioerosive dynamics of limestone scaffolds in shallow tropical waters, but the main interest in them lies in their ability to harbor highly complex communities of micro- and macro-organisms. The overall sponge-associated microbial component is extremely diverse, bacteria alone representing up to half of the sponge biomass and occurring everywhere within their host, often as consortia. Only a very small percentage of these bacteria have been cultured in isolation. The term bacteriosponge (Reiswig, 1981) reflects the uniqueness of this prokaryotic-eukaryotic functional consortium, the holobiome being the sum total of all associated microbial components.

The demosponge microbiome occasionally includes zooxanthellae (Weisz et al., 2010) that share surface cortical layers with cyanobacteria (Li, 2009), while the eubacteria and archeae typically dominate the inner regions. Cyanobacteria can be extremely abundant within their host and provide them with competitive advantages (cyanobacteriosponges *Aphanocapsa raspaigellae* in *Tersiops hoshinota*, König et al., 2006). Some unicellular strains are widely distributed across sponge hosts and reef localities, e.g. *Synechococcus spongiarum* which is regarded as a generalist (Erwin & Thacker, 2008). Sponge-specific clades of filamentous cyanobacteria, on the other hand, suggest unique coevolutionary histories (Thacker & Starnes, 2003; Hill et al., 2006) which provide greater benefit to their hosts (Thacker, 2005). Symbiotic cyanobacteria, situated both intercellularly and intracellularly, have been reported in a large variety of marine sponges (Thajuddin et al., 2005; Usher, 2008).

4.3.2 Sponges with different life strategies host different bacterial populations

Demosponge microbiologists have contrasted species that form dense and phylogenetically complex microbial communities (HMA or high microbial abundance) against those that contain only few and less diverse microbes of essentially non-specific types (LMA or low microbial abundance), representing two different basic life strategies (Weisz et al., 2008). HMA sponges typically have a reduced aquiferous system with lower pumping rates, while LMA sponges have highly porous tissues with high pumping rates that enable rapid uptake of small particulate organic matter, yet the nitrification/denitrification rates remain comparable (Schlappy et al., 2010). Calcareous sponges have been comparatively less studied as regards their bacterial flora. Quévrain et al. (2009) and Roué et al. (2010) found that two North Atlantic calcareous species had a stable bacterial population throughout the year.

4.3.3 Specificity, host selectivity and vertical transmission to offspring

Bacterial associations may range from non-specific commensalism to species-specific symbiosis, with some intracellular or even endonuclear examples which are more common in protistan hosts than in metazoans in which they are regarded as signs of a very ancient association (see Hoyos, 2010). In sponges (Vacelet, 1970; Friedrich et al., 1999), such intranuclear examples are found in the genus *Aplysina*. More commonly, the existence of core-consortia of bacteria which are found in distant conspecific sponge populations and all year around indicates a high degree of functional specificity and species selectivity. Vertical transmission of "essential" bacteria from parent to offspring via the eggs or the larvae in, respectively, oviparous and ovoviviparous species have been investigated by Webster et al. (2010b) in major poriferan clades: Demospongiae, Homoscleromorpha, Calcarea and Hexactinellida (Ereskovsky, 2011), also including cyanobacteria (Usher et al., 2001).

4.3.4 Other components of the sponge holobiont

Archaea are found in a wide variety of sponges, and their populations have been characterized in several taxa, e.g. *Axinella*, either as clade-specific mutualists or not, with some sponges lacking them altogether (Holmes & Blanch, 2007). This understudied component is primarily involved in ammonia oxidation, but the co-production of bioactive metabolites is not excluded.

In addition, sponges are host to microscopic algae, viruses, yeast and fungi, which add to the biodiversity and, at least for the fungi, to the chemodiversity of the holobiome (e.g. König et al., 2006). Sponges are host to a strain of *Aspergillus sydowii*, a fungus which has been identified as a causative agent of epidemics that affect gorgonian corals. The authors of the study (Ein-Gil et al., 2009) postulate that sponges may act as reservoirs of potential marine pathogens, in the same way as the bacterial populations associated with turf and macroalgae are considered as potential sources of infection to scleractinian corals which are host to specific microbiomes (Barott et al., 2009).

Polychaetes, prosobranchs, ophiuroids and crustaceans are also commonly found in association with sponges, and the disappearance of the latter would automatically reduce the specialized associated invertebrate biodiversity.

4.4 The sponge holobiont – Functional aspects

The sponge microbiome is a prime example of natural chemodiversity, occupying an extensive range of functions in primary (C and N cycling - see e.g Li, 2009) and secondary metabolism (thousands of original molecules of various classes and modes of action) which has been studied worldwide by natural product chemists.

4.4.1 Functional "primary" aspects of symbiosis

The biochemical nature of sponge-microbe symbioses is largely unknown, and ideally requires investigations at single strain microniche consortium and whole microbiome levels (Kamke et al., 2010) for a given host to obtain a better insight into the functional dynamics of the holobiont. Several basic (primary metabolism) sponge-associated microbial processes have been described, including:

i. nitrification, the oxidation of ammonia (NH_3) to nitrite (NO_2-) and subsequently to nitrate (NO_3-) for energy purposes, both steps being carried out by two different bacterial groups: AOB or ammonia oxidizing bacteria and archaea, and NOB or nitrite oxidizing bacteria (Bayer et al., 2008),
ii. nitrogen fixation which appears important in nutrient poor reef environments (Mohamed et al., 2008). Recently Hoffman et al. (2009) described the complex nitrogen cycle of the sponge *Geodia barretti* and speculated upon the possible role of marine sponges as nitrogen sinks,
iii. photosynthesis with cyanobacteria (Arillo et al., 1993; Li., 2009) and even zooxanthellae that appear to resist elevated temperatures (but not light) better than coral

zooxanthellae (Schönberg et al., 2008) and when present actively contribute to carbon supply to the host (Weisz et al., 2010),

iv. methane oxidation (Vacelet et al., 1996) and sulfate reduction (Hoffmann et al., 2005).

4.4.2 Communication "secondary" aspects of symbiosis

So-called secondary metabolites may be produced and released according to age and reproductive status, but also as a response to abiotic and biotic stresses, against predation and microbial infection, for resource defense against competitors, etc. The communication chemistry of soft-bodied sessile invertebrates can indeed be regarded as a vocabulary of molecular words, and its transcriptomics can be equated to proper syntax, in order to respond as exactly and as economically as possible to an identifiable conflict. According to their mode of action, these molecules can be volatile (short MW halocarbons), surface or tissue bound, or mucus-borne. The participation of the microbiome to the biosynthesis of sponge metabolites has been established in a number of cases in natural conditions, but cultivated individual strains or functional consortia of interest may not express the desired phenotype (production of a specific molecule), or may not be cultivatable outside their host. Aside from possible applications in human welfare, bacteria provide prime examples of prokaryote-metazoan coevolution which have endured an estimated 600 million years of existence and survived major biogeoclimatic changes.

The sponge mesohyl provides a broad variety of ecological microniches that host bacterial consortia (Thiel et al., 2007), with varying degrees of dependence to the host, while cortical regions tend to be dominated by cyanobacteria (Li, 2009). Both components are known to be involved in the synthesis of bioactive secondary metabolites which are naturally produced (or prompted) in response to microbial pathogens (antibiotics), space competitors (allelopathic substances), epibionts (antifouling molecules), and predators (antifeedants, intoxicants, serine protease inhibitors). This chemical arsenal, together with the presence of structural (sharp mineral sclerites, or tough spongy texture) and visual (warning or cryptic colors and patterns) defenses, are necessary to the survival of these non-motile and often exposed invertebrates. Most classes of so-called secondary metabolites are represented, making sponges a treasure trove for the discovery of new drugs. Here we are concerned with the global chemodiversity aspect and the reader is prompted to consult updated reviews (for example in the dedicated issues of *Natural Products Reports* since 1977), or texts such as Kornprobst (2010b) that provide a user-friendly review of sponge-derived metabolites, addressing their possible biosynthetic origins and their potential applications. Metagenomic screening to identify key polyketide synthase (PKS) and non-ribosomal peptide synthetase (NRPS) genes, and new cloning and biosynthetic expression strategies may provide a sustainable method to obtain new pharmaceuticals derived from the uncultured bacterial symbionts, e.g. with cyanobacteria (Li, 2009). Novel culturing techniques (e.g. Selvin et al., 2009), including co-culturing of microorganisms modulating the proliferation (through quorum sensing) and the expression of strains of interest are now actively investigated (Dusane et al., 2011).

5. Reef-building corals

5.1 Scleractinian corals as reef architects

Scleractinian corals present a rich fossil record dating back approximately 240 million years, i.e. they appeared much more recently than macroalgal lineages (1 billion years) and also than sponges (600 million years).

About half of the 1,300 scleractinian coral species are reef-building, largely colonial, zooxanthellate (hermatypic) and occurring in the clear, shallow and oligotrophic waters of the tropics. The other half of the order is largely solitary and azooxanthellate, occurring in all regions of the oceans, including the greatest depths (Budd et al., 2010). The reef-building corals function as primary ecosystem engineers, constructing the framework that serves as a habitat for all other coral reef-associated organisms (Wild et al., 2011). Scleractinians are actively engaged in the production and transformation of mineral and organic materials. Coral limestone structures are broken down by bioeroding organisms and abiotic processes into sand, itself acting as a natural biocatalytic filter for the cycling of organic matter by resident heterotrophic microorganisms (Wild et al., 2005).

5.2 The coral holobiont: An example of 3-way functional integration

The term holobiont, sometimes collectively defined as biota engaged in a host-symbiont partnership (Santiago-Vázquez et al., 2006), has been borrowed by coral researchers to conveniently include the coral host and all of its associated interactive life forms (reviewed in Rohwer, 2010). This includes tissue-associated symbiotic photosynthetic microalgae, surface and mucus-dwelling bacteria and archaea (Siboni et al., 2008) and recently investigated viruses (Vega Thurber, 2008). Endolithic algae and fungi that bore into the mineral skeleton (Wegley et al., 2004) may be included in this definition as permanent associates (as in Bourne et al., 2009). Very recently a true symbiotic relationship has been described between the acroporid *Acropora muricata* and the hydrozoan *Zanclea margaritae* (Pantos & Bythell, 2010). "Mobile" associates (crustaceans, mollusks, polychaetes etc.) provided they have developed a specific niche or trophic preference with the host and developed a cryptic or aposematic appearance as a result, should logically be included in this definition in that the host's disappearance would probably signify a loss of this additional biodiversity.

The coral holobiont is now regarded as a functional unit by scientists who are interested in physiology, pathology, biochemistry and environmental issues of reef-buiding anthozoans. A review of the functional microbiota associated with corals is provided by Laming (2010). Fig. 1 illustrates the typical coral holobiont with its associated macro- and micro-organisms.

5.3 The coral holobiont on autotrophic, heterotrophic and mixotrophic feeding modes

Mixotrophic organisms can functionally combine different modes of nutrition: (i) by using photosynthesis for inorganic carbon fixation; and (ii) by taking up organic sources. Coral polyps are diploblastic and hence have no mesoderm-derived digestive tract or specialized respiratory organ. Nutrient and energy requirements of the whole colony must depend (i) on direct diffusion of dissolved gases and simple organic molecules across the polyp body wall, (ii) on "assisted metabolism" with pseudo-respiratory and pseudo-digestive functions in association, respectively, with symbiotic macroalgae sequestered in the endoderm, and mucus-bound bacterial consortia, (iii) during the night, on heterotrophy, i.e. ingestion of bacteria and planktonic particles that are digested in the coelomic cavity, the organic products being further broken down by other bacteria. During the day, polyps function in autotrophic mode, i.e. relying on oxygen production and carbon photosynthates provided by the symbiotic zooxanthellae. Other commensal members of the "extended" holobiont, i.e. crustaceans, echinoderms, polychaete worms, mollusks, etc., live mostly off the food particles trapped in the coral mucus, or as parasites.

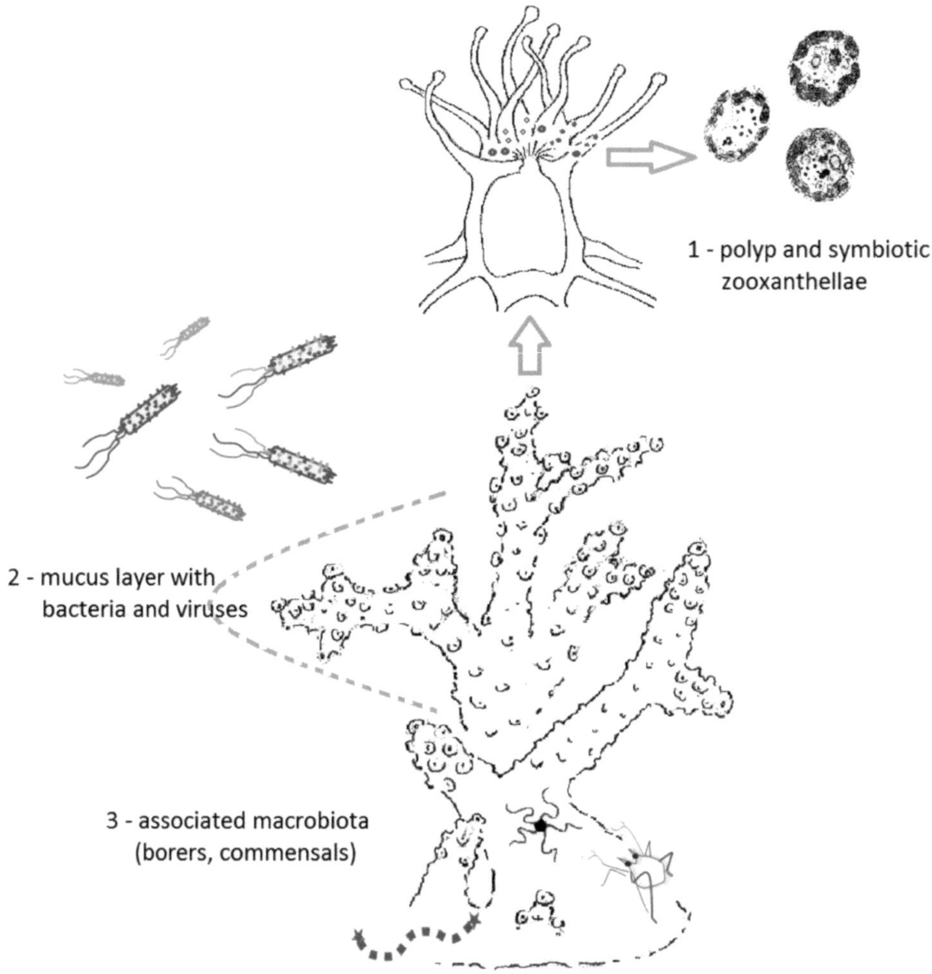

Fig. 1. Schematic representation of a typical coral holobiont.
1- The polyp endoderm harbors symbiotic dinoflagellates (various clades of the genus *Symbiodinium*), i.e. the zooxanthellae; 2- polyps produce copious amounts of mucus, a multifunctional interface with the outside world, in which bacteria and viruses are trapped together with food particles; 3- associated organisms that have a permanent (non-mobile forms) or an obligate (mobile forms) relationship with the host.

5.3.1 Direct assimilation of DOM (dissolved organic matter)

This mode of nutrient acquisition is commonplace in benthic diploblastic organisms. It has also been experimentally demonstrated in sponges (de Goeij et al., 2008), while plankton capture is the most commonly described method for filter feeders in open waters, while POM (particulate organic matter) feeding is generally regarded as assimilation of POM bacterial degradation products. To the holobiont, bacteria are as important in nitrogen fixation as zooxanthellae are in carbon fixation. Corals are opportunists in their nitrogen acquisition modes, as its sources are capable of fluctuating rapidly in an oligotrophic environment. Polyps are capable of absorbing dissolved amino acids non-discriminately through a photosynthesis-enhanced (i.e. daylight) mechanism, thus providing about one quarter of the polyp's nitrogen requirements (Grover et al., 2008), while dissolved ammonia, nitrate and urea make up for nearly all of the remaining needs. In another study (Fitzgerald & Szmant., 1997) consider that corals can "synthesize" 16 of the 20 amino acids (including the eight regarded as essential), meaning that bacteria metabolize them from degrading mucus and that the coral readily absorbs the amino acids. Agostini et al. (2009) have shown that a resident bacteria produce B12 vitamin that is taken up by the coral in the coelenteron. The zooxanthellae are fertilized with the inorganic nitrogenous and phosphate excreta of the polyp as well as obtaining a sheltered niche in exchange for their essential role in functional maintenance of the holobiont. This feature of acclimatization (nitrogen capture) is considered important in a fluctuating environment (Gates & Edmunds, 1999). Finally, scleractinians are potentially capable of modulating their epibiotic bacterial community by producing antibacterial compounds as part of a constitutive defense mechanism, as demonstrated in alcyonarians by Harder et al., (2003), and a feature shared by other marine diploblastic invertebrates investigated for their pharmacological potential. The detection of QS signals in coral-associated microbiota (Goldberg et al., 2011) might explain the role of the bacterial component in the holobiont dynamics. Bacterial colonization may indeed be modulated at strain-level by molecules produced by the coral host and interfering with intraspecific (Gram - AHL and Gram + AIP) or interspecific (AI-2) QS inducers.

5.3.2 Respiration and translocation of photosynthates

The carbon requirements for respiration in branching and foliose corals are essentially provided by carbon dioxide fixation by the algal symbiont (Muscatine & Cernichiari, 1969) during daytime photosynthesis. Under normal conditions subsurface branched corals acquire up to 95% of their energy requirements by coupling their primary metabolism with that of their symbiotic zooxanthellae, which perform carbon fixation and provide other shared benefits, e.g. mucus production.

5.3.3 Food capture and extracoelenteric digestion

Food capture of nano- and macro-plankton, especially bacteria and zooplankton, affords the heterotrophic alternative to most corals, i.e. it helps with acquisition of other elements that are essential to the functioning of the holobiont, including that of the bacterial compartment. In addition, heterotrophy enables calcification by providing the elements of the organic matrix upon which carbonate biomineralization is initiated. Furthermore, feeding rate is enhanced in bleached corals. Switching between autotrophic and heterotrophic modes of feeding has been investigated by Houlbrèque & Ferrier-Pagès (2009).
Evidence of coelenteric digestion in hermatypic corals has been long been documented (Boschma, 1925) and extracoelenteric digestion has been suggested in some corals to be a strategy to defend space from competing neighbours (Lang, 1990).

5.3.4 Mucus and mixotrophic metabolism

Coral mucus represents a major metabolic investment on behalf of the coral (up to half of its carbon budget), and it actively contributes to the carbon budget of coral reef systems by bringing photosynthates and trapped organic matter to lagoon sediment where heterotrophic nutrient recycling operates efficiently (Wild et al., 2004). To the colony, mucus is used primarily for protection against silting (layers are sloughed off regularly) and against dessication at low tide, as well as acting as natural sunscreens since they concentrate UV-absorbing mycosporin-like amino acids (MAAs) produced by the zooxanthellae (Dunlap & Schick, 1998). Mucus is also used to entrap food particles (plankton and POM) which are then drawn into the endodermal sac by the tentacles. Mucus is also an obvious source of carbon to many bacteria that degrade it and use it as a dispersion vector, somewhat like bacterial biofilm streamers (Allers et al., 2008). Solitary corals (e.g. *Fungia*) may even asphyxiate and poison their competitors by emitting thick mucus layers that cover neighboring colonies. A comprehensive review on the nature and multiple roles of coral mucus has been proposed by Brown & Bythell (2005), who hypothesize that the intraspecific composition of coral mucus is not homogenous and that different types of mucus may be produced according to needs and environmental stressors. Chemically, coral mucus is a polysaccharide-protein-lipid complex secreted by corals at their surface (ectodermal goblet cells), and there are as many different types of mucus as there are types of corals. Carbohydrate photosynthates are produced at a fast rate as the major component of the mucus which is essential to the whole colony as an exchange medium. In addition to serving as food to polyps in oligotrophic waters, mucus-associated bacterial consortia appear to be an essential component for the recycling of waste and for the biosynthesis of molecules that are useful as osmocompatibles, ROS (reactive oxygen species) detoxicants, anti UV sunscreens, biogeochemical sulphur cycling, etc.

5.4 Biomineralization in reef-building corals

In corals and in mollusks, extracellular mineral structures develop upon a matrix derived from secretory products of calicoblastic tissues. This phosphate-rich 3-D matrix is typically made up of acidic proteins, carbohydrates and glycoproteins, and is genetically programmed to perform essential regulating and/or organizing functions that will result in the formation of composite biominerals (Weiner & Dove, 2003). The latter are formed as outer membranes release the cationic elements that are taken up intracellularly and delivered / diffused to the matrix where they may be orderly arranged with organic components (e.g. in oyster nacre). There has been some debate about the role of photosynthesis in the calcification process of corals, but light is certainly not a prerequisite for the calcification process. There is substantial evidence for a diurnal cycle in the coral calcification and skeleton-building process during which the types of crystals deposited, their distribution about the skeletal surface and the overall rate of calcification changes (Cohen & McConnaugh, 2003). Central to these changes is the bimodal functioning of the light-sensitive Ca^{++} - ATPase pump that has the dual role of transporting cations into the calcifying space while removing protons. On site mineralization is probably initiated at sulfated sites of exopolysaccharides, with crystal growth possibly "guided" by organic processes at the interface with the calicoblastic epithelium (Cohen & Mc Connaugh, 2003). An up-to-date treatment on the subject of cellular mechanisms of scleractinian coral calcification is provided by Reyes-Bermudez (2009).

5.5 Scleractinian responses to environmental changes

In contrast to the aforementioned "regular" and transient biotic and abiotic events that may contribute to the classical Darwinian view of evolution and biodiversity, the multifaceted consequences of the rise in atmospheric carbon dioxide (aggravated by industrialization) on coral reefs are now estimated to be the primary cause of massive losses of reef-associated biodiversity within the next decades.

Drastic changes in the "regular" holobiont microbiodiversity, gradual loss of physiological functions and cell/tissue damage beyond recovery, are generally regarded as an aggravation of naturally existing challenges of the holobiont system, due to an unfavorable biogeochemical evolution of the environment.

Research now focuses on how ocean warming and acidification can lead to rapid coral mortality and affect biodiversity quantitatively and qualitatively, especially in human-impacted areas. Coral bleaching, i.e. loss of the symbiotic microalgal partner, skeletal demineralization, and also effects on larval and juvenile stages and disease susceptibility to pathogens, are the central themes of most of the reef-building coral literature published since 1998. Parallel to stress and mortality studies, investigators report examples of species, communities, habitats or geographical areas that actually resist or acclimatize to e.g. repeated bleaching, whether generated by biotic or abiotic factors.

5.5.1 Scleractinians and thermal stress

Bleaching (discoloration) is due to the rupture of the resident *Symbiodinium* algae with their coral host, due to mortality, loss of pigments or expulsion (Brown, 1997). Decrease in uptake and endodermal sequestration of photosymbionts may be a response to mutual perception of toxic levels of reactive oxygen species (ROS) during enhanced light-induced photosynthesis (Vidal-Dupiol et al., 2009). Coral holobiont resilience to thermal stress was extensively discussed by Coles & Brown (2003) in the light of twenty years of worldwide short- and long-term observations on various types of corals. Reef organisms are stenotolerant to heat, with tolerance limits usually not exceeding 31-32°C for branched acroporids before significant mortality occurs (Berkelmans & Willis, 1999), a range similar to that of sponge tolerance limits in the same locality of the Australian Great Barrier Reef (Webster et al., 2008). Other localities have selected for coral communities that are adapted to their climatic regime. In the Red Sea and the Arabian Gulf, the ambient summer seawater temperature reaches 34-36°C. In general, increases of 1-3°C above mean long-term annual maximum temperature have consistently induced coral bleaching.

Recovery is very much species-specific in shallow water scleractinians. Skeletal drillings of hundreds of years old massive forms (e.g. *Porites*) provide records of severe climatic events which colonies have survived (Cleveland et al., 2004). Physiological and cytological adjustments concern the protection of the light-harvesting complex from solar photosynthesis-active radiation (PAR) and ultraviolet radiation (UVA and UVB), and against overheating. Excessive PAR over-stimulates photosynthesis with production of cytotoxic reactive oxygen species (ROS), and UV radiations are genotoxic, while infrared radiations (IR) cause thermal stress.

Mechanisms allowing acclimatization to elevated seawater temperatures include activation of the xanthophyll cycle as a photoprotective defence (dissipation of PAR as heat), induction of heat-shock proteins and antioxidant enzymes, and the production of mycosporin-like amino acid sunscreens. The photoprotective role of GFP-like fluorescent pigments in scleractinian polyps has been established both in the visible light range (dissipation of PAR through fluorescence and light diffraction), and in the UV range (transformation of UVA radiation into longer-range non actinic fluorescence), in order to protect the algal symbiont chlorophyll and peridin from photo-oxidative damage (Salih et al., 2000). The peroxidase-mediated production of reactive oxygenated species (ROS) is an initial response against abiotic stress and microbial pathogens, also observed in corals. Palmer et al. (2010) have shown that components of the immune system of corals respond to oxidative stress by activating phenoloxidases (PO) and laccase activities on-site, with the production of melanin (in addition to GFP-like pigments). Akin to algal phlorotannins, melanin and other putative antioxidants play a role in wound healing and oxidative stress mitigation, as well as pathogen encapsulation (Mydlarz & Palmer, 2011). Indeed, PO activity levels, melanin abundance and distribution and GFP-like pigments may serve as useful descriptors of susceptibility to environmental changes in shallow water scleractinian corals (Palmer et al. 2010). Overall, long-lived massive coral growth forms appear better equipped to withstand transient stress than short-lived and more delicate branched forms (Palmer et al., 2011).

A shift in the composition of resident *Symbiodinium microadriaticum* zooxanthellae from stenotolerant to thermoteolerant clades may occur following short bleaching episodes. Indeed, this "upgrading" is the basis of the adaptive bleaching hypothesis or ABH (Kinzie et al., 2001), according to which the bleached basibiont can select a clade better suited to a momentary environmental disturbance.

Another strategy against bleaching is to increase heterotrophy as a source of food energy to compensate for loss of light-derived photosynthates, in species equipped with functional food capture devices.

5.5.2 Scleractinians corals and seawater acidification

The increase in pCO_2 in the water column leads to an increase in total dissolved CO_2 at the expense of carbonate ions [CO_3^{2-}]. The reduction in the latter, at constant seawater calcium concentration [Ca^{2+}], consequently results in the decrease of the saturation state of aragonite (Ω_{arag}), the polymorph of $CaCO_3$ produced by coral calcification (Wild et al., 2011). Aragonite super-saturation is necessary for efficient accretion in most scleractinians, and a lowered external Ω_{arag} will impede the calcification rate in internal fluids and crystal formation at the calicoblastic/seawater interface, translating into weaker skeletons and general deterioration in reef habitat construction. By the year 2100, Ω_{arag} is expected to decrease from an average value of 4 (necessary for calcium accretion) to less than 3 (insufficient for most shallow-water branching corals). Together with natural bioerosion, decalcification may breakdown entire reef constructions in time. Along with decalcification, the lowering of seawater pH causes a shift in the bacterial microbiodiversity, in favor of the most resilient strains, some of which are typically associated with stressed and diseased corals, e.g. Vibrionaceae and Alteromonadaceae (Meron et al., 2011).

5.5.3 Scleractinian corals, anoxia and eutrophication

Shallow water colonial corals thrive in pristine oligotrophic waters, with a finely tuned carbon and nitrogen metabolism as long as the associated microalgal and bacterial flora remains unaffected.

Nutrient enrichment generally occurs close to human settlements and is therefore likely to first affect fringing reefs. Experiments using media artificially enriched with inorganic ions (PO_4^- and NO_3^-) have shown that the release of dissolved organic carbon and of dissolved organic nitrogen actually decreased relative to tissue surface area in the test species *Montipora digitata* thus affecting not only the efficiency of the photobiosis, but also the mucus-feeding bacteria (Tanaka et al., 2010). However, the counterbalancing combination of eutrophication and seawater warming generates stress to the coral colony, setting the stage for a change in the metagenomic profile of bacteria from functional and diverse to pathogenic and restricted. Webster & Hill (2007) have reviewed ways into which the microbial life of the Great Barrier Reef ecosystems may be impacted by global warming. Along these lines, Ainsworth et al. (2009) have predicted a "microbial perspective" for coral reefs in the next decades, and proposed the addition of a metagenomic component in building predictive models.

6. Conclusion: Towards an integrative approach

Since mild episodes of adverse climatic and biotic conditions regularly occur, it can be assumed that coral reefs are quite resilient with respect to biodiversity losses. Today, human activities are having a steadily increasing impact on climate (global warming) and on biodiversity (overexploitation and toxic waste) and although accurate predictions cannot be made, it is assumed that massive losses of marine biodiversity associated with coral reefs will occur within the next decades. The capacity of organisms to acclimatize and that of communities to maintain their biodiversities is largely unknown, and estimates are based on isolated laboratory experiments in which tolerance limits to individual stressors are measured. The spread of microbial diseases from the shallow (0-30 m) to the mesophotic (30 - 200 m) zones has been discussed by Olson & Kellogg (2010) who prompt investigations on the algal, sponge and coral holobionts from the hitherto neglected deeper reef communities. At the holobiont level, it is important to detect early signs of stress (e.g. coral bleaching due to loss of zooxanthellae) and to identify the critical stages beyond which permanent damage results (e.g. tissue necrosis, various *band diseases* due to microbial pathogens). Intraspecific variations in the profiles of bacterial communities associated with corals may show up between distant localities (Littman et al., 2009), suggesting that environmental conditions may significantly influence the dominance profiles of a few strains, possibly reflecting on the health status of the host species. At the community and ecosystem levels, it is important to evaluate the extent of damage and whether recolonization is possible if the primary cause can be removed (e.g. by classifying an impacted area as a protected zone). Along the same lines, it should now be possible to devise molecular-based methods that can accurately estimate a shift in bacterial biodiversity from standard to pathogenic in a given sample, to estimate stress levels, or to compare metabolomic signatures between unaffected and impacted specimens. Comparable methods already exist in medical microbiology, and in cancerology in which specific markers are sought in addition to metabolomic profiling of biological fluids.

6.1 The common fate of biodiversity and chemodiversity

Biodiversity emerged as the first unicellular forms of life appeared, presumably as chemoautotrophs in hydrogeothermal environments. The original biogeochemistry was quite different to what it is today, and different mineral-water-air equilibria must have been attained several times since the emergence of life, e.g. basic Precambrian vs. acidic modern ocean (David & Alm, 2011). As biodiversity and chemodiversity increased, their influence on global biogeochemistry gradually increased (Falkowski et al., 2008). Cyanobacteria are thought to have enriched the atmosphere in oxygen to the point that through various endosymbiosis scenarios a "compromise" was reached that allowed living entities to use oxygen generated by photosynthesis for respiration while controlling its toxicity. Taming the generation of highly reactive radical oxygen species by using oxidative stresses as useful warning signals of an environmental aggression is an example of this adaptation to aerobic life. Competition between an ever-increasing number of species and ever reducing availability of resources fueled a new form of chemodiversity, that of secondary or communication metabolites. Thus, the biochemistry involved in physiological adaptation and inter-organism communication links biodiversity and chemodiversity. This simplified picture does not account for the contribution of microorganisms to both primary and secondary metabolisms of the host they live with. In aquatic environments, attempts to integrate all biotic communities participating in the same functional dynamics gave rise to the hologenome theory and to the holobiont concept. When addressing environmental issues, and especially stress responses, this new biome approach can benefit from the fast evolving molecular and genomic fields of metabolic fingerprinting and microbial metagenomics, in the quest for suitable analytical tools. Recent culture-independent approaches have revealed an unsuspected quantity of microbial diversity, not only bacterial (Sogin et al., 2006), but also eukaryotic (Dawson & Hagen 2009), the "rare biosphere", representing a potential reservoir of genes and genomes that marine life may rely on to adapt to new environmental conditions.

6.2 Intellectual diversity should work for, not against natural diversity

Man has been impacting his environment with extensive use of metals since the onset of the 19th century industrial era. Artificial chemodiversity followed, due to the development of the petroleum chemistry, synthetic polymers and new pharmaceuticals. Unfortunately, the industry has shortsightedly released into the environment thousands of novel organic molecules without proper evaluation of allergenic and carcinogenic effects on human health, both by direct exposure or through accumulation in terrestrial and aquatic food chains. Degradability and speciation of industrial and household waste have been neglected, with destruction of entire fragile ecosystems e.g. rivers and ponds. Biodiversity has also been "directed" with an increasing number of selected crop plants and farm animals and now with genetically modified versions which can arguably hybridize and displace natural conspecifics. The coming of age of bioresponsibility vs. exploration may soon become a necessity, i.e. what can be collectively termed as *intello-diversity* must be applied to the uptake of acceptable practices for the maintenance of a balanced natural equilibrium while making resources lasting and renewable.

6.3 How can scientists mediate environmental conflicts and help political action

At present, research concentrates on a handful of "convenient" model reef species and on how their metabolism reacts to experimental disturbances. This approach is of limited use outside the confines of the laboratory. To be applicable to real-life situations, i.e. mining pollution in an otherwise undisturbed environment, multi-scale studies must be undertaken. In the laboratory, stress studies must not only concentrate on the transcriptomics of stress-associated genes of the host organism, but should also include assessment of accompanying changes of the microbiodiversity which is likely to reveal a profile change from functional to pathogenic, and from autotrophic to heterotrophic (Littman et al., 2011). New molecular tools could be developed to address all types of environmental issues, ranging from specific toxicity to geoclimatic changes, including urban or industrial pollution. For example, spotting out the "sentinel" bacteria strain in an ecologically sensitive holobiont would provide a quantifiable biomarker of an environmental disturbance. The more specific its reactivity to a specific stress, the better. The logic behind this approach is similar to extracting the chemical profile of a marine invertebrate from an impacted environment and comparing it with the profile database of a conspecific from a pristine environment, and hopefully identifying a chemical marker that would act as a signature (as in clinical analyses of body fluids and the search of specific markers of cancer). The association of biological and chemical profiling and signature marking into a standardized procedure would undoubtedly help industry deciders to evaluate the importance of environmental impacts and to take appropriate action. At an intermediate scale, in-situ monitoring of key environmental parameters in mesocosms allows modeling that is useful for decision-making in creating natural reserves. Respecting protected reef areas should be a mature political decision based on sound scientific investigations. For example, active nickel mining and ore processing in the immediate vicinity of coral reefs protected under the UNESCO's World Heritage list in New Caledonia generates difficult choices between short-term economic profit and long-term and durable exploitation of reef resources.

Finally, accessible information through the media and adapted and participative school programs should be encouraged in order to modify the management of reef resources to take into account sustainable and durable applications.

6.4 Biodiversity beyond ecosystem resilience – What's next?

Joseph Connell's summary of his landmark Science paper of 1978 emphasizes the necessity to maintain some kind of equilibrium between human activities and natural environments, a condition without which very large clusters of biodiversity may be irremediably lost:

"*The commonly observed high diversity of trees in tropical rain forests and corals on tropical reefs is a nonequilibrium state which, if not disturbed further, will progress toward a low-diversity equilibrium community. This may not happen if gradual changes in climate favor different species. If equilibrium is reached, a lesser degree of diversity may be sustained by niche diversification or by a compensatory mortality that favors inferior competitors. However, tropical forests and reefs are subject to severe disturbances often enough that equilibrium may never be attained.*" Joseph Connell (1978)

Life would then have to re-create itself in unforeseeable evolutionary scenarios, counting on the inventiveness and adaptability of the microbial world where man's inventiveness will have failed.

7. Acknowledgment

I am very grateful to Dr. Ian Probert for his critical reading of the manuscript and for checking the English, and to Prof. Jean-Michel Kornprobst for useful suggestions.

8. References

Ainsworth, T.D.; Vega Thurber, R. & Gates, R.D. (2009). The future of coral reefs: a microbial perspective. *Trends in Ecology and Evolution* Vol.25 No.4, pp. 233-240, ISSN: 0169-5347

Agostini, S.; Suzuki, Y.; Casareto, B.E.; Nakano, Y.; Hidaka, M. & Badrun, N. (2009). Coral symbiotic complex: Hypothesis through vitamin B12 for a new evaluation. *Galaxea, Journal of Coral Reef Studies* Vol. 11, pp. 1-11, ISSN: 1883-0838

Alegado, R.A., Kontnik, R.; Fairclough, S.R.; Zuzow, R.; Clardy, J. & King, N. (2010). A bacterial sphingolipid induces choanoflagellate colony development. 3rd Conference on Beneficial Microbes. Oct 25-29, 2010 Miami Florida. *American Society for Microbiology.*

Allers, E.; Niesner, C.; Wild, C. & Pernthaller, J. (2008). Microbes enriched in seawater after addition of coral mucus. *Appl. And Environ. Microbiol.*, Vol.74, No10, pp. 3274-3278, ISSN: 0099-2240

Arillo, A.; Bavestrello, G.; Burlando, B. & Sarà, M. (1993). Metabolic integration between symbiotic cyanobacteria and sponges: a possible mechanism. *Marine Biology*, Vol.117, pp. 159-162, ISSN: 0025-3162

Azam, F. & Malfatti, F. (2007). Microbial structuring of marine ecosystems. *Nature Reviews Microbiology*, Vol.5, pp. 782-791, ISSN: 1740-1526

Ball, S.; Hollingsworth, A..M.; Humble, J.; Leblanc C.; Potin P. & Mc Figgans G. (2010). Spectroscopic studies of molecular iodine emitted into the gas phase by seaweed. *Atmos. Chem. Phys.*,Vol.10, pp. 6237–6254, ISSN: 1680-7316

Barott, K.; Smith, J.; Dinsdale E.; Hatay, M.; Sandin S. & Rohwer F. (2009). Hyperspectral and physiological analyses of coral-algal interactions. *PLoS ONE*, Vol.4, No4, e8043, pp. 1-9, ISSN: 1932-6203

Bayer, K.; Schmitt, S. & Hentschel, U. (2008). Physiology, phylogeny and *in situ* evidence for bacterial and archaeal nitrifiers in the marine sponge *Aplysina aerophoba. Environ. Microbiol.*, Vol.10, No.11, pp. 2942-2921, ISSN: 14622912

Bensoussan, N. & Gattuso, J-P. (2007). Community primary production and calcification in a NW Mediterranean ecosystem dominated by calcareous microalgae. *Mar. Ecol. Prog. Ser.* Vol.334, pp.37-45, ISSN: 0171-8630

Bentov, S.; Brownlee, C. & Erez, J. (2009). The role of seawater endocytosis in the biomineralization process in calcareous foraminifera. *PNAS,* Vol.106, No51, pp.21500-21504, *ISSN: 0027-8424*

Berkelmans ; R. & Willis, B.L. (1999). Seasonal and local spatial patterns in the upper thermal limits of corals on the inshore Central Great Barrier Reef. *Coral Reefs*, Vol.18, pp. 219-228, ISSN: 0722-4028

Blackford, J.C. (2010). Predicting the impacts of ocean acidification: challenges from an ecosystem perspective. *Journal of Marine Systems*,Vol. 81, pp.12–18, ISSN: 0924-7963

Borchardt, S. A.; Allain, E. J.; Michels J. J.; Stearns, G. W.; Kelly, R. F. & McCoy, W. F. (2001) Reaction of acylated homoserine lactone bacterial signaling molecules with oxidized halogen antimicrobials. *Appl. Environ. Microbiol.* Vol.67, pp. 3174-3179, ISSN: 0099-2240

Boschma, H. (1925). On the feeding reactions and digestion in the coral polyp Astrangia danae, with notes on its symbiosis with zooxanthellae. *Biol. Bull.* Vol.49, pp. 407-439, ISSN: 0006-3185

Bourne, D.G.; Garren, M.; Work, T.M.; Rosenberg, E.; Smith, G.W. & Harvell, C.D. (2009). Microbial disease and the coral holobiont. *Trends in Microbiology*, Vol.17 ,No.12, pp. 554-562, ISSN: 0966-842X

Brawley, S.H. & Adey, W.H. (1977). Territorial behavior of threespot damselfish (*Eupomacentrus planifrons*) increases reef algal biomass and productivity. *Env. Biol. Fish.*,Vol2, No1, pp. 45-51, ISSN: 0378-1909

Brown , B.E. (1997). Coral bleaching: causes and consequences. *Coral Reefs*, Vol.16, Suppl. S129-S138, ISSN: 0722-4028

Brown , B.E. & Bithell, J.C. (2005). Perspectives on mucus secretion in coral reefs. *Mar. Ecol. Prog. Ser.*, Vol. 296, pp. 291, ISSN: 0171-8630

Budd, A.F.; Romano, S. L., Smith, N.D. & Barbeitos, M.S. (2010). Rethinking the phylogeny of scleractinian corals: a review of morphological and molecular data. *Integrative and Comparative Biology*, volume 50, number 3, pp. 411–427, ISSN: 1540-7063

Burkepile, D.E. & Hay, M.E. (2010). Impact of herbivore identity on algal succession and coral growth on a Caribbean reef. *PLoS ONE*, Vol.5, No1, e8963, ISSN: 1932-6203

Butler, A. & Carter-Franklin, J.N. (2004). The role of vanadium bromoperoxidase in the biosynthesis of halogenated marine natural products. *Nat. Prod. Rep.*, Vol.21, pp. 181-188, ISSN: 0265-0568

Carballo, J.L. & Avila, E. (2004). Population dynamics of a mutualistic interaction between the sponge *Haliclona caerulea* and the red alga *Jania adherens. Mar. Ecol. Prog. Ser.*,Vol.279, pp. 93-104, ISSN: 0171-8630

Carlton, J.T. & Geller, J.B. (1993). Ecological roulette: The global transport of non-indigenous marine organisms. *Science*, Vol.261, pp. 78–82, ISSN: 0036-8075

Carlton, J.T. (1996). Marine bioinvasions: the alteration of marine ecosystems by nonindigenous species. *Oceanography*, Vol.9, No1, pp. 36-43, ISSN: 1042-8275

Cesar, H.S.J. (2002). Coral reefs, their functions, threats and economic value. http://hdl.handle.net/1834/557

Cesar, H.J.S.; Burke, L. & Pet-Soede, L. (2003). The Economics of Worldwide Coral Reef Degradation. Cesar Environmental Economics Consulting, Arnhem, and WWF-Netherlands, Zeist, The Netherlands. 23pp. Online at: http://assets.panda. org/downloads/cesardegradationreport100203.pdf

Chateau-Degat, M-L; Chinain, M.; Cerf, N., Gingras, S., Hubert . & Dewailly, E. (2005). Seawater temperature, *Gambierdiscus spp.* variability and incidence of ciguatera poisoning in French Polynesia. *Harmful Algae*, Vol.4, pp. 1053-1062, ISSN: 1568-9883

Chisholm, J.R.M. & Barnes, D.J. (1998). Anomalies in coral reef community metabolism and their potential importance in the reef CO2 source-sink debate. *PNAS*, Vol.95, pp.6566-6569, *ISSN: 0027-8424*

Clark, D; Lamare, M. & Barker, M. (2009). Response of sea urchin pluteus larvae (Echinodermata: Echinoidea) to reduced seawater pH: a comparison among a tropical, temperate, and a polar species. *Marine Biology*, Vol.156, pp.1125–1137, ISSN: 0025-3162

Cleveland, R.O.; Cohen, A.; Roy, R.A.; Singh, H. & Szabo, T.L. (2004). Imaging Coral II: using ultrasound to image coral skeleton. *Subsurface Sensing Technologies and Applications*,Vol. 5, No. 1, pp. 43-61, ISSN: 1566-0184

Cock, J.M.; Sterck, L.; Rouzé, P.; Scornet, D.; Allen, A. et al. (2010). The *Ectocarpus* genome and the independent evolution of multicellularity in brown algae. *Nature*, Vol.465, pp. 617-621, ISSN: 0028-0836

Cohen, A.L.. & McConnaughey, T.A. (2003). Geochemical perspectives on coral mineralization. In *Biomineralization*. Eds. P.M. Dove, S. Weiner and J.J. De Yoreo. Mineralogical Society of America, Washington, D.C., v. 54, p. 152-186

Coles, S.L. & Brown, B. (2003). Coral bleaching - capacity for acclimatization and adaptation. *Advances in Marine Biology*, Vol.46, pp. 143-183, ISSN: 0065-2881

Colin, S.; Deniaud, E.; Jam,M.; Descamps, V.; Chevolot, Y.; Kervarec, N.; Yvin, J.-C.; Barbeyron, T.; Michel, G. & Kloareg, B. (2006). Cloning and biochemical characterization of the fucanase FcnA: definition of a novel glycoside hydrolase family specific for sulfated fucans. *Glycobiology*, Vol.16, pp. 1021–1032, ISSN: 0959-6658

Connell, J.H. (1978). Diversity in tropical rain forests and coral reef high diversity of trees and corals is maintained only in a nonequilibrium state. *Science*, New Series, Vol.199, No 4335, pp. 1302-1310, ISSN: 0036-8075

Cosse, A.; Potin, P. & Leblanc, C. (2009). Patterns of gene expression induced by oligoguluronates reveal conserved and environment-specific molecular defense responses in the brown alga *Laminaria digitata. New Phytologist*, Vol. 182, pp. 239–250, ISSN: 0028-646X

Crabbe, M.J.C.; Martinez, E.; Garcia, C; Chub, J.; Castro, L. & Guy, J. (2008). Growth modelling indicates hurricanes and severe storms are linked to low coral recruitment in the Caribbean. *Mar. Environ. Res.*,Vol.65, pp. 364-368, ISSN: 0141-1136

David, L.A. & Alm, E.J. (2011). Rapid evolutionary innovation during an Archean genetic expansion. *Nature*, Vol.469, pp. 93-96, ISSN: 0028-0836

Dawson, S. & Hagen, K.D. (2009). Mapping the protistan 'rare biosphere'. *Journal of Biology*, Vol.8, http://jbiol.com/content/8/12/105

de Goeij, J.M.; Moodley, L.; Houtekamer, M.; Carballeira, N.M. & van Duyl, F.C. (2008). Tracing 13C-enriched dissolved and particulate organic carbon in the bacteria containing coral reef sponge *Halisarca caerulea*: evidence for DOM feeding. *Limnol. Oceanogr.*, Vol.53, No4, pp. 1376-1386, ISSN: 0024-3590

Dinsdale, E.A.; Pantos, O.; Smriga, S.; Edwards, R.A.; Angly, F. et al. (2008). Microbial ecology of four coral atolls in the Northern Line Islands. PLoS ONE Vol.3, No2: e1584. doi:10.1371/journal.pone.0001584

Donner, S.D. & Potere, D. (2007). The inequity of the global threat to coral reefs. *BioScience*, Vol.57, No.3, pp. 314-215, doi:10.1641/B570302, ISSN 0006-3568

Dusane, D.H.; Matkar, P.; Venugopalan, V.P.; Kumar, A.R. & Zinjarde, S.S. (2011). Cross-species induction of antimicrobial compounds, biosurfactants and quorum-sensing inhibitors in tropical marine epibiotic bacteria by pathogens and biofouling microorganisms. *Curr Microbiol.*,Vol.62, pp. 974–980, ISSN: 1369-5274

Dunlap, W. & Schick, J.M. (1998). Ultraviolet radiation-absorbing mycosporine-like amino acids in coral reef organisms: a biochemical and environmental perspective *J. Phycol.* 34, 418–430, ISSN: 0022-3646

EPOCA European project (http://epoca-project.eu/)

Ein-Gill, N.; Ilan, M.; Carmeli, S.; Smith, G.W.; Pawlik, J.R. & Yarder, O. (2009). Presence of *Aspergillus sydowii*, a pathogen of gorgonian sea fans in the marine sponge *Spongia obscura. The ISME Journal,* Vol.3, pp. 752–755, ISSN: 1751-7362

Erba, E.; Bottini, C.; Weissert, H.J. & Keller, C.S. (2010). Calcareous nannoplankton response to surface-water acidification around oceanic anoxic event 1a. *Science,* Vol.329, p.428, ISSN: 0036-8075

Ereskovsky, A.V. (2011). Vertical transmission of bacteria in sponges. 1st Int. Symp. on Sponge Microbiology, March 21-22, Würzburg, Germany.

Erwin, P.M., & Thacker, R.W. (2008). Cryptic diversity of the symbiotic cyanobacterium *Synechococcus spongiarum* among sponge hosts. *Molecular Ecology*, Vol.27, pp. 2937-2947, ISSN: 0962-1083

Fabricius, K. & De'ath, G. (2001). Environmental factors associated with the spatial distribution of crustose coralline algae on the Great Barrier Reef. *Coral Reefs*,Vol.19, pp. 303-309, ISSN: 0722-4028

Falkowski, P. (2008). The microbial engines that drive earth's biogeochemical cycles. *Science*, Vol.320, pp. 1034-1039, ISSN: 0036-8075

Fitzgeral, L.M. & Szmant, A.M. (1997). Biosynthesis of ' essential ' amino acids by scleractinian corals. *Biochem. J.* Vol.322, pp. 213-221, ISSN: 0264-6021

Fraune, S. & Bosch, T.C.G. (2010). Why bacteria matter in animal development and evolution. *Bioessays*, Vol.32, pp. 571–580. ISSN: 0265-9247

Friedrich, A.B.; Merkert, H.; Fendert, T.; Hacker, J.; Proksch, P. & Hentschel, U. (1999). Microbial diversity in the marine sponge *Aplysina cavernicola* (formerly *Verongia cavernicola*) analyzed by fluorescence *in situ* hybridization (FISH). *Marine Biology*, Vol.134, pp. 461-470, ISSN: 0025-3162

Fusetani, N. (2011) Antifouling marine natural products. *Nat. Prod. Rep.*, vol.28, pp. 400-410, ISSN: 0265-0568

Garderes, J.; Henry, J.; Wiens, M. & Le Pennec, G. (2011). Bacteria-sponge molecular cross-talk: role of quorum-sensing molecules. 1st Int. Symp. on Sponge Microbiology, March 21-22, Würzburg, Germany.

Gates, R.D. & Edmunds, P.J. (1999). The physiological mechanisms of acclimatization in tropical reef corals. *Amer. Zool.*, Vol. 39, pp.30-43, ISSN: 0003-1569

Gattuso, J.P.; Pichon, M. & Franckignoulle, M. (1995). Biological control of air-sea CO_2 fluxes: effect of photosynthetic and calcifying marine organisms and ecosystems. *Mar. Ecol. Prog. Ser.*, Vol.129, pp. 307-312, *ISSN:* 0171-8630

Gauna, C.; Parodi, E.R. & Caceres, E.J. (2009). Epi-endophytic symbiosis between *Laminariocolax aecidioides* (Ectocarpales, Phaeophyceae) and *Undaria pinnatifida* (Laminariales, Phaeophyceae) growing on Argentinian coasts. *J. Appl. Phycol.* , Vol.21, pp. 11–18, ISSN: 0921-8971

Geng, H. & Belas, R. (2010). Molecular mechanisms underlying *Roseobacter*-phytoplankton symbioses *Current Opinions in Biotechnology*, Vol. 21, pp. 332-338, ISSN: 0958-1669

Genovese, G.; Tedone, L., Hamann, M.T. & Morabito, M. (2009). The Mediterranean red alga *Asparagopsis*: a source of compounds against *Leishmania Marine Drugs*,Vol.7, pp. 361-366, ISSN: 1660-3397

Goecke, F.; Labes A.; Wiese, J. & Imhoff J.F. (2010) Chemical interactions between marine macroalgae and bacteria – a review. *Mar. Ecol. Prog. Ser.*, Vol.409, pp. 267–300, ISSN*:* 0171-8630

Goldberg, K; Eltzov, E.; Shnit-Orland, N.; Marks, R.S. & Kushmaro, A. (2011). Characterization of quorum sensing signals in coral-associated bacteria. *Microbiol. Ecol.* Vol.61, pp. 783-792, ISSN: 0095-3628

Grover, R.; Maguer, J-F.; Allemand, D. & Ferrier-Pagès, C. (2008). Uptake of dissolved free amino acids by the scleractinian coral *Stylophora pistillata. The Journal of Experimental Biology*, Vol.211,pp. 860-865, ISSN: 0022-0949

Hadfield, M.G. (2011). Biofilms and marine invertebrate larvae: what bacteria produce that larvae use as settlement sites. *Annu. Rev. Mar. Sci.,* Vol.3, pp. 453–70, ISSN: 1941-1405

Harder , T.; Lau, S.C.K.; Dobretsov, S.; Fang, T.K. & Qian, P-Y. (2003). A distinctive epibiotic bacterial community on the soft coral *Dendronephthya sp.* and antibacterial activity of coral tissue extracts suggest a chemical mechanism against bacterial epibiosis. *FEMS Microbiology Ecology* , Vol.43, pp. 337-347, ISSN: 0168-6496

Hall-Spencer, J.M.; Rodolfo-Metalpa, R.; Martin, S.; Ransome, E. Fine, M. et al. (2008). Volcanic carbon dioxide vents show ecosystem effects of ocean acidification. *Nature,* Vol 454, pp.96-99, *ISSN*: 0028-0836

Hay, M.E. (1997). The ecology and evolution of seaweed-herbivore interactions on coral reefs. *Coral Reefs*, Vol. 16, Suppl.: S67 - S76, ISSN: 0722-4028

Hill, M.; Hill, A.; Lopez, N. & Harriott, O. (2006) Sponge-specific bacterial symbionts in the Caribbean sponge, *Chondrilla nucula* (*Demospongiae, Chondrosida*). *Marine Biology,* Vol.148, pp. 1221–1230, ISSN: 0025-3162

Hoffmann, F.; Larsen, O.; Thiel, V.; Rapp, H.T.; Pape, T.; Michaelis, W. & Reitner, J. (2005). An anaerobic world in sponges. *Geomicrobiology Journal,* Vol.22, pp. 1–10, ISSN: 0149-0451

Hoffman, F.; Radax, R.; Woebken, D.; Holtappels, M.; Lavik, G. et al. (2009). Complex nitrogen cycle in the sponge *Geodia barretti*. *Environ. Microbiol.,* Vol.11, No.9, pp. 2229-2243 ISSN: 14622912

Holmes, B. & Blanch, H. (2007). Genus-specific associations of marine sponges with group I crenarchaeotes. *Marine Biology,* Vol.150: pp. 759-772, ISSN: 0025-3162

Hooper, J.N.A. & Van Soest, R.W.M. (2002). Systema Porifera – a guide to the classification of sponges. Kluwer, Academic/Plenum Publishers , vol. 1, pp. 1-1101, vol. 2, pp. 1103-1708, ISBN: 0-306-47260-0

Houlbrèque, F. & Ferrier-Pagès, C. (2009). Heterotrophy in tropical scleractinians corals. *Biological Reviews*, Vol.84, pp. 1-17, ISSN: 1464-7931

Hoyos, L.R. (2010). Bacterial-invertebrate symbioses: from asphalt cold seep to shallow waters. PhD. Dissertation. Max-Planck Institut für Marine Mikrobiologie in Bremen, 200 pages.

Huggett, M.J.; Williamson, J.E.; De Nys, R.; Kjelleberg, S. & Steinberg, P.D. (2006). Larval settlement of the common Australian sea urchin *Heliocidaris erythrogramma* in response to bacteria from the surface of coralline algae. *Oecologia* , Vol.149, pp. 604-619, ISSN: 0029-8549

Jackson, J.B.C. & Buss, L. (1975). Allelopathy and spatial competition among coral reef invertebrates. *PNAS*, Vol.72, No12, pp. 5160-5163, *ISSN: 0027-8424*

Javaux, E.J.; Knoll, A.H. & Walter, M.R. (2004). TEM evidence for eukaryotic diversity in mid-Proterozoic oceans. *Geobiology*, Vol.2, pp. 121–132, ISSN: 1472-4669

Johnson, C.R.; Muir, D.G. & Reysenbach, A.L. (1991). Characteristic bacteria associated with surfaces of coralline algae: a hypothesis for bacterial induction of marine invertebrate larvae. *Mar. Ecol. Prog. Ser.,* Vol.74, pp. 281-294, ISSN: 0171-8630

Johnson, C.R. & Sutton, D.C. (1994). Bacteria on the surface of crustose coralline algae induce metamorphosis of the crown-of-thorns starfish *Acanthaster planci. Marine Biology,*Vol.120, pp. 305-310, ISSN: 0025-3162

Joint, I.; Tait, K. & Wheeler, G. (2007). Cross-kingdom signalling: exploitation of bacterial quorum sensing molecules by the green seaweed *Ulva*. *Phil. Trans. R. Soc. B* 2007, Vol.362, pp. 1223-1233, ISSN: 0962-8436

Kamke, J.; Taylor, M.W. & Schmitt, S. (2010). Activity profiles for marine sponge-associated bacteria obtained by 16S rRNA vs 16S rRNA gene comparisons. *The ISME Journal*, Vol.4, pp. 498-508, ISSN: 1751-7362

Kang, S-H. & Lee K-W. (2002). Phylogenetic relationships between *Ulva conglobata* and *U. pertusa* from Jeju Island inferred from nrDNA ITS 2 sequences. *Algae*, Vol.17, No2, pp. 75-81, ISSN: 1226-2617

Keeling, P. J. (2010). The endosymbiotic origin, diversification and fate of plastids. *Phil. Trans. R. Soc. B*, Vol. 365, pp. 729-748, ISSN: 0962-8436

Kim, D.E.; Lee, E.Y. & Kim H.S. (2009). Cloning and Characterization of Alginate Lyase from a Marine Bacterium *Streptomyces sp.* ALG-5. *Mar Biotechnol*, Vol.11, pp. 10–16, ISSN: 1436-2228

King, N.; Westbrook, J; Young, S.L.; Kuo, A.; Abedin, M. et al. (2008). The genome of the choanoflagellate *Monosiga brevicollis* and the origin of metazoans. *Nature*, Vol.451, pp. 783-788, ISSN: 0028-0836

Kinzie, R.A. ; Takayama, M. ; Santos, S. & Crofforth, M.A. (2001) The adaptive bleaching hypothesis: experimental test of critical assumptions. *Biol. Bull.* Vol.200, pp. 51–58, ISSN: 0006-3185

Kitamura, M.; Koyama, T.; Nakano, Y. & Uemura, D. (2007). Characterization of a natural inducer of coral larval metamorphosis. *J. Exp. Mar. Biol. Ecol.*,Vol.340, No1, pp. 96-102, ISSN: 0022-0981

Kladi, M.; Vagias, C. & Roussis, V. (2004) Volatile halogenated metabolites from marine red algae. *Phytochemistry Reviews* 3, 337–366, ISSN: 1568-7767

Klein, J. & Verlaque, M. (2008). The *Caulerpa racemosa* invasion: a critical review. *Marine Pollution Bulletin*, Vol.56, pp. 205–225, ISSN: 0025-326X

König, G.M.; Kehraus. S.; Seibert, S.F.; Abdel-Lateff, A. & Müller, D. (2006) Natural Products from Marine Organisms and Their Associated Microbes. *ChemBioChem.*, Vol.7, pp. 229 - 238, ISSN: 1439-4227

Kornprobst, J-M. (2010a). Algae. Vol.1, pp. 251-440. In: Encyclopedia of Marine natural Products. Wiley-Blackwell Verlag GmbH & Co. KGaA, Weinheim, Germany, 1594 pp. (3 vols); ISBN; 978-3-527-32703-4

Kornprobst, J-M. (2010b). Porifera (sponges). Vol. 2. In: Encyclopedia of marine natural products. Wiley-Blackwell, Verlag GmbH & Co. KGaA, Weinheim, Germany, 1594 pp. (3 vols); ISBN 978-3-527-32703-4

Kuntz, N.M.; Kline, D.I.; Sandin, S.A. & Rohwer, F. (2005). Pathologies and mortality rates caused by organic carbon and nutrient stressors in three Caribbean coral species. *Mar. Ecol. Prog. Ser.*, Vol.294, pp. 173-180, ISSN: 0171-8630

La Barre, S.C. & Haras D. (2007). Rencontres avec les bactéries marines. *J. Soc. Biol.*, Vol.201, No.3, pp. 281-289, ISSN: 1295-0661

La Barre, S.; Potin, P.; Leblanc, C. & Delage, L. (2010).The halogenated metabolism of brown algae (Phaeophyta) – its biological importance and its environmental significance. *Marine Drugs*. 2010; 8(4), 988-1010, ISSN: 1660-3397

Lachnit T.; Wahl, M. & Harder, T. (2010). Isolated thallus-associated compounds from the macroalga *Fucus vesiculosus* mediate bacterial surface colonization in the field similar to that on the natural alga. *Biofouling*, Vol.26, No.3, pp. 247-255, ISSN: 0892-7014

Lam, C. & Harder, T. (2007). Marine macroalgae affect abundance and community richness of bacterioplancton in close proximity. *J. Phycol.*, Vol. 43, pp. 874–881, ISSN: 0022-3646

Laming, K. (2010). Beneficial roles of microorganisms in the coral reef ecosystem. *The Plymouth Student Scientist*, Vol.3, No.2, pp. 289-299, ISSN: 1754-2383 (online)

Lang, J.C. & Chornesky, E.A., 1990. Competition between scleractinian reef corals - a review of mechanisms and effects. In: Dubinsky, Z. (Ed.). Ecosystems of the World, Vol. 25. Elsevier, Amsterdam, pp. 209–252.

Largo, D.B.; Fukami, K. & Nishijima, T. (1999). Time-dependent attachment mechanism of bacterial pathogen during ice-ice infection in *Kappaphycus alvarezii* (Gigartinales, Rhodophyta). *Journal of Applied Phycology*, Vol.11, pp. 129–136, ISSN: 0921-8971

Latham, H. (2008). Temperature stress-induced bleaching of the coralline alga *Corallina officinalis*: a role for the enzyme bromoperoxidase. *Bioscience Horizons*, Vol.1, No.2, pp. 104-113, ISSN: 1754-7431

Leggatt, W.; Ainsworth, T.; Bythell, J.; Dove, S.; Gates, R. et al. (2007). The hologenome theory disregards the coral holobiont. *Nature Rev. Microbiol. Online correspondance*, doi:10.1038/nrmicro1635C1, ISSN*: 1740-1526*

Li, Z. (2009). Advances in symbiotic cyanobacteria. In: Handbook on Cyanobacteria, eds. PM Gault & HJ Marler, Nova Science Publishers Inc., chapter 16, pp. 1-10

Littman, R.A., Willis B. L., Pfeffer, C. & Bourne, D. G. (2009). Diversities of coral-associated bacteria differ with location, but not species, for three acroporid corals on the Great Barrier Reef. *FEMS Microbiol. Ecol.*, Vol. 68, pp. 152-153, ISSN: 0168-6496

Littman R.; Willis, B.L. & Bourne, D.G. (2011). Metagenomic analysis of the coral holobiont during a natural bleaching event on the Great Barrier Reef *Environmental Microbiology Reports* ISSN: 1758-2229 (online), DOI: 10.1111/j.1758-2229.2010.00234.x

Lough, J.M. (2008). 10th anniversary review: a changing climate for coral reefs. *Journal of Environmental Monitoring*, Vol.10, No.1, pp 1-148, ISSN 1464-0325

Loughlane, C.J.; McIvor, L.M.; Rindi, F.; Stengel, D.B. & Guiry, M. (2008). Morphology, rbcL phylogeny and distribution of distromatic *Ulva* (Ulvophyceae, Chlorophyta) in Ireland and southern Britain. *Phycologia*,Vol.47, No4, pp. 416-429, ISSN: 0031-8884

Maldonaldo, M. (2004). Choanoflagellates, choanocytes and animal multicellularity. *Invertebrate Biology*, Vol.123, No1, pp. 1-22, ISSN: 1077-8306

Manefield M.; Rasmussen T.B.; Henzter M.; Andersen J.B.; Steinberg P.; Kjelleberg S. & Givskov M. (2002). Halogenated furanones inhibit quorum sensing through accelerated LuxR turnover. *Microbiology*, Vol.148, pp. 1119–1127, ISSN: 1350-0872

Manning, G.; Young, S.L.; Miller, W.T.; Zhai, Y. (2008). The protist, *Monosiga brevicollis*, has a tyrosine kinase signaling network more elaborate and diverse than found in any known metazoan. *PNAS*, Vol.105, No28, pp. 9674-9679, ISSN*: 0027-8424*

Maragos, J.E.; Crosby, M.P. & McManus, J.W. (1996). Coral reefs and coral reef biodiversity: a critical and threatened relationship. *Oceanography*, Vol.9, No1, pp. 83-99, ISSN: 1042-8275

Martone, P.T.; Alyono, M. & Stites, S. (2010). Bleaching of an intertidal coralline alga: untangling the effects of light, temperature, and desiccation. *Mar. Ecol. Prog. Ser.*, Vol.416, pp. 57-67, ISSN: 0171-8630

Marubini, F.; Ferrier-Pagès, C.; Furia, P. & Allemand, D. (2008). Coral calcification responds to seawater acidification: a working hypothesis towards a physiological mechanism. *Coral Reefs*, Vol.27, pp.491-499, ISSN: 0722-4028

Matsuo, Y.; Imagawa, H.; Nishizawa, M. & Shizuri, Y. (2005). Isolation of an algal morphogenesis inducer from a marine bacterium. *Science*, Vol.307, p. 1598, ISSN: 0036-8075

McConnell, O.J. & Fenical, W. (1977) Halogen chemistry of the red alga *Asparagopsis*. *Phytochemistry*, Vol.16, pp. 367-369, ISSN: 0031-9422

Meron, D.; Atias, M.; Kruh, L.I.; Elifantz, H.; Mintz, D.; Fine, M. & Banin, E. (2011). The impact of reduced pH on the microbial community of the coral *Acropora eurystoma*. *The ISME Journal*, Vol. 5, pp. 51-60,

Moberg, F. & Folke, C. (1999). Ecological goods and services of coral reef ecosystems. *Ecological Economics*, Vol.29, pp. 215-233, ISSN 0921-8009

Mohamed, N.M.; Colman, A.S.; Tai, Y. & Hill, R.T. (2008). Diversity and expression of nitrogen fixation genes in bacterial symbionts of marine sponges. *Environ. Microbiol.*, Vol.10, No.11, 2910-2955, ISSN: 14622912

Mouchka, M.E.; Hewson, I. & Harvell D. (2010). Coral-associated bacterial assemblages: current knowledge and the potential for climate-driven impacts. *Integrative and Comparative Biology*, Vol.50, No.4, pp. 662–674, ISSN: 1540-7063

Muscatine, L. & Cerniciari, E. (1969). Assimilation of photosynthetic products of zooxanthellae by a reef coral. *Biol. Bull.* Vol.137, pp. 506-523, ISSN: 0006-3185

Mydlarz; L.D. & Palmer, C.A. (2011). The presence of multiple phenoloxidases in Caribbean reef-building corals. *Comparative Biochemistry and Physiology,* Part A Vol. 159, pp. 372–378, ISSN: 1096-4959

Nedelcu, A.M.; Miles, I.H.; Fagir, A.M. & Karol, K. (2008). Adaptive eukaryote-to-eukaryote lateral gene transfer: stress-related genes of algal origin in the closest unicellular relatives to animals. *J. Evol. Biol.*, Vol.21, pp. 1852-1860, ISSN: 1420-9101

Negri, A.P.; Webster, N.S.; Hill, R.T. & Heyward, A.J. (2001). Metamorphosis of broadcast spawning corals in response to bacteria isolated from crustose algae. *Mar. Ecol. Prog. Ser.*,Vol.223, pp. 121-131, ISSN: 0171-8630

Nishizawa, M.; Iyenaga, T.; Kurisaki, T.; Yamamoto H.; Sharfuddin M. et al. (2007). Total synthesis and morphogenesis-inducing activity of (±)-thallusin and its analogues. *Tetrahedron Letters*, Vol.48, pp. 4229–4233, ISSN: 0040-4039

Olson, J. & Kellogg, C.A. (2010). Microbial ecologyof corals, sponges and algae in mesophotic coral environments. *FEMS Microbiol. Ecol.*, Vol. 73, pp. 17-30, ISSN: 0168-6496

Palmer, C.A., Bythell, J.C. & Willis, B.L. (2010). Levels of immunity parameters underpin bleaching and disease susceptibility of reef corals. The FASEB Journal, Vol. 24, pp. 1935 - 1946, ISSN: 0892-6638

Palmer, C.A., Bythell, J.C. & Willis, B.L. (2011). A comparative study of phenoloxidase activity in diseased and bleached colonies of the coral *Acropora millepora*. *Developmental and Comparative Immunology,* Vol. 35, pp. 1096-1099, ISSN: 0145- 305X

Pantos, O. & Bythell, J.C. (2010). A novel reef coral symbiosis. *Coral Reefs*, Vol. 29, pp. 761-770

Paul, V.J.; Ritson-Williams, R. & Sharp, K. (2011). Marine chemical ecology in benthic environments. *Nat. Prod. Rep.*, Vol.28, pp. 345-387, ISSN: 0265-0568

Quévrain, E.; Roué, M.; Domart-Coulon, I.; Ereskovski, A.; Perez, T. & Bourguet-Kondracki, M-L. (2009) Novel natural parabens produced by a *Microbulbifer* bacterium in its calcareous sponge host *Leuconia niveae. Environ. Microbiol.*, Vol.11, No6, pp. 1527-1539, ISSN: 1462 2912

Rao, D.; Webb, S. & Kjelleberg, S. (2006). Microbial colonization and competition on the marine alga *Ulva australis. Appl. And Environ. Microbiol.*, Vol.72, No8, pp. 5547-5555, ISSN: 0099-2240

Rasher, D.B. & Hay, M.E. (2010a). Seaweed allelopathy degrades the resilience and function of coral reefs. *Communicative & Integrative Biology*, Vol.3, No6, pp. 564-566, ISSN: 1942-0889

Rasher, D.B. & Hay, M.E. (2010b). Chemically rich seaweeds poison corals when not controlled by herbivores. *PNAS*, Vol.107, No.21, pp. 9683-9688, ISSN: *0027-8424*

Reiswig, H.M. (1971). Particle feeding in natural populations of three marine demosponges. *Biol Bull.*, Vol.141, pp. 568–591, ISSN: 0006-3185

Reitzel, A.M.; Sullivan, J.C.; Traylor-Knowles N. & Finnerty, J.R. (2008). Genomic survey of candidate stress-response genes in the estuarine anemone *Nematostella vectensis. Biol. Bull.* Vol.21 pp.: 233-254, ISSN: 0006-3185

Reyes-Bermudez, A. (2009). Cellular mechanisms of coral calcification. PhD thesis, James Cook University of North Queensland, 145pp with figures. (http://eprints.jcu.edu.au/8084)

Rohwer, F. (2010). Coral reefs in the microbial seas. Plaid Press, USA, 201 pages, ISBN: 978-0-9827012-0-1.

Rosenberg, E.; Koren, O.; Reshef, L.; Efrony, R. & Zilber-Rosenberg, I. (2007)a. The role of microorganisms in coral health, disease and evolution. *Nature Rev. Microbiol.*, Vol.5, pp. 355–362, ISSN: *1740-1526*

Rosenberg E;, Koren O.; Reshef L; Efrony R. & Zilber-Rosenberg, I. (2007)b. The hologenome theory disregards the coral holobiont: reply from Rosenberg et al. *Nature Rev. Microbiol. Online correspondance.* doi:10.1038/nrmicro1635C2, ISSN: *1740-1526*

Roué, M.; Domart-Coulon, I.; Ereskovski, A.; Djediat, C.; Perez, T. & Bourguet-Kondracki, M-L (2010). Cellular localization of clathridimine, an antimicrobial 2-aminoimidazole alkaloid produced by the mediterranean calcareous sponge *Clathrina clathrus. J. Nat. Prod.*, Vol.73, No.7, pp. 1277-1282, ISSN: 0163-3864

Salih, A.; Larkum, T.; Cox, G.; Kühl, M. & Hoegh-Guldberg, O. (2000). Fluorescent pigments in corals are photoprotective. *Nature*, Vol.408, pp. 850-853, ISSN: 0028-0836

Sammarco, P.W.; Hallock, P.; Lang, J.C. & LeGore, R.S. (2007). Roundtable discussion groups summary papers: environmental bio-indicators in coral reef ecosystems: the need to align research, monitoring, and environmental Regulation. *Environmental Bioindicators,*Vol.2, pp. 35-46, ISSN: 1555-5275

Sammarco, P.W. & Strychar, K.B. (2009). Effects of climate change/global warming on coral reefs: adaptation/exaptation in corals, evolution in zooxanthellae, and biogeographic shifts. *Environmental Bioindicators*, Vol.4, pp. 9–45, ISSN: 1555-5275

Santiago-Vasquez L.Z.; Ranzer, L.K. & Kerr, R.G. (2006). Comparison of two total RNA extraction protocols using the marine gorgonian coral *Pseudopterogorgia elisabethae* and its symbiont *Symbiodinium* sp. *Electronic Journal of Biotechnology,* Vol.9, No.5, Issue of October 15, pp. 598-603, ISSN: 0717-3458

Schlappy, M-L.; Schöttner, S.I.; Lavik, G.; Kuypers, M.M.; de Beer, D. & Hoffmann, F. (2010). Evidence of nitrification and denitrification in high and low microbial abundance sponges. *Marine Biology,* Vol.157, pp. 593-602, ISSN: 0025-3162

Schönberg, C.H.; Suwa, R.; Hidaka, M. & Loh, W.K.W. (2008). Sponge and coral zooxanthellae in heat and light: preliminary results of photochemical efficiency monitored with pulse amplitude modulated fluorometry. *Marine Ecology,* Vol.29, pp. 247-258, ISSN: 0173-9565

Selvin, J.; Gandhimathi, R.; Kiran, G.S.; Priya, S.S.; Ravji, T.R. & Hema, T.A. (2009). Culturable heterotrophic bacteria from the marine sponge *Dendrilla nigra*: isolation and phylogenetic diversity of actinobacteria. *Helgol. Mar. Res.*, Vol. 63, pp. 239–247, ISSN: 1438-3888

Siboni, N.; Ben-Dov, E.; Sivan, A. & Kushmaro, A. (2008). Coral-associated ammonium oxidizing Crenarchaeota and their role in the coral holobiont nitrogen cycle. *Proceedings of the 11th International Coral Reef Symposium,* Ft. Lauderdale, Florida, 7-11 July 2008, pp. 252-256, doi:10.1111/j.1462-2920.2007.01383.x

Smyrniotopoulos, V.; Vagias, C.; Rahman, M.M.; Gibbons, S. & Roussis V. (2010). Structure and antibacterial activity of brominated diterpenes from the red alga *Sphaerococcus coronopifolius. Chemistry and Biodiversity,* Vol. 7, pp. 186-195, ISSN: 1612-1872

Sogin, M; Morrison, H.G. ; Huber, J.A.; Welch, D.M.; Huse, S.M. et al. (2006). Microbial diversity in the deep sea and the underexplored "rare biosphere". *PNAS,* Vol. 103, No.32, pp. 12115-12120, ISSN: 0027-8424

Sotka, E.E. & Hay, M.E. (2009). Effects of herbivores, nutrient enrichment, and their interactions on macroalgal proliferation and coral growth. *Coral Reefs,* Vol.28, pp. 555–568, ISSN: 0722-4028

Srivastava, M.; Simakov, O.; Chapman, J.; Fahey, B.; Gauthier, M.A. et al. (2010). The *Amphimedon queenslandica* genome and the evolution of animal complexity. *Nature,* Vol.466, pp. 720-727, ISSN: 0028-0836

Stanley, S.M. & Hardie, L.A. (1998). Secular oscillations in the carbonate mineralogy of reef-building and sediment-producing organisms driven by tectonically forced shifts in seawater chemistry. *Palaeogeography, Palaeoclimatology, Palaeoecology,* Vol.144, pp. 3–19, ISSN: 0031-0182

Stat, M. & Gates, R.D. (2011). Clade DSymbiodinium in Scleractinian Corals: A "Nugget" of Hope, a Selfish Opportunist, an Ominous Sign, or All of the Above? *Journal of Marine Biology* Vol. 2011, Article ID 730715, 9 pages, doi:10.1155/2011/730715

Steneck, R.S. & Martone, P.T. (2007). "Calcified algae." In: *Encyclopedia of Tidepools,* eds. M.W. Denny & S.D. Gaines. pp 21-24. University of California Press, ISBN: Berkeley. ISBN: 3-7653-9271-5

Stewart, P.; Soonklang, N.; Stewart, M.J.; Wanichanon, C.; Hanna PJ, et al.(2008). Larval settlement of the tropical abalone, *Haliotis asinina* Linnaeus, using natural and artificial chemical inducers. *Aquaculture Research,* Vol.39, No.11, pp. 1181-1189, ISSN: 1355-557X

Suggett DJ & Smith DJ (2011). Interpreting the sign of coral bleaching as friend vs. foe. *Global Change Biology*, Vol.17, pp. 45-55, ISSN: 1354-1013

Talmage, S.C. & Glober, C.J. (2010). Effects of past, present, and future ocean carbon dioxide concentrations on the growth and survival of larval shellfish. *PNAS,* Vol.107, No40, pp. 17246-17251, ISSN: 0027-*8424*

Tanaka, Y.; Ogawa, H. & Miyajima, T. (2010). Effects of nutrient enrichment on the release of dissolved organic carbon and nitrogen by the scleractinian coral *Montipora digitata. Coral Reefs*, Vol.29, pp. 675-682, ISSN: 0722-4028

Taris, N.; Comtet, T., Stolba, R.; Lasbleiz, R.; Pechenik, J.A. & Viard F. (2010). Experimental induction of larval metamorphosis by a naturally-produced halogenated compound (dibromomethane) in the invasive mollusk *Crepidula fornicata* (L.). *J. Exp. Mar. Biol. Ecol.*, Vol.293, pp. 71-77, ISSN: 0022-0981

Teo, S.E.; Ho, C.L.; Teoh, S. & Rahim, R.A. (2009). Transcriptomic analysis of *Gracillaria changii* (Rhodophyta) in response to hyper- and hypo- osmotic stresses. *Journal of Phycology*, Vol.45, No5, pp. 1093-1099, ISSN: 0022-3646

Thacker, R.W. & Starnes, S. (2003). Host specificity of the symbiotic cyanobacterium *Oscillatoria spongeliae* in marine sponges, *Dysidea* spp. *Marine Biology*, Vol.142, pp. 643-648, ISSN: 0025-3162

Thacker, R.W. (2005). Impacts of shading on sponge-cyanobacteria symbioses: a comparison between host-specific and generalist associations. *Integrative and Comparative Biology*, Vol.45, pp. 369-376, ISSN: 1540-7063

Thajuddin, N. & Subramanian, G. (2005). Cyanobacterial biodiversity and potential applications in biotechnology. *Current Science*, Vol.89, No1, pp. 47-57, ISSN: 0011-3891

Thiel, V.; Leininger, S.; Schmaljohann, R.; Brümmer, F. & Imhoff, J.S. (2007). Sponge-specific bacterial associations of the Mediterranean sponge *Chondrilla nucula* (Demospongiae, Tetractinomorpha). *Microbial Ecology*, Vol.54, pp. 101-111, ISSN: 0095-3628

Tujula, N.A.; Crocetti, G.R.; Burke, C.; Thomas, T.; Holmström, C. & Kjelleberg; S. (2010). Variability and abundance of the epiphytic bacterial community associated with a green marine *Ulvacean* alga. *The ISME Journa*l, Vol.4, pp. 301–311, ISSN: 1751-7362

Turque, A.S.; Batista, D.; Silveira, C.B.; Cardoso, A.M.; Viera, R.P. et al. (2010). Environmental shaping of sponge-associated archaeal communities. *PLoS ONE*, Vol.5, No12, e15774, pp. 1-10, ISSN: 1932-6203

Usher, K.M.; Kuo, J.; Fromont, J. & Sutton, D. (2001) Vertical transmission of cyanobacterial symbionts in the marine sponge *Chondrilla australiensis* (Demospongiae). *Hydrobiologia*, Vol.461, pp. 15–23, ISSN: 0018-8158

Usher, K.M. (2008). The ecology and phylogeny of cyanobacterial symbionts in sponges. *Marine Ecology*, Vol.29, pp. 178-192, ISSN: 0173-9565

Vacelet, J. (1970) Description de cellules à bactéries intranucléaires chez des éponges *Verongia*. *J. Microscopie*, Vol.9, No.3, pp. 333-346, ISSN: 1365-2818

Vacelet, J.; Fiala-Medioni, A.; Fisher, C.R. & Boury-Esnault, N. (1996). Symbiosis between methane-oxidizing bacteria and a deep-sea carnivorous cladorhizid sponge. *Mar. Ecol. Prog. Ser.*, Vol.145, pp. 77-85., ISSN: 0171-8630

Vairappan, C.S. (2006). Seasonal occurrences of epiphytic algae on the commercially cultivated red alga *Kappaphycus alvarezii* (Solieriaceae, Gigartinales, Rhodophyta). *Journal of Applied Phycology*, Vol.18, pp. 611–617, ISSN: 0921-8971

Vega Thurber, R.L.; Barott, K.L.; Hall D.; Liu, H.; Rodriguez-Mueller, B. et al. (2008). Metagenomic analysis indicates that stressors induce production of herpes-like viruses in the coral *Porites compressa. PNAS*, Vol.105, No.47, pp. 18413-18418, ISSN: *0027-8424*

Vega Thurber, R.L.; Willner-Hall, D.; Rodriguez-Mueller, B.; Desnues, C.; Edwards, R.A. et al. (2009). Metagenomic analysis of stressed coral holobionts. *Environmental Microbiology* Vol.11, No.8, pp. 2148-2163, ISSN 1462-2920

Veron, J.E. (2008). Mass extinctions and ocean acidification: biological constraints on geological dilemmas. *Coral Reefs*, Vol.27, pp. 459–472, ISSN: 0722-4028

Vidal-Dupiol, J.; Adjeroud, M.; Roger, E.; Foure, L.; Duval, D. et al. (2009). Coral bleaching under thermal stress: putative involvement of host/symbiont recognition mechanisms. *BMC Physiology*, Vol.9, No.14, doi:10.1186/1472-6793-9-14

Wang, G.; Shuai L.; Li Y.; Lin W.; Zhao X. & Duian D. (2008). Phylogenetic analysis of epiphytic marine bacteria on Hole-Rotten diseased sporophytes of *Laminaria japonica. J. Appl. Phycol.*, Vol.20, pp. 403–409, ISSN: 0921-8971

Webster, N.; Webb, R.I.; Ridd, M.J.; Hill, R.T., Negri, A.P. (2001). The effects of copper on the microbial community of a coral reef sponge. *Environ. Microbiol.*, Vol.3, No1, pp. 19-31, ISSN: 14622912

Webster, N.A.; Smith, L.D.; Heyward, A.J.; Watts, J.E.M.; Webb, R.I. et al. (2004). Metamorphosis of a scleractinian coral in response to bacterial biofilm. *Appl. Environ. Microbiol.* Vol.70 No2, pp. 1213-1221, ISSN: 0099-2240

Webster NS and Hill RT (2007) Chapter 5 Vulnerability of marine microbes on the Great Barrier Reef to climate change. pp. 97-120. In: Johnson JE and Marshall PA (eds) Climate change and the Great Barrier Reef: A vulnerability assessment. Great Barrier Reef Marine Park Authority and the Australian Greenhouse Office, Department of the Environment and Water Resources. 818 p.

Webster, N.S.; Cobb, R.E. & Negri, A.P. (2008). Temperature threshold for bacterial symbiosis with a sponge. *The ISME Journal*,Vol.2, pp. 830-842, ISSN: 1751-7362

Webster, N.S.; Soo, R.; Cobb, R. & Negri, A.P. (2010)a. Elevated seawater temperature causes a microbial shift on crustose coralline algae with implications for the recruitment of coral larvae. *The ISME Journal* (2010), pp. 1-12, ISSN: 1751-7362

Webster, N.S.; Taylor, M.W.; Benham, F.; Lücker, S.; Rattei, T. et al. (2010b) Deep sequencing reveals exceptional diversity and modes of transmission for bacterial sponge symbionts. *Environ. Microbiol.*, Vol.12, No8, pp. 2070-2082, ISSN: 1462-2912

Webster, N. (2011). Sponge symbionts: sentinels of marine ecosystem health. 1st Int. Symp. on Sponge Microbiology, March 21-22, Würzburg, Germany.

Wegley, L.; Yu, Y.; Breitbart, M.; Casas, V.; Kline, D. & Rohwer, F. (2004). Coral-associated archaea. *Mar. Ecol. Prog. Ser.*, Vol.273, pp. 89-96, ISSN: 0171-8630

Wegley, L.; Edwards, R.; Rodriguez-Brito, B.; Liu, H. & Rohwer, F. (2007). Metagenomic analysis of the microbial community associated with the coral *Porites astreoides. Environmental Microbiology* (Vol.9, No11 pp. 2707–2719, ISSN: 1462-2912

Weinberger, F. (2007). Pathogen-induced defense and innate immunity in macroalgae. *Biol. Bull.*, Vol.213, pp. 290-302, ISSN 1062-3590

Weiner, S. & Dove, P. (2003). An overview of biomineralization processes and the problem of vital effect. In: Biomineralization. Reviews in Mineralogy & Geochemistry. Eds. Dove, P.M., DeYoreo, J.J. and Weiner, S. Vol. 54, pp. 1-29

Weisz, J.B.; Massaro, A.J.; Ramsby, B.D. & Hill, M. (2010). Zooxanthellar symbionts shape host sponge trophic status through translocation of carbon. *Biol. Bull.*, Vol.219, pp. 189–197, ISSN: 0006-3185

Weisz, J.H.; Lindquist, N. & Martens, C.S. (2008). Do associated microbial abundances impact marine demosponge pumping rates and tissue densities? *Oecologia*, Vol.155, pp. 367-376, ISSN: 0029-8549

Welsh, D.T.; Viaroli, P.; Hamilton, W.D. & Lenton, T.M. (1999). Is DMSP synthesis in chlorophycean macro-algae linked to aerial dispersal? *Ethology Ecology & Evolution*, Vol.11, pp. 265-278, ISSN: 0394-9370

Wild, C.; Huettel, M.; Klueter,A.; Kremb, S.; Rasheed, M.Y.M. & Jorgensen B.B. (2004). Coral mucus functions as an energy carrier and particle trap in the reef ecosystem. *Nature*, Vol.428, pp. 66-70, ISNN: 0028-0836

Wild C., Rasheed M., Jantzen C., Cook P., Struck U., Huettel M. & Boetius A. (2005). Benthic metabolism and degradation of natural particulate organic matter in carbonate and silicate reef sands of the northern Red Sea. *Mar. Ecol. Prog. Ser.*, Vol. 298, pp. 69–78, ISSN: 0171-8630

Wild, C.; Hoegh-Guldberg, O.; Naumann, M.S.; Colombo-Pallotta, F.; Ateweberhan, M. et al. (2011). Climate changes impede scleractinian corals as primary reef ecosystem engineers. *Marine and Freshwater Research*, Vol.62, pp. 205–215, ISSN: 1323-1650

Williams, E.A.; Craigie, A; Yates, A. & Degnan, S.M. (2008). Articulated coralline algae of the genus *Amphiroa* are highly effective natural inducers of settlement in the tropical abalone *Haliotis asinine. Biol. Bull.*, Vol.215, pp. 98-107, ISSN: 0006-3185

Wong, K.K.W.; Lane, A.C.; Leung, P.T.W. & Thiyagarajan, V. (2011, in press). Response of larval barnacle proteome to CO2-driven seawater acidification. *Comp. Biochem. Physiol. Part D*, 12 pages, in press.

Woods Hole Oceanographic Institute: http://www.whoi.edu/OCB-OA/FAQs/)

Woolard, F.X.; Moore, R.E. & Roller, P.P. (1979). Halogenated acetic and acrylic acids from the red alga *Asparagopsis taxiformis Phytochemistry* 18 (4), Vol.18, No4, pp. 617-620, ISSN: 0031-9422

5

Biogeography of Platberg, Eastern Free State, South Africa: Links with Afromontane Regions and South African Biomes

Robert F. Brand[1], L.R. Brown[1] and P.J. du Preez[2]
[1]*Applied Behavioural Ecology and Ecosystem Research Unit, University of South Africa*
[2]*Department of Plant Sciences, University of the Free State, Bloemfontein*
South Africa

1. Introduction

This chapter comprises a vegetation analysis of Platberg, eastern Free State, South Africa (Figure 1). Platberg is an inselberg which has high botanical diversity, with associated species richness, and high numbers of endemic taxa only found at high altitudes over 2 000 m.a.s.l. indicative of montain flora.

Inselbergs are one of the most striking and persistent landform types in Africa (Goudie 1996) and are defined as an isolated hill, knob, koppie or small mountain, which stands alone and rises abruptly, island-like, from the surrounding terrain (Sarthou & Villiers 1998). In the eastern Free State, South Africa, rising abruptly from the flat terrain of the Karoo sediments, are a series of more than 20 prominent inselbergs all over 2000 m high. These flat-topped, steep-sided inselbergs stretch north from the Qwa-Qwa scarp, which constitutes the northern endpoint of the Maluti-Drakensberg. Geologically, these inselbergs are an extension of the Maluti Drakensberg, and occur along a line prescribed by the Great Escarpment (King 1963) at 1800 m. They form a discrete island-like archipelago stretching over 200 km, connecting the main Lesotho-Maluti-Drakensberg with the Mpumalanga-Drakensberg, and the Highveld Mountains to the north.

Inselbergs are formed from igneous or sedimentary rocks, capped by more resistant strata. The remnants of the resistant igneous capping form the distinct flat-topped mesas, buttes or table mountains. Platberg (Figure 2) and all the other inselbergs are the structurally controlled remnants of such weathering processes (King 1963; Moon & Dardis 1992). As prominent landscape features, geomorhologically they are formed by extensive subsurface decay along joints. The subsequently loosened material is removed by water, wind and gravity, with thawing and freezing assisting the process (Moon & Dardis 1992). The accompanying erosional detritus, rocks, boulders and gravel, form the steep slopes to the inselbergs, which are slowly buried in the rock debris (Figure 3). It is the packing of this weathered material, which provides the numerous ecological niches for the development of biodiverse plant communities characterised by inselbergs (Sarthou & Villiers 1998).

The unique, high altitude conditions found above 2 000 m, lead to high levels of endemism in organisms; bryophytes, plants and animals (Hillard & Burtt 1987; Van Wyk & Smith 2001; Carbutt & Edwards 2006; Mucina & Rutherford 2006). This is due to the compression of climatic life zones over a short distance that makes mountains hot spots for biological diversity (Körner 2003). Inselbergs may be regarded as analogous with an archipelago of islands in an 'ocean' of low-level vegetation types which act as an isolation factor (Taylor 1996; MacArthur & Wilson 2001).

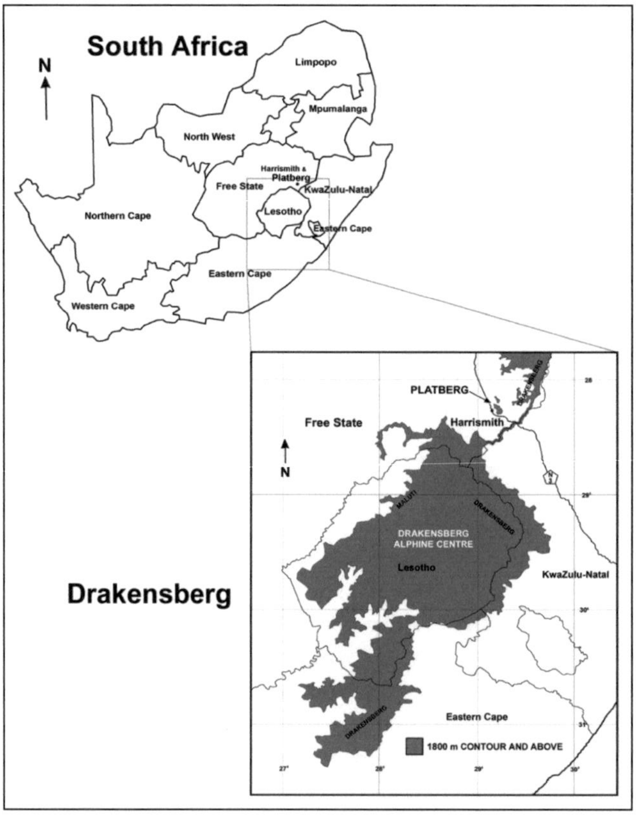

Fig. 1. Platberg, Eastern Free State, South Africa (Brand 2008).

This in turn precludes plant species with less mobile seed dispersal mechanisms from propagating over wide ranges, and allows for high levels of endemism to develop (Taylor 1996). High levels of endemism mean that a large proportion of the available gene pool is unique to that site, and inselbergs and mountains therefore have an important role to play in the maintenance of genetic diversity (Taylor 1996; Mucina & Rutherford 2006).

Inselbergs are differentiated from their surroundings by harsher edaphic and microclimatic conditions (Porembski & Barthlott 1995), consequently they host distinctive species, forming unique phytosociological associations, which differ from the lowland vegetation matrix in which they are embedded (Porembski & Brown 1995; Porembski et al., 1996, 1997, 1998; Sarthou & Villiers 1998; Parmentier et al., 2006). The vegetation map of Mucina & Rutherford (2006) shows Platberg as an inselberg embedded in the lowland grassland but also as a high altitude vegetation unit designated as the Gd 8 Lesotho Highland Basalt Grassland, part of the Drakensberg Grasslands predominantly associated with the broader Drakensberg Alpine Centre (DAC). Prior to this study no field data was available to prove this hypothesis. These studies provides empirical data on a detailed vegetation level that shows strong floristic and plant community similarities with Gd 8 Lesotho Highland Basalt Grassland, and demonstrates Platberg is an extension of the same phytochoria as the DAC.

Fig. 2. The contact between the dark dolerite cap and the lighter sandstone is a distinct feature of Platberg.

Fig. 3. Mobile boulder beds on Platberg provide habitat for species and vegetation unique to inselbergs and the Afromontane region.

1.1 African phytochoria and species richness

The phytochoria of Africa have been formally classified by White (1983), who defined eighteen major phytochoria for Africa. The two phytochoria of interest for this work as are the *Afromontane* archipelago-like regional centre of endemism (VIII), and embedded within the *Afromontane* phytochoria is the *Afroalpine* archipelago-like centre of extreme floristic impoverishment (IX). The basis of White's work was to produce a vegetation map for Africa, using two criteria: the physiognomy of the most extensive vegetation type, and floristic composition. This produced a useful, broad scale map of regional biodiversity, but did not have the resolution to show species richness on the local or community level.

A map showing continental wide African sites of high biodiversity using both the taxon-based approach and the geographical or inventory-based approach are presented by Mutke et al., (2001). This Global Information System (GIS) approach to map African phytodiversity shows good correlation with climatic, edaphic and biotic parameters for African phytodiversity for the Drakensburg/Natal Area - analogous to White's *Afromontane* archipelago-like regional centre of endemism. This study has improved on White's broad scale map, but being a desktop study still does not provide detailed information of species richness on the local community scale.

Regional mapping of the biodiversity by Van Wyk and Smith (2001) and Mucina & Rutherford (2006) provides a more detailed pattern of plant diversity, which correlates well with the geological map of the African land surface (White 1983; Hillard & Burtt 1987). The South African Drakensberg is shown within this archipelago-like regional endemic centre, which Hillard & Burtt (1987) called the *Eastern Mountain Region* (EMR), first used by Phillips in 1917 (Carbutt & Edwards 2004).

The use of the EMR has not been followed as it shows a broader, topographically less well-differentiated area and is a loose correlation with the more precise geographical/ topographical designation of the Drakensberg Alpine Centre (DAC) (Van Wyk & Smith 2001). However, what Hillard & Burtt (1987) where alluding to, in using the term 'EMR', was to differentiate the Drakensberg and surrounding high altitude areas from the rest of the continental wide *Afromontane* region due to its unique and rich floristic composition. Biogeographically, the DAC and Platberg also show relatively strong floristic affinities with the Cape Floristic Region (Linder 2003; Carbutt & Edwards 2004, 2006) or the Fynbos Biome as defined by Mucina & Rutherford (2006). Other biogeographical links are shared with Nama-Karoo Biome found in the drier interior of the sub-continent and further north the sub tropical African and Eurasian flora (Mutke et al., 2001). It is within this broader Afromontane biogeographical context, including being one of an 'island-like' archipelago of inselbergs that Platberg's biological diversity is considered.

2. Methods

The field-derived data was analysed using phytosociological principles. The statistical analysis of vegetation and environmental data, which underpins phytosociology, provides a measure for biodiversity that is incorporated in the concepts of species richness and evenness or relative abundance. A total of 393 relevés where analysed for the entire Platberg. The scope of the study was to sample vegetation plots above the 1 800 m contour in order to work within the limits set by Killick (1978a) who regarded the region in the Drakensberg above the 1 800 m as a distinct floristic region - the Afroalpine Region. The topography of the plain in which Platberg is situated, is relatively flat, rising abruptly at the 1 900 m contour, this being the start of the footslopes, which was used as the lower limit set for sampling.

Additionally, the PRECIS (National Herbarium Pretoria **[PRE]** **C**omputer **I**nformation **S**ystem) data from the South African National Biodiversity Institute (SANBI), Pretoria, was used to compare species with the Platberg data. The PRECIS list is compiled from field collections and plotted on a grid square frame with each grid covering 30 x 30 km^2. Even though the PRECIS data covers the same grid square as Platberg, it is mostly flatland under 1 800 m lower than the footslopes of Platberg which start at 1 900 m. The comparison was done to reveal correlations and connections, which exist with Platberg and the vegetation of the lower lying regions, in which the vegetation of Platberg is embedded. The disadvantage of the PRECIS list is that very few of the common species are normally recorded which will result in there being a lower number of species per hectare.

3. Results

The vegetation on Platberg is dominated by grasses and shows an intergrade of floristic associations and habitat features in common with several other major vegetation types of the Grassland, Fynbos, Afrotemperate Forest and Nana-Karoo Biomes. Fynbos as well as succulents, particularly Mesembryanthemaceae from the Nama-Karoo Biome grow on Platberg. These floristic elements also extend to the DAC. Woody shrubs and forest remnants grow in the specalised ecological niches in sheltered gullies, boulder beds and rocky terrain. Numerous wetlands occur, forming distinct hygrophilous communities. Geophytes and forbs add significantly to the biodiversity of all vegetation types and grow in all habitats. Only two Gymnosperms occur, the indigenous, forest emergent tree *Podocarpus latifolius*, and the exotic *Pinus patula*, established in timber plantations at lower altitudes, which have now invaded and are replacing the indigenous vegetation on the cool southern slopes. Pterdophyte diversity is relatively low, comprising a total of 16 species, 10 genera and 8 families. Ferns are widespread and occur throughout all habitats, on all aspects - hot northern, cool southern and at all altitudes. Three of the ferns, *Dryopteris dracomontana, Mohria rigida* and *Polystichum dracomontanum* are endemic to the DAC. All but one species (*Pellaea calomelanos*) occur at altitude throughout the Afromontane region. The exotic bracken, *Pteridium aquilinum* occurs on the lower footslopes with *Searsia pyroides* subsp *gracilis*. The prostrate fern, *Selaginella caffrorum* is a mat forming species, which forms unique communities that contribute to inselberg vegetation structure. *Selaginella* communities are found on open, sheet rock on Platberg, Korannaberg, Thaba Nchu and other inselbergs as well in the DAC. *Afrotrilepis pilosa* mats occurring on granite inselbergs of west Africa are physiognomic equivalents.

Of the 974 species, in the PRECIS database, collected in the 2828AC Harrismith grid, about 670 (68.8%) species occur on Platberg above 1 900m. The rest (31.2%) occur on the surrounding lowlands. The 670 species found on Platberg were correlated with DAC species lists of Van Zinderen Bakker (1973), Killick (1963, 1978a, 1979b), Hill (1996), Low & Rebelo (1996), Carbutt & Edwards (2004, 2006), Hoare & Bredenkamp (2001), (Moffett 2001), Smith & Van Wyk (2001) and Mucina & Rutherford (2006). A strong genus level correlation of over 80% was found between Platberg and the DAC, which includes exotic angiosperm taxa such as *Pinus, Acacia,* Bidens, Tagetes, etc. (Carbutt & Edwards 2004).

Of the 670 species recorded on Platberg, several species are new records for the Free State and represent range extensions, or have only been collected once before (Brand et al., 2010). One of these species *Struthiola angustiloba*, a rare KwaZulu-Natal species was collected on Platberg. Two Asteraceae species also collected on Platberg, *Helichrysum harveyanum* and *H. truncatum* are Northwest/Gauteng and Mpumalanga endemics, giving new range extensions of 400-500 km south to Platberg. These new range extensions are not totally unexpected, given that mountain chains act as routes of migration (Körner 2003) and with similar altitude and ecological conditions, distance is not a critical factor (400-500 km), but similarity in live zones determines speciation, rarity and endemism (Körner 2003).

Of the 670 vascular plants recorded on Platberg, there are 305 genera in 96 families (Table 1). Of these 27 are endemic or near-endemic species also found in the Drakensberg Alpine Centre

(DAC). Only 22 alien plants occur of which most are annual Dicotyledons (Brand et al., 2010), which is a good indicator of limited human influence on the vegetation of the Platberg plateau.

	Families	Genera	Species
Pteridophytes	8	10	16
Gymnosperms	2	2	2
Angiosperms	86	293	652
(Monocotyledons)	(23)	(104)	(214)
(Dicotyledons)	(63)	(189)	(438)
Totals	**96**	**305**	**670**

Table 1. Floristic composition for Platberg (Brand et al., 2010)

4. Affinities with other regions

Platberg shares many climatic, edaphic and biotic similarities with the Drakensburg Alpine Centre (DAC). Biogeographically and chorologically it falls within the DAC and shares many of the same plant species and endemic taxa (Brand et al., 2010). Biogeographical and botanically the DAC is seen as a transition zone, a migratory pathway and repository for taxa of diverse regions and biomes (Killick 1963, 1978a, 1978b, Hillard & Burtt 1987, Carbutt & Edwards 2004, Mucina & Rutherford 2006). The DAC has some 2800 species of vascular plants of which 929 are endemic or near-endemic angiosperms in an area of 40 000 km^2 (Carbutt & Edwards 2004, 2006). The Global Information System (GIS) approach of Mutke et al. (2001) correlates well with these figures, showing a plant species richness > 3000 species per 10 000 km^2.

The two largest plant families on Platberg are Asteraceae (40 genera, 126 species, comprising 18.8% of the flora), and Poaceae (39 genera, 73 species, 10.9% of the flora). This pattern of high diversity of Asteraceae is common to the DAC where it is represented by 65 genera and 430 species and 17% of the total flora (Carbutt & Edwards 2004). The pattern is the same for Poaceae where, in the DAC it is represented by 86 genera and 267 species and contributes 11% to the total species flora (Carbutt & Edwards 2004). The Grassland Biome with the DAC, Platberg in its centre, has provided idea habitat for the rapid spread of Asteraceae, as well as the development and radiation of grasses (Carbutt & Edwards 2004).

Floristic composition for Platberg and the DAC shows Asteraceae as the largest family (Brand et al., 2010). This trend is the same for the Cape flora, with significant correlation within the top 12–20 families (Goldblatt & Manning 2000). The second most specious family on Platberg is Poaceae (39 genera, 73 species, 10.9% of the flora) followed by Cyperaceae (18 genera, 39 species, 5.8% of total flora), which reflects a similar floral composition to the DAC (Carbutt & Edwards 2004, 2006; Mucina & Rutherford 2006). This trend with the same ranking for the top 3-10 families plus rations of floristic compositions is also found with the more grassy regions to the arid western interior and higher altitude, wetter northern areas of South Africa (Kooij et al., 1990; Du Preez & Bredenkamp 1991; Fuls 1993; Eckhardt et al. 1993, 1995; Malan 1998).

The ranking of Poaceae and Cyperaceae as the second and third richest families on Platberg and the DAC, which is in contrast to the Cape flora, where Poaceae and Cyperaceae are poorly represented with Restionaceae filling the environmental and floristic position of the Poaceae (Goldblatt & Manning 2000; Brand et al., 2010). For the Cape fynbos Fabaceae is the second largest family followed by Iridaceae, Aizoaceae, Ericaceae and Scrophulariaceae

(Goldblatt & Manning 2000). With the exception of Ericaceae, these families for significant floral structure and species composition represented in the top 7 families for Platberg and the DAC (Brand et al., 2010).

The fynbos vegetation elements found on Platberg and the DAC show close affinities with similar fynbos of the Cape Floral Region (CFR). These fynbos elements are characterised by *Passerina, Cliffortia, Metalasia* and *Muraltia* species which all exhibit narrow, but extensive ranges, located on depaupered soils of the Clarens Sandstone Formation. There are two distinct fynbos vegetation types found on Platberg and the DAC. They are described as the Gd 9 Drakensberg-Amathole Afromontane Fynbos, and the Gm 24 Northern Escarpment Afromontane Fynbos (Mucina & Rutherford 2006).

4.1 Vegetation composition and structure: Grasses and forbs using C_3, C_4 and CAM pathways

To understand how rising CO_2 levels with increased temperatures may affect plant species composition and vegetation structure, an analysis of plant communities using CAM, C_4 and C_3 metabolic pathways was done. At high temperatures C_4 outcompete C_3 plants (Retallack 2001), it would be predicted that high altitude vegetation using the C_3 pathway would show a reduction of range, an upward shift in distribution and a loss of species. Most of the grasses comprising the Grassland Biome occur at altitudes below 1 200 m and consequently use the C_4 pathway, while most montane grasses, trees and shrubs use the C_3 pathway. Asteraceae, also use the C_3 pathway while Wetland plants, ferns, Gymnosperms and succulents use Crassulacean Acid Metabolism (CAM). The analysis of the plant families on Platberg using C_3, C_4 and Crassulacean Acid Metabolism (CAM) are from Brand et al., (2010).

The phytosociological analysis for Platberg found 39 distinct plant communities (Brand et al., 2008, 2009, 2010) all of which contain a combination of Asteraceae Poaceae, and Cyperaceae. Only 7 plant communities are dominated by woody/shrubs and exclusively use the C_3 pathway. The remaining 32 that plant communities are structure by species using C_4 pathways, CAM or high altitude grasses which use the C_3 pathway. Excluding the woody/shrub communities 82% of the formally classified plant communities use C_3, C_4 or CAM pathways (Table 2).

Plant Family	Pathway	Gen. No.	Gen %	Spp. No	Species (%)
Asteraceae	most C_4/ C_3	40	13.1	126	18.8
Poaceae	mixed C_4 C_3	39	12.8	73	10.9
Cyperaceae	C_4	18	5.9	39	5.8
Crassulaceae	CAM	3	1.0	13	1.9
Caryophyllaceae	C_4	9	1.2	4	1.3
Euphorbiaceae	C_4	3	1.0	8	1.2
Brassicaceae	C_4	4	1.3	4	0.6
Amaranthaceae	C_4	2	0,7	5	0.7
Mesembryanthe-maceae	CAM	2	0.7	4	0.6
Chenopodiaceae	C_4	1	0.3	1	0.1
Restionaceae	C_4	1	0.3	1	0.1
Totals		**122**	**44.6%**	**278**	**42%**

Table 2. Platberg vegetation using C_3, C_4 or CAM pathways

On a species level, succulent families using CAM pathways total 278 species and 122 genera **(Table 1).** This accounts for 44.6% of the total genera and 42% of the total species found on Platberg.

4.2 Grass diversity

A total of 73 grass species where collected on Platberg with 29 species using C_3 metabolism, and 30 species using C_4 metabolism, the remaining 14 species use mixed C_3/C_4 metabolism (Figure 4). The C_3/C_4 split is almost equal however, an examination of grass community structure on Platberg shows grasses using C_4 metabolism dominate.

Only one grass *Helictotrichon longifolium* and the mountain bamboo *Thamnocalamus tessellatus*, uses C_3 metabolism, with *Digitaria monodactyla* and *Pennisetum sphacelatum* showing mixed C_3/C_4 metabolism, the majority of the communities and species show the use of C_4 metabolism, which represents 83% of the grass community structure on Platberg.

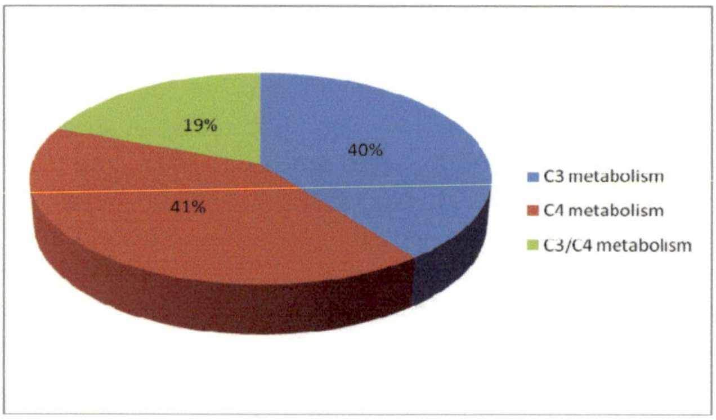

Fig. 4. Ratio of grass metabolic pathways on Platberg.

5. Discussion

5.1 Forest Biome affinities

The entire forest biome in South Africa is composed of a number of fragmented forest patches, either continuous as in the Southern Afrotemperate Forests of the Tsitsikamma region of the western Cape, fragmented Southern Mistbelt Forest for the Eastern Cape and Northern Afrotemperate Forests in the Drakensberg, or contiguous Northern Mistbelt mountain patches in Mpumalanga and Limpopo Provinces, or disjunct, scattered Northern Afrotemperate Forests growing in the Magaliesberg in the North Western Province (Von Maltitz 2003, Mucina & Rutherford 2006).

These relictual elements of the Afrotemperate Forests (von Maltitz 2003) are all regarded as depaupered or 'biogeographically' eroded as well as previously linked as recently as the Pleistocene > 2 My BP (Scott et al., 2003). Today they are fragmented and are found as refugia in inselbergs (Brand et al., 2009) or relictual scarp situations on elevated ranges (Von Maltitz 2003; Mucina & Rutherford 2006). Afrotemperate forests have also retreated into kloofs (narrow, incised canyons) or have lost numerous Afrotemperate species through episodes of climate change, encroachment by Grassland and Savanna Biomes, or reduction by fire and grazing. Fire in particular is important in shaping the modern appearance and distribution of the forest (Mucina & Rutherford 2006), patches of which are found below

Van Reenen's Pass 60km east of Platberg, Royal Natal National Park to the west and the main Drakensberg at Qwa-Qwa to the south (Figure 5).
The highest altitude Afrotemperate *Podocarpus* forest in South Africa grows on Nelsons Kop at 2 230 m (Von Maltitz 2003), an inselberg about 20 km northeast of Platberg, and as with the forest patch on Platberg, it is located on the cool south-eastern side, below the basalt on the Clarens Formation sandstones. Other Northern Afrotemperate Forest affinities are found to the south, in the Fynbos Biome, and the Savanna Biome mountains to the north (von Maltitz 2003; Mucina & Rutherford 2006).

Fig. 5. *Podocarpus* forest patches below the Clarens Formation sandstone confined to fire protected gullies of the Drakensberg and Great Escarpment.

5.2 Fynbos Biome affinities

Fynbos on Platberg occurs in 13 different communities, which can be grouped into two communities (Brand 2008). The sclerophyllous vegetation is characterised by *Passerina montana*, a fynbos taxa endemic to the Cape (Figure 6). These fynbos communities have a species richness varying between 14 - 54 species per 30 m^2, with an average of 28.34 species per 30 m^2. This is lower compared with the grassland vegetation, which has 11 - 54 per 30 m^2 with an average of 32 per 30 m^2, which gives a moderate species diversity index. A minimum 16 fynbos genera comprising 22 species occur on Platberg. The fynbos is located in two distinct habitats: the lower altitude zone at 2000 m growing on the mineral poor soils of the Cave Sandstone of the Clarens Formation, and the higher altitude zone at 2200 m on the rocky, basaltic, mineral rich rim of the plateau just below the exposed summit grassland (Brand et al., 2008).
Fynbos and the Gm 24 Northern Escarpment Afromontane Fynbos (Mucina & Rutherford 2006). This afromontane fynbos community is found on most of inselbergs including the Floristically, and structurally, the *Passerina montana* fynbos-like shrubland elements found on Platberg conform with the Gd 6 Drakensberg-Amathole Afromontane Korannaberg 200km west, and a similar altitude of 2000-2200 m in the Drakensberg. For Mucina & Rutherford (2006) it is structurally and floristically different from the other Afromontane fynbos species rich community found at higher altitudes and embedded in the Gd 8 Lesotho Highveld Basalt Grassland.

The Afromontane fynbos genera found on Platberg are Passerina, Cliffortia, Erica, Euryops, Helichrysum, Macowania, Metalasia, Muraltia, Pentaschistis, Ischyrolepis, Schoenoxiphium and Watsonia, all of which are endemic taxa typically found in the Cape Floristic Region (Goldblatt & Manning 2000).

The grass-like genus *Restio* is a major component of the Cape flora not found on Platberg or the broader Drakensberg, however, *Ischyrolepis schoenoides,* is found on Platberg and other inselbergs in the Free State (Du Preez & Bredenkamp 1991; Malan 1998) and could be regarded as an ecological equivalent for *Restio,* as its growth form and habitat is similar. Phylogenetically, these two genera are closely related (Haaksma & Linder 2000) and Germishuizen et al., (2006) lists *Restio schoenoides, Kunth* (1) as a synonym for *Ischyrolepis schoenoides* (Kunth) H.P. Linder (Brand et al. 2010). The remnants of fynbos vegetation on Platberg may be relicts from cooler periods when more extensive fynbos migrated over lower altitudes starting during the Late Pleistocene and evident up to the last Glacial Maximum (Scott et al., 1997).

Fig. 6. Fynbos *Passerina montana* shrub community on Clarens Sandstone on Platberg.

Pollen taken from sites at Clarens, a town in the semi-arid interior of the Free State, 200km west of Platberg, and the Rose Cave site, near Ladybrand, 500km to the west of Platberg, located at the extreme western footslopes of the DAC, show a similar pattern for fynbos genera *Protea* and *Cliffortia* which were more abundant and are typical of upland vegetation.

5.3 Nama-Karoo Biome affinities

Limited affinities are found with the Nana-Karoo flora on Platberg. These are mostly succulents from the Mesembryanthemaceae (2 genera, 4 species) and low shrubs, *Chrysocoms ciliata, Felicia filifolia, F. muricata* the most representative, found on the warm north and west sides of Platberg (Brand et al., 2010), and may represent previous climatic conditions of hotter, drier periods with less seasonal fluctuation (Scott 1988). Succulent evolution and distribution is not favoured by high rainfall and freezing temperatures extended over days (Smith et al., 1998; van Wyk & Smith 2001), which would be the environmental limiting factors responsible for the low numbers of succulents found at altitude (Körner 2003) on Platberg and DAC.

5.4 Grassland Biome affinities

Grasses, the family Poaceae, is the single most important plant family for humanity (Gibbs-Russell 1991). It is distributed over all seven continents and is the fifth largest plant family on earth and the second largest family for the DAC. Globally, Poaceae comprise some 770 genera and 9 700 species (Gibbs-Russell 1991). Southern Africa there are approximately 194 genera and 967 species of which 329 are endemic (Carbutt & Edwards 2004, 2006). For the Drakensberg Alpine Centre there are 86 genera and 267 species, of which 22 endemic or near-endemic genera comprising 39 endemic or near-endemic species (Carbutt & Edwards 2004, 2006). Grasses, particularly C_4 pathway users, are better than plant families in stripping and utilising CO_2 from the atmosphere. Grasses store the Carbon below ground in roots and soils (Retallack 2001), and with the Savanna Biome are enormous sinks for Carbon (Table 3).

Country/Region	Vegetation	Mean C (kg m $^{-2}$)
Tanzania:		
Serengeti	Tall grassland	51.4
Serengeti	Dry woodland	12.8
India:		
Uttar Pradesh	Terai grassland	23.5
Uttar Pradesh	Monsoon forest	3.9
U.S.A.:		
Iowa	Tall prairie	14.9
Illinois	Oak forest	8.2

Table 3. Mean Carbon (C) content in soils under Modern Grasslands and adjacent Woodlands (modified from Retallack 2001)

Platberg and the DAC are within the Grassland Biome, and as such are dominated by grasses. Grassland is a complex mix of graminoids, forbs and geophytes (Mucina & Rutherford 2006). Geophytes are abundant on Platberg, with Amaryllidaceae, Asphodelaceae, Hyacinthaceae, Hypoxidaceae and Iridaceae growing throughout the grassland, fynbos, woody/shrub and wetland communities. These geophyte families are also important fynbos components (Goldblatt & Manning 2000) and are prominent members in the Succulent and Nama Karoo Biomes (Van Wyk & Smith 2001; Pond et al., 2002; Mucina & Rutherford 2006).

On a regional scale, strong floristic affinities exist between Platberg and the Drakensberg grasslands to the south (Bester 1998), which includes Qwa-Qwa (Moffett 2001), the Golden Gate Highlands National Park (Roberts 1969; Kay et al., 1993), the central Cathedral Peak area, and southern Drakensberg (Killick 1963, 1978a, 1978b; Van Zinderen Bakker 1973; Hillard & Burtt 1987), the Stormberg and Eastern Cape Drakensberg (Hill 1996; Carbutt & Edwards 2004), as well as Korannaberg to the western interior of the Free State (Du Preez & Bredenkamp 1991; Du Preez et al., 1991).

The development of high biodiversity and species richness of Poaceae across the Grassland Biome has a number of explanations; these include a combination of weather (Mean Frost Days, and minimum daily temperature) and moisture availability (Mutke et al., 2001), soils, and the effects of fire and grazing (Seabloom & Richards 2003). Moisture availability is an important factor in determining species richness (Mutke et al., 2001) particularly for grasses, maximum growth occurs for up to four days after rain (Cavagnora 1988), which allows grasses to outcompete geophytes and forbs with similar morphology (Mucina & Rutherford 2006). The Drakensberg and Platberg have much higher precipitation than the surrounding areas 700-2400 mm (Mucina & Rutherford 2006), this provides for more moisture availability at higher, cooler altitudes. The lower regions of the Grassland Biome have lower rainfall 454 mm average (Mucina & Rutherford 2006), and are more humid. This moisture availability divides the grassland into high altitude Moist grassland dominated by species using C_3 metabolism and low, altitude Dry grassland dominated by species using C_4 metabolism.

Platberg on the cool southern and eastern sides in particular, provide in current times, a similar habitat and climatic conditions reminiscent of both Holocene and Pleistocene with the grassland on the plateau a mix of upland C_3 grasses form cooler times, mixed with C_4 grasses from corresponding warmer times (Scott & Vogel 2000). On Platberg, the predominance of C_4 grasses indicates that it falls within the core of the Grassland Biome, with a species composition similar to that dominated by C_4 grasses from the supertribe Andropogonodae which includes *Andropogon, Trachypogon, Heteropogon, Cymbopogon, Diheteropogon, Monocymbium, Tristachya, Schizachyrium, Themeda* and *Hyparrhenia* (Gibbs-Russell et al., 1991; Mucina & Rutherford 2006). The abundant C_3 grass *Helictotrichon longifolium,* found on the open plateau area of Platberg, would suggest a link with the predominant high altitude Drakensberg grasses dominated by C_3 grasses. The other abundant C_3 grass on Platberg is from the smallest of the five Grass subfamilies, Bambusoideae, the mountain bamboo *Thamnocalamus tessellatus.* On Platberg *Thamnocalamus tessellatus* forms dense stands, which grow on the cool, moist, sheltered south slopes of Platberg. These stands occur below the vertical cliffs at about 2000 m, and in some of the gullies which drain the seasonal streams as low as 1980 m (Brand et al., 2009). The *Thamnocalamus tessellatus* vegetation community is dominated by this monotypic, endemic genus, which is a species-poor community with limited presence of the low trees *Buddleja loricata, Searsia divaricata* and *Leucosidea sericea* (Brand et al., 2009). Shading out of competition and the dense rhizomatous, root system inhibits growth of other species.

In Africa, Bambusoideae are mainly tropical species confined to the humid forest shade where *Arundinaria alpina* grows in dense stands on mountains between 2 130 m and 3 200 m (White 1983). In South Africa *Thamnocalamus tessellatus* (previously named *Arundinaria tessellatus*) is confined predominantly at high altitudes of 2 700 m in the Stormberg and Drakensberg, but may be found as low as 1450 m (Pooley 2003) on the Ngeli inselberg in KwaZulu/Natal. Where it does occur in South Africa, *Thamnocalamus tessellatus* has a limited range, composed of disjunct populations - it only occurs again in the Himalayas (Pooley 2003).

The almost total dominance of C_4 grasses on Platberg is somewhat anomalous as it would have been predicted that for Platberg, with its relatively high altitude of 2 350 m with snow, frost and freezing temperatures, high rainfall between 700 - 1 200 mm per annum and close proximity to the Drakensberg, would be dominated or at least have a higher cover/abundance of C_3 grasses. However, despite the cold conditions on Platberg, its altitude may be below the limit for continuous cold for extended periods, and thus below the altitude at which C_3 grasses metabolic pathway predominates (Pitterman & Sage 2000; Sage 2001) as with the higher elevations at 3000 m for the Drakensberg (Hillard & Burtt 1987; Mucina & Rutherford 2006). The grassland structure and composition on Platberg may also be a reflection of palaeocological conditions, which started in the Miocene some 20 million years ago (Scott et al., 1997).

5.5 Wetland affinities

Wetlands form distinct and unique vegetation communities embedded in all eight Biomes in South Africa as well as through out the mountains and associated phytochoria of the Afromontane Region. Consequently, in South Africa, wetlands have been assigned the formal vegetation designation of Azonal units (Mucina & Rutherford 2006). Wetlands on Platberg are embedded in the grassland as playas or pans with semi-permanent, or permanent open water (Figure 7), which is the key factor that determines the common species shared by wetlands (Stock et al., 2004). A total of 13 naturally occupying, different wetland types where identified with 188 species including five alien (weeds) plants. The wetlands are not particularly species rich, with an average of 13.56 species per 30 m^2 (the lowest for all vegetation types), ranging from 7–29 species per sample plot. This is lower than found for other high altitude wetlands by Fuls (1993) and Malan (1998) who recorded 21 species per plot. Wetland vegetation is dominated by a single species or two to three species mostly hydrophilic grasses, sedges or juncales, which contributes to a low diversity index (Burgoyne et al., 2000). Sedge dominated wetlands occur on inselbergs in West Africa (Porembski & Brown 1995) and are a feature of most, if not all inselbergs (Parmentier et al., 2006). Very few species are endemic to high altitude wetlands, the majority also occurring in low altitude fresh water wetlands (Collins 2005). This inflates the total species richness in wetlands as numerous species will also occur in association with, but not exclusively to wetlands, the status of these species associated with wetlands (wetland indicator status), are classified into 5 categories. Due to the geology on Platberg; it is capped with lava, no vernal pools occur. Vernal pools are abundant on the lower altitude inselbergs (Korannaberg, Thaba Nchu and Thaba Patswa), in the arid interior and north of Platberg where the more resistant igneous capping has been lost and the softer sandstone exposed weathering to form vernal pools. Vernal pools are species poor with a low biodiversity, but contain Obligate Wetland, and Facultative wetland species with a high proportion of endemics (Du Preez & Bredenkamp 1991; Mucina & Rutherford 2006).

6. Phytogeographic comparison and biodiversity of Platberg

The concept of islands having high numbers of endemics (MacArthur & Wilson 2001) associated; generally with low species richness (Linden 2003) is typical of island fauna and flora as well as African mountain regions (Kingdom 1989). Platberg, as an inselberg 'island' does not fully reflect this association; it has high numbers of endemics, 27 (Brand et al., 2010), but has high species richness with a total of 670 taxa.

On a global scale the United Kingdom (Whales, Scotland, Northern Ireland and England), covers 312 000 km^2 (a surface area four-orders-of-magnitude larger than Platberg which is 30 km^2) with a total vascular plant species count of approximately 1400 of which none are endemic (Preston et al., 2002). In comparison Platberg has 670 species with the 27 endemic. It is problematic trying to compare species richness (alpha-diversity), with turnover between habitats (beta-diversity) and turnover between floras from one landscape to another (gamma-diversity), (Cowling et al., 1992). Similarities in edaphic conditions must take into account and environmental differences minimised to be able to match the differences in habitat gradients (Cowling et al., 1992). For Platberg it is relatively easy to match alpha-diversity with sites elsewhere. However, this becomes more problematic with beta and gamma diversity as different researcher use different plot sizes for similar biomes. For grassland Perkins et al., (1999a) uses 25^2 m, for Platberg plot size is 30 m^2 while Eckhardt et al. (1993, 1995), Fuls (1993) and Kay (1993) use 100 m^2 plot sizes.

Fig. 7. High altitude freshwater wetlands embedded in grassland on Platberg summit plateau.

Phytogeographic comparison for the vegetation of Platberg and Korannaberg is shown in Table 5 (Brand et al., 2010). There is a high degree of similarity for the Angiosperm flora for Platberg and Korannaberg with the larger numbers for Korannaberg possibly the result of the much larger area, (MacArthur & Wilson 2001, Cowling et al., 2002), which, as surface area increases, more species can be supported (Linder 2003, Gröger & Barthlott 1996). The higher species numbers for Korannaberg could be due to its position, which has strong Afromontane, Nana Karoo and Savanna Biome floristic influences (Du Preez & Bredenkamp 1991, Du Preez et al. 1991).

Site	Families	Genera	Species	Area (km^2)
Platberg this survey 2007	96	305	670	30.0
Korannaberg (Du Preez 1991)	115	385	767	129.52
Golden Gate National Park (Kay 1983)	N/A	N/A	566	47.9
Kammanasie Nature Reserve (Clever et al. 2005)	> 76	> 229	481	494.30
Bourke's Luck, Mpumalanga (Brown et al. 2005)	73	176	263	5.244

Table 5. A summary of Platberg and Korannaberg floras other sites, (modified from Brand et al., 2010)

However, climate and topography are only part of the explanation for high biodiversity and it is rather actual diversity of plant communities, which provides a measure of habitat diversity, (Cowling & Lombard 2002) and is a reflection of response to climate change (Scott et al., 1997; Mucina & Rutherford 2006).

For the Ivory Coast inselbergs, species composition was highest for Poaceae, then Cyperaceae, Scrophulariaceae and Lentibulariaceae. Species number and diversity were closely related to rainfall patterns for inselberg habitats (Porembski & Barthlott 1995). An examination of *Afrotrilepis pilosa* vegetation mats, shows the most species rich families are

Cyperaceae (3 species), Poaceae (2), Orchidaceae (2) and Rubiaceae (2) with a total of only 15 species found on 100 relevés. For the *Selaginella caffrorum* mats on Platberg Poaceae was the most species rich family (9), second was Asteraceae (8), Cyperaceae with 4 and Crassulaceae with 3 species with a total of 100 species in 4 relevés giving an average of 25 species. Geophytes are abundant (6) with none occurring on the *Afrotrilepis pilosa* mats and few occurring on the Ivory Coast inselbergs. The high geophytes numbers is a trend seen in the DAC and an influence from the geophyte rich Cape flora (Goldblatt & Manning 2001).

Porembski et al., (1996) found that on the vegetation of small size inselbergs in West African rain forests, 66 species of vascular plants occurred in 29 families, with the numbers showing good correlation with inselberg size. The largest Families are Poaceae, Cyperaceae, with Acanthaeceae, Commelinaceae and Malvaceae. Biogeographic affinities show species distribution widespread for Sudano-Zambezian elements, which are widespread for vegetation on inselbergs in the Comoé National Park in the Ivory Coast with Poaceae, Cyperaceae and Fabaceae comprising the most species rich families out of a total of 216 species found on 18 inselbergs (Porembski & Brown 1995); with most species found not being restricted to the inselbergs but in other habitats, savanna, marshes or waste ground.

Inselbergs host a comparatively high amount of endemic species, 86 on Venezuelan inselbergs (Gröger & Barthlot 1996). Endemic angiosperms in the Drakensberg number 410, on Platberg, endemics number 27, three of which are Pterdophytes (Brand et al., 2010). Additionally, Table 6 shows a comparison of the most abundant inselberg family level vegetation.

Platberg South Africa (Brand et al., 2010)	**Zimbabwe Inselbergs (Seine et al., 1998)**	**Ivory Coast Inselbergs Porembski et al., 1996)**	**Venezuela Inselbergs (Gröger & Barthlot 1996)**
1. Asteraceae 126	Poaceae 53	Poaceae 72	Cyperaceae 40
2. Poaceae 73	Cyperaceae 40	Cyperaceae 68	Rubiaceae 40
3. Cyperaceae 39	Fabaceae 40	Fabaceae 62	Melastomataceae 36
4. Fabaceae 33	Asteraceae 33	Scrophulariaceae 23	Orchidaceae 33
5. Scrophulariaceae 28	Scrophulariaceae 22	Rubiaceae 22	Poaceae 31
6. Hyacinthaceae 21	Euphorbiaceae 18	Orchidaceae 15	Bromeliaceae 20
7. Orchidaceae 16	Rubiaceae 18	Commelinaceae 14	Apocynaceae 18
8. Iridaceae 16	Lamiaceae 14	Lentibulaceae 14	Caesalpiniaceae 18
9. Geraniaceae 15	Adiantaceae 12	Caesalpiniaceae 13	Fabaceae 17
10. Crassulaceae 13	Acanthaceae 10	Euphorbiaceae 13	Euphorbiaceae 15

Table 6. A comparison of the 10 largest plant families for African and South American inselbergs (numbers represent species per family).

It is obvious that some families have better representation on inselbergs than others. An immediate difference is Poaceae, which is fifth largest for Venezuela while it is first for the Ivory Coast and Zimbabwe, and second for Platberg and the DAC (Brand et al., 2010), with other species rich families of Cyperaceae, Asteraceae, Fabaceae, Scrophulariaceae and Orchidaceae. The trend for Platberg species rich flora is comparable to the DAC (Brand et al., 2010). The floristic trends shown on these disparate sites, all show high abundance for

Asteraceae, Poaceae and Fabaceae which may represent the pattern of evolutionary radiation seen for these groups during the Eocene as well as associated ungulate grazers and predatory fauna (Retallack 2001). Cyperaceae is another important group, and its exploitation of environmental niches and evolution, may be parallel with Grasses, but may show a some-what different route. Cyperaceae are a widespread family, but generally have a lesser cover abundance throughout their range compared with grasses (Du Preez & Bredenkamp 1991; Kooij 1990; Perkins et al., 1999a; Bester 1998). Cyperaceae (like Poaceae) seems to have evolved C_4 photosynthetic pathways several times (Clayton 1975; Ferrier 2002; Stock et al., 2004) and even though they form components of grasslands and are commonly associated with wetlands and other seasonally moist areas, they also occur in arid regions (Gordon-Grey 1995; Jürgens 1997; Stock et al., 2004). Unlike Grasses (Gibbs-Russell et al., 1991) for Cyperaceae in southern Africa, no strong relationship is found such as altitude and rainfall (Stock et al., 2004). Their development may be less dependent on the mechanisms responsible for grass radiation, it may be that co-evolution of grasses and ungulates, climate change and CO_2 level changes, is a reflection of composition found in the DAC, Platberg and the inselbergs compared in Table 3, which show a general pattern for the top most specious families to include Asteraceae, Poaceae, Cyperaceae and Fabaceae.

For Venezuela inselberg flora, Asteraceae is significantly under represented which is a reflection of the South American regional flora, which occurs in high rainfall areas (Gröger & Barthlot 1996). This is the same trend shown for Ivory Coast inselbergs, moisture availability is the common environmental factor (Gröger & Barthlot 1996). Asteraceae is not well represented in west Africa, unlike East Africa where in Zimbabwe, the DAC and Platberg significant high levels of Asteraceae are found. This trend is representative of the Asteraceae for arid and semi-arid regions, including the CFR (Jürgens 1997; Linder 2003, Mucina & Rutherford 2006) where high levels of endemic species (63.2%) are found.

7. Factors influencing inselberg flora

7.1 Hybridisation

For the DAC, Hillard and Burtt (1987) list 21 hybrid taxa of which three (two *Senecio* crossed species and one *Cephalaria*) are exclusive to high altitude DAC records, and do not occur on Platberg, while the *Protea roupelliae* x *P. subvestita* cross has both species growing in the vicinity of Platberg. Of the remaining 17 hybrids it could be that some or all of them may occur on Platberg. This level of taxonomic expertise must wait for future detailed analysis. Hillard & Burtt (1987) do not offer an explanation for the occurrence of these 21-recorded hybrids (it is possible there are more). However, hybridisation breaks down species boundaries. Linden (2003) reports for the Cape flora, where species co-occur, such as *Moraea* hybrids, these are frequently found. This is the case also for *Romulea* with artificial hybrids cultivated for *Sparaxis, Watsonia* and *Ixia* of which *Watsonia* and *Morea* occur on Platberg (Linden 2003). Most of these Cape hybrids are sterile, however, hybridisation, sterile or not, is an environmental response to selection forces, which, in the case of Cape flora, limit gene flow (Linden 2003).

For the higher areas of the DAC, and Platberg as an inselberg, a degree of geographical isolation occurs which allows for edaphic and microclimatic, as well as larger, longer climatic and geological processes, to provide for geographical isolation of species (MacArthur & Wilson 2001; Porembski & Brown 1995; Gröger & Barthlott 1996; Linden 2003). This isolation has allowed for speciation and to quote from Linden (2003, page 623):

" ... factors that allow (and drive?) speciation may be very similar to those that allow species to co-exist. These factors could be expected to include those that limit gene flow between sister species, as well as those that result in differential selection. Reproductive isolation is needed to prevent the ecological specialisation from being lost...."

This may offer a partial explanation for the high numbers of hybrids recorded, as well as the speciation and high numbers of endemics or near-endemics observed within the DAC and by extension, Platberg. Hybrids and rare species do not contribute to community patterns (beta-diversity), but are important for maintenance of biological diversity (Mutke et al. 2001; Cowling & Lombard 2002).

7.2 Pollination: Birds, insects and wind

7.2.1 Birds

For the central and southern Drakensberg, Hillard & Burtt (1987) have recorded bird pollination for *Protea, Kniphofia, Watsonia, Pelargonium schlecterii, Sutherlandea montana, Halleria lucida* and *Leonotis*. Most *Protea* species have a specalised feeding connection with Sugar birds *(Promerops* spp.) and Sunbirds (*Nectariania* species) detailed by Kingdom (1989) as well as the Cape Canary (*Serinus leucopterus*) and a beetle specialists which also feeds on Protea nectar. Except for *Protea*, all the other plant species occur on Platberg. This form of pollination, where a single pollinator will visit different, but close-by species with similar flower morphology and physiology, may also be responsible for the phytosociological associations of some plants. On Platberg, such community assemblage corresponds to the *Leonotis ocymifolia–Watsonia lepida* and *Halleria lucida* plant associations (Brand et al., 2009). Limited empirical research has been done to establish the connection between bird/plant community associations and is a subject for future investigation.

7.2.2 Insects

Hillard and Burtt (1987) have also described pollination by bees of highly specalised plants; Orchidaceae, and Scrophulariaceae. In the Fynbos Biome closely related plant species could have very different pollinators, with pollinator selection playing a central role in speciation and species richness (Linden 2003). Hillard & Burtt (1987) and Linden (2003) discuss the possible link between insect pollinators and floral guilds (phytosociological communities), with the possibility that such guilds are the results of insect/plant interaction and association.

Orchidaceae is a species rich family in the Cape flora and comprise 3.3% of the entire flora (Goldblatt & Manning 2000), 5.2% of the angiosperm DAC flora (Carbutt & Edwards 2004) and 2.7% of Platberg flora. Many orchid species are unique to select habitats. It may be that the process of pollination as well as pollinator selection (Hillard & Burtt 1987, Linden 2003) may play a significant part in accounting for the high numbers of Orchidaceae found in the Cape flora, on the Drakensberg and at Platberg.

7.2.3 Wind

On Platberg, Asteraceae has the greatest numbers of genera and species, however grasses are the most dominant plant group with Cyperaceae the third most abundant. Wind dispersal is an important mechanism of pollination for grasses (Gibbs-Russell 1991) and for sedges (Gordon-Grey 1995). Wind pollination is also important for *Cliffortia* and *Anthospermum* fynbos species (Hillard & Burtt 1987), both of which are abundant on

Platberg as well as the DAC. This long-distance dispersal by wind allows for crossing mountain valleys, from one isolated peak or inselberg to another.

Pollination by wind allows for gene flow between disparate areas (Cowling & Lombard 2002), which have different geology, soils, moisture availability and climate (Burke 2001, 2002; Linder 2003). The effects of wind to influences floristic composition and species richness over long distances and connect different regions is seen on the granite inselbergs in West African (Porembski et al., 1996), Namibian (Burke 2001), and Mulanje (Burrows & Willis 2005), basalt lavas of Platberg (Brand et al., 2010), the Drakensberg (Mucina & Rutherford 2006), and quartzite of the Cape Fold Mountains (Linden 2002).

8. Conclusion

Platberg is a centre of significant biological diversity, with high species richness, vegetation and ecosystem complexity and a centre of genetic diversity and variation. It occurs as an island in the Grassland 'sea' and shares inselberg floral richness and endemism which can be tracked via the Afromontane archipelago-like string of inselbergs and mountains which stretch north through the Chimanimani Mountains, into Malawi, the East African Arc of Mountains via Tanzania and north through Ethiopia into Eurasia. It also shows a western tract via the Congo, Ivory Coast and Cameroon inselbergs and mountains.

The current vegetation patterns on Platberg reflect changes in palaeo-environmental cycles of cooling and warming which, since Miocene times have had the greatest influence on the texture and composition of plants and speciation on Platberg. Floristically and choriologically, Platberg has a grass composition showing transition between C_3 high altitude grasses and C_4 grasses, the latter favouring hotter temperatures, lower altitude and lower rainfall. It also has a significant composition of plant families using CAM, C_3, C_4 and metabolism. This floristic composition is shared by the DAC. Current trends towards climate change will alter the composition and species numbers of grasses as well as other plant families, which use these metabolic pathways.

Floristic and compositional connections extend up the Great Escarpment via a series of inselbergs as well along sheltered gullies into the Lebombo Mountains, and west via the Magaliesberg (Figure 8). Affinities are also shown in the Chimanimani Mountains, with attenuation in the Mulanje Massif and Nyika Plateau. Floristic connections are also shared with other inselbergs in west and central Africa and show extension along the Afro montane mountain region.

Climate changes have directly influenced the evolution, structure and composition of the vegetation of southern Africa, the main driving force possibly being orbital (Croll/Milankovitch) cycles (Kutzbach 1976; Bennett 1999; Muller & MacDonald 1999; Linder 2003). The most significant, relatively recent changes started in the Palaeocene (65 million years BP) with the Gondwana break-up. This started a cycle of erosion and uplift of the interior of southern Africa. Cycles of glaciations and interglaciation, cool wet, hot dry periods, influenced the environment and forced high speciation (Bennett 1999), still evident in the high species richness exhibited today in the flora of the DAC, Platberg and the Cape flora. At the start of the Quaternary (2.4 million years BP) the DAC and Cape Flora were much as they are today, but showed range extensions and contractions due to the longer, cooler, wetter 100 000 year glaciations, and the shorter, hotter, drier 10 000 year interglaciation periods.

Fire and the grazing by herbivores also helped shape the vegetation, with the greatest influence on the Grassland Biome. These cumulative effects, geological processes, climate

change, CO_2 level fluctuations, fire and grazing are all responsible for the present day species richness and diversity recorded on Platberg and its parent vegetation of the Drakensberg. The Grassland Biome in South Africa is second largest of all eight Biomes (354 593.501 km^2) after the Savanna Biome (412 544.091 km^2, Mucina & Rutherford 2006), with both Biomes providing an enormous sink for Carbon as well as climate amelioration.
Between 30–55% of the Grassland biome has already been transformed with only 5.5% protection. Grassland is under significant threat of continual transformation, the most severe is ploughing, which disrupts the soil and releases not only moisture, but also nutrients and Carbon (Mucina & Rutherford 2006).

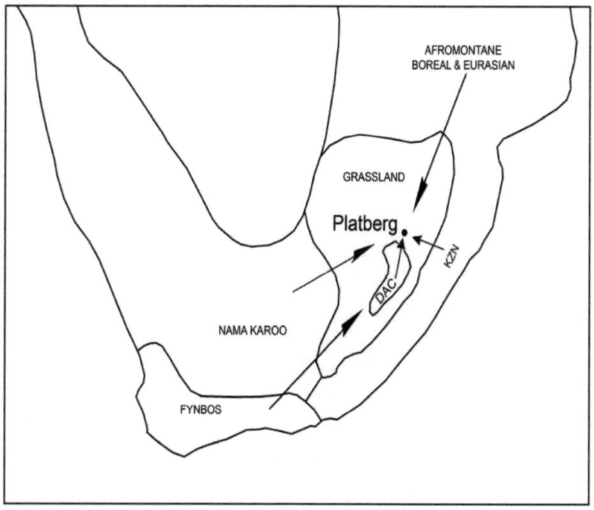

Fig. 8. Platberg showing tracks of the major floristic influences (modified from Rutherford & Westfall 1994).

Undisturbed grassland provides a significant Carbon sink, and should thus be given priority status for conservation, which should include both the above ground and belowground grass and soil ecosystems (Retallack 2001).
Historically in South Africa, areas for conservation were not selected primarily for ecological reasons until the mid 1970's (and then only in the Cape), but were rather based on other criteria, such as national strategically protection of the watershed around dams; fragmented forest areas to protect the water sources higher up in the mountains, or politically determined boundaries such as the Kruger National Park. It was only between 1971 and 1982 in the Cape, that a few wilderness areas were established for scientific research in natural ecosystems, aesthetic values they engendered and physical and spiritual opportunities they afforded (Rebelo 1992). Biogeographical considerations such as fynbos being 'islands' surrounded by forest were used as well as critical plant population size and habitat. These parameters plus total land surface were selected in determining conservation areas. It was found that for mountain areas, the minimum statuary size for reserves should be at least 10 000 ha (Rebelo 1992). Human population growth and increased urbanisation are seen as two major threats to natural ecosystems, with the implications for conservation of the effects of unrestricted human population growth being considered (Rebelo 1992; Ferrier 2002).

The scale of human impact on the biogeographical distribution of plants and animals will only continue to grow globally, but specifically in Africa, with the continued existence of many plant and animal species more dependant on human social, rather than on physical environmental factors (Meadows 1996; Anderson et al., 2001; Mutke et al., 2001). The African landscape, in particular East Africa, has had a long history of influence by humans, i.e. the Savanna and Grassland Biomes (Meadows 1996). Globally, and in the context of the use of resources and climate change in Africa, the human species now exerts an influence on all other species of organism, which in its scale and intensity is critical in the future evolution of the African flora and fauna (Kingdom 1989; Meadows 1996; Anderson 2001; Mucina & Rutherford 2006).

The conservation value of Platberg is significant: is a link between high altitude C_3 grasses and lowland C_4 grasses. It is also the best-preserved continuous high altitude grassland within the Grassland Biome and has escaped agricultural activity which giving the site a unique status, and leaves enormous scope for future research (Smith & Young 1987). Platberg is an excellent natural history laboratory, which connects the present day environment with geological, palaeo-environmental and evolutionary processes from modern Holocene to Jurassic times (Scott et al.1997; Anderson 2001; Linder 2003; Mucina & Rutherford 2006).

9. Acknowledgements

National Geographic (Grant number 7920-05) for the generous funds without which the fieldwork and the study would not have been possible.

Snow on Platberg, Figure 4. Kind permission of Theunis Bekker.

Figure 8, Wetlands on Platberg is courtesy of Nacelle Collins.

10. References

Anderson, J.M. (2001). Management for the New Millennium; How will biodiversity benefit? In: *Towards Gondwana Alive*, Anderson, N. (Ed.), pp. 134-135, National Botanical Institute, Pretoria

Bennett, K.C., (1999). Milankovitch cycles and their effects on species in ecological and evolutionary time. *Paleobiology*, Vol. 16, No. 1, pp.11-21

Bester, S.P. (1998). Vegetation and Flora of the Southern Drakensberg Escarpment and adjacent areas. MSc Thesis, University of Pretoria, Pretoria

Brand, R.F.; du Preez, P.J. & Brown, L.R., (2008). A floristic description of the Afromontane Fynbos communities on Platberg, Eastern Free State, South Africa. *Koedoe*, Vol. 50, pp. 32-41

Brand, R.F.; du Preez, P.J. & Brown, L.R., (2009). A classification and description of the shrubland vegetation on Platberg, Eastern Free State, South Africa. *Koedoe* Vol. 51, No. 1, Art. #696,11 pages. DOI: 10.4102/ koedoe.v51i1.696

Brand, R.F.; du Preez, P.J. & Brown, L.R. (2010). A Floristic Analysis of the vegetation of Platberg, Eastern Free State, South Africa. *Koedoe* Vol. 52, No. 1, Art. #710, 11 pages, DOI: 10.4102/koedoe.v52i1.710.

Brown, L.R., Marais, AJ., Henzi, S.P. & Barrett, L. 2005. Vegetation classification as the basis for baboon management in the Bourke's Luck Section of the Blyde Canyon Nature Reserve, Mpumalanga. *Koedoe* 48(2): 71-92

Burgoyne, P.M.; Bredenkamp, G.J. & Van Rooyen, N. (2000). Wetland vegetation in the North-eastern Highveld, Mpumalanga, South Africa. *Bothalia* Vol. 30, No. 2, pp. 187–200

Burke, A. (2001). Determinants of inselberg floras in arid Nama Karoo landscapes. *Journal of Biogeography*, Vol. 28, pp. 1211–1220

Burke, A. (2002). Plant Communities of a Central Namib Inselbnerg Landscape. *Journal of Biogeography*, Vol. 13, No. 4, pp. 483–492

Burrows, J. & Willis, C. (2005). *Plants of the Nyika Plateau: an account of the vegetation of the Nyika National Parks of Malawi and Zambia*, Report No. 31, SABONET, Southern African Botanical Diversity Network, Pretoria

Carbutt, C. & Edwards, T.J. (2004). The Flora of the Drakensberg Alpine Centre. *Edinburgh Journal of Botany*, Vol. 60, No. 3, pp. 581-607

Carbutt, C. & Edwards, T.J. (2006). The endemic and near-endemic angiosperms of the Drakensberg Alpine Centre. *South African Journal of Botany*, Vol. 72, pp. 105–132

Cavagnora, J.B., (1988). Distribution of C_3 and C_4 grasses at different altitudes in a temperate arid region of Argentina. *Oecologia*, Vol. 76, pp. 273-277

Clayton, W.D. (1976). The Chorology of African Mountain Grasses. *Kew Bulletin*, Vol. 31, No. 2, pp. 273-288

Cleaver G, Brown LR & Bredenkamp GJ. 2005. The phytosociology of the Vermaaks, Marnewicks and Buffelsklip Valleys of the Kammanassie Nature Reserve, Western Cape. *Koedoe* 48(1): 1-16

Collins, N. B. (2005). *Wetlands: the basics and some more.* Free State Department of Tourism, Environmental and Economic Affairs, Bloemfontein

Cowling, R.M.; Holmes, P.M & Rebelo, A.G. (1992). Plant Diversity and Endemism. In: *The Ecology of Fynbos: Nutrients, Fire and Diversity*, Cowling, (Ed.), pp. 62-112, Oxford University Press, Cape Town

Cowling, R.M., & Lombard, A. T. (2002). Heterogeneity, speciation/extinction history and climate: explaining regional plant diversity patterns in the Cape Floristic Region. *Diversity and Distributions*, Vol. 8, pp. 163–179

Du Preez, P. J., 1991, 'A syntaxonomical and Synecological Study of the Vegetation of the South-Eastern Orange Free State and related areas with special reference to Korannaberg', Unpublished PhD Thesis, University of the Orange Free State, Bloemfontein.

Du Preez, P. J. & Bredenkamp, G. J. (1991). Vegetation classes of the southern and eastern Orange Free State, (Republic of South Africa) and the highlands of Lesotho. *Navorsing van die Nasionale Museum Bloemfontein*, Vol. 7, pp. 477–526

Du Preez, P. J.; Bredenkamp, G. J. & Venter, H.J.T. (1991). The syntaxonomy and synecology of the forests in the eastern Orange Free State, South Africa. *South African Journal of Botany*, Vol. 57, pp. 198–206

Eckhardt, H.C. Van Rooyen, N. & Bredenkamp, G.J. (1993). The phytosociology of the thicket and woodland vegetation of the north-eastern Orange Free State. *South African Journal of Botany*, Vol. 59, 4, pp. 401 - 409

Eckhardt, H.C. Van Rooyen, N. & Bredenkamp, G.J. (1995). The grassland communities of the slopes and plains of the north-eastern Orange Free State. *Phytocoenologia*, Vol. 25, 1, pp. 1-21

Ferrier, S. (2002). Mapping Spatial Pattern in Biodiversity for Regionl Conservation Planning: Where to from Here? *Systematic Biology*. Vol. 51, No. 2, pp. 331-363

Fuls, E.R. (1993). *Vegetation Ecology of the northern Orange Free State*. Ph D Thesis, University of Pretoria, Pretoria

Germishuizen, G. & Meyer, N.L. (2003). *Plants of southern Africa: an annotated checklist.* Strelitzia 14, National Botanical Institute, Pretoria

Gibbs-Russell, G.E.; Watson, L.; Koekemoer, M.; Smook, L.; Barker, N.P.; Anderson H.M. & Dallwitz, M.J. (1991). *Grasses of Southern Africa.* Memoirs of the Botanical Survey of South Africa, No. 58, Botanical Research Institute of South Africa, Pretoria

Goldblatt, P. & Manning, J. (2000). *Cape Plants. A conspectus of the Cape Flora of South Africa.* Strelitzia 9, Missouri Botanical Gardens USA & National Botanical Institute, Cape Town, South Africa

Gröger, A., & Barthlott, W. (1996). Biogeography and diversity of the inselbergs (Laja) vegetation of southern Venezuelan. *Biodiversity Letters,* Vol. 3, pp. 165–179

Gordon-Gray, K.D. (1995). *Cyperaceae in Natal.* Strelitzia 2, National Botanical Institute, Pretoria

Goudie, S. (1996). The Geomorphology of the Seasonal Tropics, In: In: *The Physical Geography of Africa,* Adams, W.M., Goudie, A.S. & Orme. A.R., (Eds.), 148-172, Oxford University Press, UK

Haaksma, E. D. & Linder, H.P. (2000). *Restios of the Fynbos.* The Botanical Society of South Africa, Cape Town

Hill, T.R. (1996). Description, Classification and Ordination of the Dominant Vegetation Communities, Cathedral Peak, KwaZulu-Natal Drakensberg. *South African Journal of Botany,* Vol. 62, pp. 263-269

Hilliard, O.M. & Burtt, B. L. (1987). *The Botany of Southern Natal Drakensberg.* National Botanical Gardens, Cape Town

Hoare, D.B. & Bredenkamp, G.J. (2001). Syntaxonomy and environmental gradients of the grasslands of the Stormberg/Drakensberg mountain region of the Eastern Cape, South Africa. *South African Journal of Botany,* Vol. 67, pp. 595-608

Kay. C.; Bredenkamp, G.J. & Theron, G.K. (1993). The plant communities of the Golden Gate Highlands National Park in the north-eastern Orange Free State. *South African Journal of Botany,* Vol. 59, No. 4, pp. 442–449

Killick, D.J.B. (1963). *An Account of the Plant Ecology of the Cathedral Peak Area of the Natal Drakensberg.* No. 34. Memoirs of the Botanical Survey of South Africa, Pretoria

Killick, D.J.B. (1978a). The Afro-alpine Region. In: *Biogeography and Ecology of Southern Africa,* Werger, M.J.A., (Ed.), 515-560, Junk, The Hague

Killick, D.J.B. (1978b). Notes on the Vegetation of the Sani Pass Area of the Southern Drakensberg. *Bothalia,* Vol. 12, pp. 537-542

King, L.C. 1963. *South African Scenery. A textbook of Geomorphology.* 3rd edn. Oliver and Boyd, Edinburgh

Kingdom, J. (1989). *Island Africa. The evolution of Africa's Rare Animals and Plants.* Princeton University Press, Princeton, New Jersey. ISBN 0-691-08560-9

Kooij, M.S.; Bredenkamp, G.J. &. Theron, G.K (1990). The vegetation of the north-western Orange Free State, South Africa. *Bothalia,* Vol. 20, No. 2, pp. 214-248

Körner, C. (2003). *Alpine Plant Life. Functional plant Ecology of High Mountain Ecosystems,* Springer-Verlag, Berlin

Kutzbach, J.E., (1976). The nature of climate and climatic variations, *Quaternary Research,* Vol. 6, pp. 471-480

Linder, H. P. (2003). The radiation of the Cape flora, southern Africa. *Biological Review.* Cambridge Philosophical Society, United Kingdom, Vol. 78, pp. 597–638

Low, A. B. & Rebelo, A. G. (1996). *Vegetation of South Africa, Lesotho and Swaziland.* Department of Environmental Affairs & Tourism, Pretoria

MacArthur, R.H. & Wilson, E.O. (2001). *The Theory of Island Biogeography*, Princeton University Press, Princeton

Malan, P. W. (1998). *Vegetation Ecology of the Southern Free State*. PhD Thesis, University of the Orange Free State, Bloemfontein

Meadows, M.E. (1996). Biogeography. In: *The Physical Geography of Africa,* Adams, W.M.; Goudie; A.S. & Orme, A.R. (Eds), pp. 161-172, Oxford University Press, Oxford

Moffett, R. O.; Daemane, M.E.; Pitso, T.R.; Lentsoane, R. & Taoana, T.R.N. (2001). A Checklist of the Vascular Plants of Qwa-Qwa and notes on the Flora and Vegetation of the Area. *UNIQWA Research Chronicles*, Vol. 3, No. pp. 32-83

Moon, B.P. & Dardis, G.F. (1992). *The Geomorphology of Southern Africa,* Halfway House, Southern Book Publishers, South Africa

Mucina, L, & Rutherford, M.C. (2006). *The Vegetation of South Africa, Lesotho and Swaziland,* South African National Biodiversity Institute, Strelitzia 19, Pretoria

Muller, R.A., & MacDonald, G.J. (1999). Glacial Cycles and Astronomical Forcing. *Science, New Series*, Vol. 277, No. 5324, pp. 215-218

Mutke, J.; Kier, G.; Braun, G.; Schultz, C.H.R. & Barhlott, W., (2001). Patterns of African vascular plant diversity – a GIS based analysis. *Syst. Geogr. Pl.*, Vol. 71, pp. 1125–1136

Parmentier, I.; Oumorou, M.; Porembski, S.; Lejoly, J. & Decocq, G. (2006). Ecology, distribution, and classification of xeric monocotyledonous mats on inselbergs in West Africa and Atlantic central Africa. *Phytocoenology,* Vol. 36, No. 4, pp. 547–564

Perkins, L.; Bredenkamp, G.J. & Granger, J.E. (1999a). The phytosociology of the incised river valleys and dry upland savanna of the southern KwaZulu-Natal. *South African Journal of Botany,* Vol. 65, pp. 321-330

Pitterman, J. & Sage, R.F. (2000). The response of high altitude C_4 grass *Muhlenbergia montana* (Nutt.) A.S. Hitchc. to long- and short-term chilling. *Journal of Experimental Botany*, Vol. 52, No. 357, pp. 829–838

Pooley, E. (2003). Mountain Flowers. *A field guide to the Flora of the Drakensberg and Lesotho.* The Flora Publishing Trust, Durban

Pond, U.; Beesley, B.B.; Brown, L.R. & Bezuidenhout, H. (2002). Floristic analysis of the Mountain Zebra National Park, Eastern Cape. *Koedoe,* Vol. 45, No. 1, pp. 35-57

Porembski, S. & Barthlott, G.B. (1995). A species poor tropical sedge community: *Afrotrilepis pilosa* mats on inselbergs in West Africa. *Nordic Journal of Botany,* Vol. 16, No. 3, pp. 239–246

Porembski, S., & Brown, G. (1995). The vegetation of inselbergs in the Comoé National Park (Ivory Park), *Conservatoire et Jardin Botanicues de Geneve,* Vol. 50, No. 2, pp. 351–365

Porembski, S.; Szarzynske, J.; Mund, J-P. & Barthlott, W. (1996). Biodiversity and vegetation of small-size inselbergs in a West African rain forest (Taï, Ivory Coast), *Journal of Biogeography,* Vol. 23, pp. 47–55

Porembski, S.; Fisher, E. & Biedenger, N. (1997). Vegetation of inselbergs, quartzitic outcrops and ferricretes in Rwanda and eastern Zaire (Kivu), *Bull. Jar. Bot. Nat. Plantentuin Bel.,* Vol. 66, pp. 81–95

Porembski, S.; Martinelli, G.; Ghlemüller, R. & Barthlott, W. (1998). Diversity and ecology of saxicolous vegetation mats on inselbergs in the Brazilian Atlantic rainforest, *Diversity and Distribution,* Vol. 4, pp. 107–119

Preston, C. D.; Pearman, D. A. & Dines, T. D. (2002). *New atlas of the British and Irish flora.* Oxford University Press, Oxford

Rebelo, A.G. (1992). Preservation of Biotic diversity. In: *The Ecology of Fynbos. Nutrients, Fire and Diversity,* R.M.Cowling, (Ed.), pp. 309–344, Oxford University Press, Cape Town

Retallack, G.J. (2001). Cenozoic Expansion of Grasslands and Climate Cooling. *The Journal of Geology,* Vol. 109, pp. 407–426

Roberts, B.R. (1969). The vegetation of the Golden Gate Highlands National Park. *Koedoe,* Vol. 12, pp. 15–28

Rutherford, M.C. & Westfall, R.H. (1994). *Biomes of Southern Africa: An Objective Categorization,* 2nd edn, Vol. 63. Memoirs of the Botanical Survey of South Africa, Botanical Research Institute, Pretoria

Sage, F.S. (2001). Variation of the kcat of Rubisco in C_3 and C_4 plaints and some implications of photosynthetic performance at high and low temperature. *Journal of Experimental Botany,* Vol. 53, No. 369, pp. 609–920

Sarthou, C. & Villiers, J-F. (1998). Epilithic plant communities on inselbergs in French Guiana. *Journal of Vegetation Science,* Vol. 9, pp. 847-860

Seabloom E. W. & Richards, S. A. (2003). Multiple Stable Equilibrium in Grassland mediated by Herbivore Population Dynamics and Foraging Behaviour. *Ecology,* Vol. 84, No. 11, pp. 2891-2904

Seine, R., Becker, U.; Porembski, S.; Follman, G. & Barthlott, W. (1998). Vegetation of inselbergs in Zimbabwe. *Edinburgh Journal of Botany,* Vol. 55, pp. 267-293

Scott, L.; Anderson, H.M. & Anderson, J.M. (1997). Vegetation History. *Vegetation of Southern Africa.* (Eds. Cowling, R.M. & Pierce, S.M.) pp. 62–84, Cambridge University, Cambridge

Scott, L, & Vogel, J.C. (2000). Evidence for environmental conditions during the last 20 000 years in Southern Africa from ^{13}C in fossil hyrax dung. *Global and Planetary Change,* Vol. 26, pp. 207-215

Smith, G.M.; Chesselet, P.; Van Jaarsveld, E.J.; Hartmann, H.; Hammer, S.; Van Wyk, B-E.; Burgoyne, P.; Kalk, C. & Kurzweil, H. (1998). *Mesembs of the World.* Briza Publications, Pretoria

Smith. A.P., & Young, T.P., (1987). Tropical Alpine Plant Ecology. *Ann. Rev. Ecol. Syst.* Vol. 18, pp. 137-58

Stock, W.D.; Chuba, D.K. & Verboom, G.A. (2004). Distribution of South African C_3 and C_4 species of Cyperaceae in relation to climate and phylogeny. *Austral Ecology,* Vol. 29, pp. 313–319

Taylor, D. (1996). Mountains. In: *The Physical Geography of Afric,* Adams, W.M.; Goudie, A.S. & Orme, A.R. (Eds.), 287-306, Oxford University Press, Oxford

Van Wyk, A.E. & Smith, G.F. (2001). *Regions of Floristic Endemism in Southern Africa: A Review with Emphasis on Succulents,* Umdaus Press, Pretoria

Van Zinderen Bakker Jr., E.M. (1973). Ecological Investigation of Forest Communities in the Eastern Orange Free State and the Adjacent Natal Drakensberg, *Vegetatio,* Vol. 28, pp. 299-334

Von Maltitz, G. (2003). *Classification System for South African Indigenous Forests.* Council for Scientific and Industrial Research, Department of Water Affairs and Forestry, Environmentek report ENV-P-C 2003-017, Pretoria

White, F. (1983). *The Vegetation of Africa.* A descriptive Memoir to accompany the Unesco/AETFAT/UNSO vegetation map of Africa, Natural Resources Research, Volume XX, Unesco, Paris

6

Biodiversity Loss in Freshwater Mussels: Importance, Threats, and Solutions

Trey Nobles and Yixin Zhang
Department of Biology, Texas State University, San Marcos, Texas
The United States of America

1. Introduction

The loss of biodiversity worldwide has been well documented for decades, and while much of the attention of the media and scientific community has been focused on terrestrial ecosystems, other biomes such as freshwater lakes and streams have received less consideration (Myers et al., 2000). Despite the current decade (2005-2015) being declared an International Decade for Action – 'Water for Life' by the United Nations General Assembly, freshwater ecosystems worldwide are as threatened as ever by the activities of a rapidly growing human population. Surface freshwater ecosystems only constitute 0.8% of the Earth's surface, yet they contain almost 9.5% of the Earth's known species, including as many as one-third all known vertebrate species (Balian et al., 2008; Dudgeon et al., 2006). The impact of human disturbances on this disproportionate amount of biodiversity has made the extinction rate in freshwater ecosystems equal to that of tropical rainforests (Ricciardi and Rasmussen, 1999).

1.1 Freshwater ecosystem services

Because we depend on water both as a biological necessity and for the myriad of resources and services it provides us, over half of the world's population lives within 20 km of a permanent river or lake (Small and Cohen, 1999). Direct benefits and ecosystem functions of freshwater lakes, streams, and wetlands include providing sources of water for municipal and industrial use, irrigation, hydroelectric power generation, transportation corridors, recreation, and producing fish and other resources used for food and medicine. Freshwater ecosystems also provide many indirect ecosystem services such as water filtration, buffering against storms and flooding, cycling of nutrients and organic matter through the environment, and supporting ecosystem resilience against environmental change (Aylward et al., 2005; Jackson et al., 2001). These indirect ecosystem services have very real economic values. One study valued the ecosystem services of freshwater aquatic ecosystems worldwide at $6.5 trillion USD, or 20% of all the world's ecosystem services (Costanza et al., 1997).

As human populations continue to develop aquatic resources to maximize a few of these anthropogenically beneficial services such as water storage, generation of electricity, and fish production, other environmental services that are less directly important to humans are being reduced or lost (Bennett et al. 2009). The reduction of these ecosystem functions can significantly alter an ecosystem's natural character. After more than a century of unprecedented human population growth and global economic development, we have created widespread and long-term ecological disturbances to freshwater ecosystems in almost all parts of the inhabited world (Strayer and Dudgeon, 2010).

Water Supply (Extractive)	Supply of Goods Other Than Water	Non-Extractive/Instream Benefits and Uses
•Drinking, cooking, washing, and other household uses	•Fish	•Flood control
•Manufacturing and other industrial uses	•Waterfowl	•Transportation
•Irrigation of crops, parks, etc.	•Clams and mussels	•Recreation.
•Aquaculture	•Pelts	•Dilution of pollution
	•Plant products	•Water quality protection
		•Hydroelectric generation
		•Bird and wildlife habitat
		•Soil fertilization
		•Enhanced property values
		•Non-user values

Table 1. Ecosystem services provided by freshwater lakes, rivers, and wetlands (after Postel and Carpenter, 1997).

1.2 Human disturbances to freshwater systems

Humans now capture more than 50% of the world's precipitation runoff behind dams for electricity generation and water storage, and through diversion canals for irrigation. 16% of all runoff is consumed, or not returned to the rivers after use, while the remainder of the captured runoff is returned but with altered timing, amount, and quality (Jackson et al., 2001). Water pollution, including siltation, is endemic to almost all inhabited parts of the world and is consistently ranked as one of the major threats to freshwater ecosystems (Richter et al., 1997). Pollution in aquatic ecosystems not only consists of chemical toxicants like heavy metals, industrial waste, and pesticides, but also includes excessive nutrient enrichment and pharmaceuticals and personal care products (PPCPs) (Jobling and Tyler, 2003; Smith et al., 1999). Habitat loss and habitat degradation are also major reasons for worldwide biodiversity loss in aquatic ecosystems, and are caused by a multitude of anthropogenic disturbances (Allan and Flecker, 1993; Richter 1997 et al. 1997). Many freshwater species are also being overharvested for human consumption or the pet trade. This affects mostly vertebrate species, especially fishes, but can impact invertebrate species like mussels and crustaceans as well (Dudgeon et al., 2006). The threat of global climate change is pervasive across all of the Earth's ecosystems, and is also often cited as a major threat to freshwater biodiversity (Sala et al., 2000; Strayer and Dudgeon, 2010).

All of these environmental disturbances alter the "natural" chemical, physical, and biological patterns of a system, and when those conditions are changed, both the absolute and functional biodiversity of that system can be threatened. This loss of biodiversity can in turn create a feedback loop that further alters ecosystem functioning. The theory that ecosystem services depend on the biological diversity of the system is well supported for terrestrial ecosystems (Kinzig et al., 2002; Loreau et al., 2001), and recent studies have shown that maintaining biodiversity in aquatic ecosystems is crucial to the continued functioning of ecosystems and the delivery of ecosystem services as well (Covich et al., 2004).

It is widely accepted that freshwater ecosystems worldwide are suffering from a "biodiversity crisis", with estimates of 10,000-20,000 species currently extinct or threatened

(Abell, 2001; IUCN, 2007; Strayer and Dudgeon, 2010). While almost all taxonomic groups of freshwater organisms are facing unprecedented declines, some groups are especially affected.

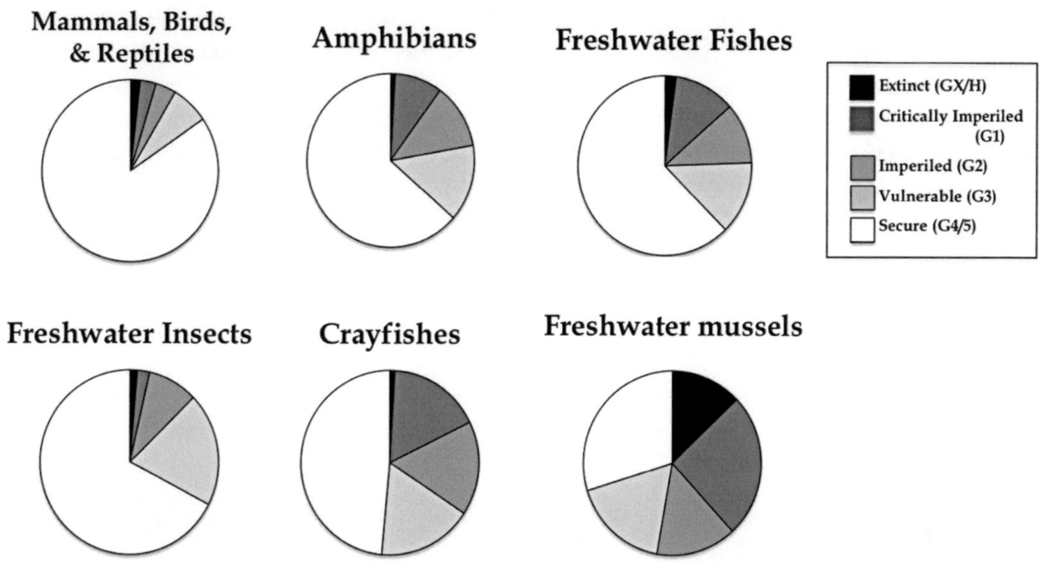

Fig. 1. Conservation status of selected groups of terrestrial and freshwater organisms using NatureServe conservation status designations (after Master et al., 2000).

2. Freshwater mussels: North America's most threatened animals

Of all groups of threatened aquatic animals, freshwater mussels (also known as unionid or pearly mussels) are the most imperiled, with 67% of North American species considered threatened (Williams et al., 1993). 35 North American freshwater mussel species have gone extinct since 1900 (Williams et al., 1993), and some scientists have estimated a 1.2% per decade extinction rate for this group, with others predicting that unless effective conservation action is taken 127 more species will become extinct over the next 100 years (Ricciardi and Rasmussen, 1999).

2.1 Classification of freshwater mussels

Freshwater mussels (order Unioniformes) belong to the subclass Paleoheterodonta, class Bivalvia, and phylum Mollusca. A total of 18 bivalve families have at least one species found in freshwater, although only about 9 have radiated to any degree there (Bogan, 1993). The order Unioniformes contains the largest number and diversity of groups with 180 out of 206 genera and 797 out of 1026 species. Within the Unioniformes, the family Unionidae is the largest, comprising nearly 80% of both the genera and species within the order (Bogan and Roe, 2008). Other important families include Hyriidae (17 genera, 83 species), Mycetopodidae (12 genera, 39 species), Sphaeriidae (8 genera, 196 species) (Bogan and Roe, 2008). As the order Unionidae is the most diverse, and has had the most research dedicated to it, we shall from here on out refer to freshwater mussels simply as Unionids, or unionid mussels.

2.2 Freshwater mussel distribution

Freshwater mussels are found on every continent with the exception of Antarctica, but reach their highest level of diversity in the Nearctic geographic region, with one-third of all species (297 recognized taxa) being found there (Bogan, 2008). The Neotropical region has 179 described species, the Oriental has 121, the Palaearctic 92, the Afrotropical 74, and the Australasian region has 29 (Bogan, 2008). Data on the conservation status of freshwater mussels globally is incomplete, with relatively strong data from only a few areas (North America, Europe, and Australia). In other areas (Africa and South America), detailed taxonomic information including the total number of species currently or historically present is lacking, which makes determining changes in species abundance and richness difficult (Bogan, 2008). There has been increased interest in the biodiversity of freshwater mussels worldwide over the last few decades, though, as scientists have realized just how rapidly this group is declining (Graf and Cummings, 2007). Hopefully this increased awareness will lead to more surveys in these understudied areas to fill in the gaps in basic knowledge that currently exist.

	Family	Genera	Species
Order Arcoida	Arcidae	1	4
Order Mytiloida	Mytilidae	3	5
Order Unioniformes	Etheriidae	1	1
	Hyriidae	17	83
	Iridinidae	6	41
	Margaritiferidae	3	12
	Mycetopodidae	12	39
	Unionidae	142	620
Order Veneroida	Cardiidae	2	5
	Corbiculidae	3	6
	Sphaeriidae	8	196
	Dreissenidae	3	5
	Solenidae	1	1
	Donacidae	2	2
	Navaculidae	1	2
Order Myoida	Corbulidae	1	1
	Erodonidae	2	2
	Teridinidae	1	1
Order Anomalodesmata	Lyonsiidae	1	1
	Total	209	1026

Table 2. Classification of freshwater mussels (6 orders and 19 families), including number of genera and species for each family (after Bogan, 2008).

2.3 Endemism and conservation

One of the major reasons for the high proportion of extinct and endangered freshwater mussels is the high degree of endemism found in this group, which is characteristic of many freshwater organisms. Endemic species have a limited geographical range, often limited to a single drainage basin or lake, and often have unique characteristics suited to that particular locale (Strayer and Dudgeon, 2010). Local rarity also puts a species at a much higher risk of extinction due to the fact that limited distribution puts most or all of a population at risk to environmental stresses simultaneously (Gaston, 1998) and also limits the ability of a population to recover through recruitment from other populations, especially in species with low dispersal ability, such as unionid mussels (Burlakova et al., 2010). One recent study showed that endemic species were critical determinants of the uniqueness of unionid communities, and as such, should be made a conservation priority (Burlakova et al., 2010).

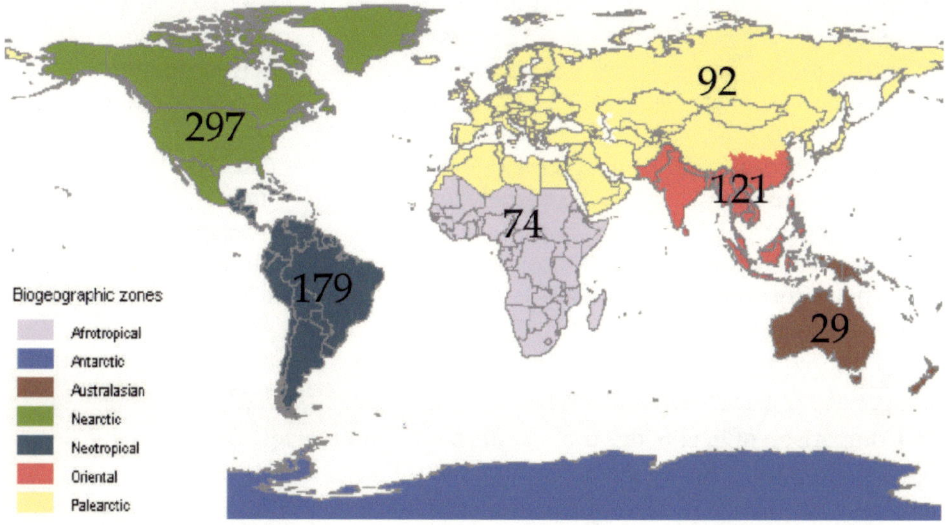

Fig. 2. Map showing the distribution of freshwater mussel species by biogeographic region.

3. Ecology and life history of freshwater mussels

Freshwater mussels are long-lived organisms, often living for decades, and some species can survive over 100 years (Bauer 1992). Typically, unionids live buried in fine substrate in unpolluted streams and rivers with benthic, sedentary, suspension-feeding lifestyles. The mussels use their exposed siphons to inhale water and use their gills to filter out fine food particles, such as bacteria, algae, and other small organic particles. Their benthic, sessile lifestyle, their obligatory dependence on fish hosts for reproduction, and their patchy distribution as a result of specific habitat requirements all contribute to their decline in the face of human disturbances. Freshwater mussels have complex life cycles with extraordinary variation in life history traits (Table 3).

3.1 Reproduction

Freshwater mussels are broadcast spawners, with males releasing sperm into the water to fertilize the eggs that are retained internally in the females' body (Wachtler et al., 2001). The defining characteristic of Unionids is their specialized larval stage known as glochidia that are released from a gravid female's modified "marsupial" gills where they developed from embryos following fertilization (McMahan and Bogan, 2001). One female mussel can produce up to 4 million or more glochidia and eject them in a sudden and synchronized action (Bauer 1987). If the glochidia are released in the proximity of a suitable host fish, they clamp onto the gills of the host, which then carries the glochidia for weeks or months until they are mature and ready to live freely on the bottom of the stream or lake. Because glochidia are heavy, short-lived, non-motile, and poorly carried in currents, facultative dispersal by fish species is necessary for the spread and maintenance of most Unionid populations (Strayer et al., 2004).

Trait	*Unionoidea*	*Sphaeriidae*
Life span range	< 6 to > 100 years	< 1 to > 5 years
Age at maturity	6 to 12 years	>0.17 to <1 year
Reproductive mode	gonochoristic	hermaphroditic
Fecundity (young/female/season)	0.2 - 17 million/female per breeding season	2 - 136/female/season
Juvenile size at release	50 - 450 μm	600 - 4150 μm
Juvenile survivorship	extremely low	high
Adult survivorship	high	intermediate
Semelparous or iteroparous	iteroparous	semelparous or iteroparous
Reproductive efforts per year	1	1-3
Non-respired energy allocated to:		
(i) growth (%)	85-98	65-96
(ii) reproduction (%)	3-15	4-35

Table 3. Comparison of life history traits of freshwater mussels (Unionoidea and Sphaeriidae) in North America (after McMahon and Bogan, 2001).

Fig. 3. a.) Fish-imitating lure of a gravid female broken-ray mussel, *Lampsilis reeveiana;* b.) Crawfish-imitating lure of the rainbow-shell mussel, *Villosa iris;* c.) Glochidia of the fluted kidneyshell, *Ptychobranchus subtentum.* Each glochidia is approximately 220 micrometers long; d.) Fish-imitating conglutinate of the Ouchita kidneyshell, Ptychobranchus occidentalis; e.) Rainbow darter, *Etheostoma caeruleum, attacking a conglutinate of* Ptychobranchus occidentalis; f.) Glochidia attached to the gills of a host fish. After attachment, the host fish's gill tissue forms a cyst around the glochidia. All photos are courtesy of Chris Barnhart (http://unionid.missouristate.edu).

In many mussel species, the gill, mantle margin, or other tissue has evolved into a lure that very realistically mimics a small minnow or invertebrate prey item used to attract a host fish. When a host fish nips at the lure, the glochidia are released into the vicinity of the fish's mouth, thus greatly increasing the odds of the glochidia attaching onto the fish's gills (Haag and Warren, 1999). Other species release large packages of glochidia called conglutinates, which often mimic prey items themselves, that rupture and release glochidia upon being bitten by potential hosts (Grabarkiewicz and Davis, 2008). These unique reproductive strategies have important implications for unionid conservation that will be discussed later in this chapter.

3.2 Feeding behavior and habitat preferences

As adults, freshwater mussels live on the surface or in the top layers of sediment; filter feeding suspended phytoplankton, bacteria, detritus, and other organic matter out of the water (Strayer et al., 2004). Juveniles often bury themselves in sediment below the surface, filtering interstitial water (Grabarkiewicz and Davis, 2008) or feeding pedally by scooping food into their mouths with their foot (Yeager et al., 1994). Unionids are highly sedentary, moving only short vertical and horizontal distances to reproduce or in response to seasonal or environmental cues (Amyot and Downing, 1998; Balfour and Smock, 1995). They are found in a wide range of habitats, from soft sediment bottoms in lakes and ponds to cobble and rock substrates in fast-moving rivers, although the majority of species are found in clear, highly oxygenated streams and rivers with sand, gravel, or cobble bottoms (Grabarkiewicz and Davis, 2008).

3.3 Small-scale spatial distribution

Freshwater mussels are often found in large multispecies aggregations known as mussel beds that can have densities of 10-100 individuals per square meter (Strayer et al., 2004). The biomass of freshwater mussels can higher than all other benthic macroinvertebrates by an order of magnitude (Layzer et al., 1993), and as a result of their large size and sheer numbers they can significantly influence both the biotic and abiotic conditions around them. Although the critical factors determining the location of mussel beds are still unclear, most researchers agree that water velocity and substrate, most notably where water velocity is low enough to limit shear stress and allow for substrate stability but high enough to prevent siltation, are strongly influential. Land use, geology, water quality, and availability of food and suitable host fish species are also strongly correlated with mussel presence/absence in other studies (Arbuckle and Downing, 2002; Newton et al., 2008; Strayer, 1983, 1999; Strayer et al., 2004). These habitat requirements result in a "patchy" distribution of mussels in riverine systems in non-continuous beds that may or may not be reproductively connected by host fish (Strayer et al., 2004).

4. The role of freshwater mussels on ecosystem functioning

As ecosystem engineers that modify their environment, freshwater mussels play many ecological roles where they are found in large numbers. These roles are a function of their life histories and behaviors, and can strongly affect both the biotic and abiotic components of the ecosystems in which they live. Loss of unionid biodiversity can result in loss of these functions and changes to the ecological regimes in those areas where mussels are in decline (Vaughn and Hakencamp, 2001).

4.1 Removing suspended particles

As suspension feeders, Unionids can remove large amounts of phytoplankton, bacteria, and inorganic nutrients from the water column, enhancing water clarity and quality (Strayer et al., 1999). When present in large numbers, they can filter an amount of water equal to or greater than that of daily stream discharge. In a study conducted in the River Spree in Germany, Welker and Walz (1998) found that freshwater mussels created a zone of "biological oligitrophication" by decreasing phytoplankton and phosphorus in the water column. Unionids can also play other important roles in nutrient cycling, such as removing pelagic nutrient resources and depositing them into nearby sediments as faeces or psuedofaeces (Roditi et al., 1997; Spooner and Vaughn, 2006). Mussels also influence nutrient cycling by serving as nutrient sinks in growing populations, or as nutrient sources in declining ones (Vaughn and Hakencamp, 2001).

4.2 Benthic influences

The presence of live mussels can increase in sediment organic matter, which has been shown to positively influence abundance and diversity of other benthic invertebrates and phytoplankton (Spooner and Vaughn, 2006). Benthic invertebrate diversity can also be increased by the presence of mussel shells (Allen and Vaughn, 2011). Other benthic organisms use the shell as habitat and flow refuges, and in large numbers, the presence of mussel shells can increase landscape-level species diversity and abundance (Gutierrez et al., 2003). The influence of Unionids on benthic communities is so great that Aldridge et al., (2006) found that the abundance of freshwater mussels successfully predicted invertebrate abundance and richness in seven lowland rivers in the UK. Mussels also act as environmental engineers, bioturbating the sediment as they move both vertically and horizontally (Allen and Vaughn, 2011). This activity can increase the depth of oxygen penetration in the sediment, homogenize sediment particle size (McCall et al., 1995), and affect the flux rates of solutes between the sediment and water column (Matisoff et al., 1985).

4.3 Ecological impacts of declining Unionid populations

Freshwater mussels are declining in both species richness and abundance, which can reduce their influence on ecosystem functioning and have multiple negative impacts on the ecosystem as a whole. If unioniddiversity declines but total abundance remains the same, these ecological functions should continue being performed if all mussel species perform these functions at equivalent rates (i.e. are functionally redundant). However, both common and rare species are in decline (Vaughn, 1997; Vaughn & Taylor, 1999), and it has been shown that some mussel species are more effective in carrying out the ecosystem functions described above (McCall et al., 1995; Vaughn et al., 2007). It is likely, therefore, that the ecological functions performed by freshwater mussels will continue to decline along with mussel populations, which can significantly impact the overall ecological functioning of freshwater systems (Vaughn, 2010).

5. Causes for the decline in freshwater mussel abundance and diversity

There are many causes for the decline in freshwater mussel biodiversity (Strayer et al., 2004; Downing et al., 2010). Dudgeon et al. (2006) describe five major contributors to the loss of freshwater biodiversity in general: over-exploitation, pollution, flow modification, exotic species invasion, and habitat degradation. These five factors are also driving the decline in freshwater mussel biodiversity and, along with the threat of global climate change, can create smaller and more isolated populations susceptible to genetic bottlenecks and burdened with extinction debts.

5.1 Commercial harvesting

Humans have gathered freshwater mussels for meat, pearls, and mother-of-pearl shells for thousands of years, although commercial harvesting on a large scale did not begin in North America until the early 19th century (Strayer et al., 2004). During this period, commercial musselers harvested untold numbers of unionids for their pearls, which were sold in domestic and international markets. Local populations of mussels were decimated following exhaustive harvesting, after which time the musselers moved on to other, previously untapped, streams (Anthony and Downing, 2001). Overharvest made marketable pearls rarer, and the pearl fishery declined near the end of the century. Around the same time, however, new manufacturing processes allowed for the production of clothing buttons from North American mussel shells, and another round of unregulated exploitation occurred that devastated many populations that had been missed by the pearl frenzy in the previous decades (Neves, 1999). As plastic buttons began to replace those made from mussel shells in the 1930s and 40s, the rising market of the Japanese cultured pearl industry sparked a new demand for mussel shells. It was found that beads of freshwater mussel shells, when placed inside saltwater pearl oysters, made superior nuclei for the formation of cultured pearls (Anthony and Downing, 2001). This most recent boom has lasted until the mid 1990's, when a combination of declining mussel stocks, increased regulation, foreign competition, and disease outbreaks in Japanese pearl oysters has significantly reduced freshwater mussel harvest in North America (Neves, 1999).

5.2 Pollution

Because mussels are such long-lived organisms, chronic exposure to pollutants can cause direct mortality or reduced fitness. This pollution can come from many different sources, such as municipal wastewater effluent, industrial waste, and agricultural and mining runoff (Bogan, 1993), and because unionids live in the sediment, the legacy effects of accumulated toxins can have long-term effects on populations (Strayer et al., 2004). Freshwater mussels can suffer direct mortality from acute or long-term exposure to high levels of organic and inorganic pollutants, and experience sublethal effects on growth, enzyme production, abnormal shell growth, reduced metabolism, and reduced fitness in general (Keller et al., 2007). Because of their complex life cycles, there are several critical life stages where unionids can be exposed to these pollutants, and each stage can have different sensitivities to them (Cope et al., 2008).

In addition to chemical toxicants, excessive sediment can also be a pollutant. Poor agricultural and forestry practices, benthic disturbance by dredging operations, runoff from construction sites, road building, urbanization, loss of riparian vegetation, erosion of stream banks, and changes in hydrologic patterns all contribute to unnaturally high amounts of fine particle sedimentation that affects mussels directly by clogging gills and feeding siphons, and indirectly by blocking light necessary for algal production (Brim Box and Mossa, 1999) and reducing visibility needed for fish hosts to find the lures of breeding female mussels (Haag et al., 1995). Siltation can also create a hardpan layer in the substrate, making it unsuitable for burrowing in (Gordon et al., 1992).

5.3 Flow alteration

Restriction or alteration of flow patterns is another major cause of mussel biodiversity loss. The construction of dams restricts the timing, frequency, and magnitude of natural flow regimes, and affects mussels by altering the stability of the substrate, the type and amount of particulate organic matter (an important food source for mussels), the temperature of the water, and water quality (Poff et al., 2007). Studies have shown decreased mussel populations below large dams, with populations increasing with increased distance downstream from dams and with increasing flow stability (Strayer, 1993; Vaughn and Taylor, 1999). Altered flow regimes after dam construction have been implicated in the extinction of several mussel species, and have resulted in the local extirpation of many more (Layzer et al., 1993). Dams also impair recruitment of juveniles by restricting access to host fish and dispersal of glochidia (Watters, 1999). Urbanization of catchment basins can also alter flow regimes by increasing the amount of impervious cover and channelizing storm runoff, causing higher, faster, and more frequent erosive storm flow events (Walsh et al., 2005). Direct withdrawals of surface and ground water for human consumption can also reduce available habitat, increase water temperatures, and impair mussels' ability to feed, respire, and reproduce (Golladay et al., 2004; Hastie et al., 2003).

5.4 Non-native organisms

Invasion of exotic species is a global phenomenon that threatens terrestrial and aquatic ecosystems alike. The zebra mussel (*Dreissena polymorpha*) and Asian clams (*Corbicula fluminea*) are the two non-native species of greatest concern in North America (although there is some debate over the impact of *C. fluminea*) (Strayer et al., 1999). *D. polymorpha* is highly invasive and fecund, and will attach to any solid substance including the shells of living Unionid mussels. They can occur in densities greater than 750,000 individuals/m^2 , with veliger (their pelagic larvae) densities reaching 400/liter of water (Leach, 1993). Zebra mussels spread rapidly, and one group of researchers has noted a 4-8 year delay from time of introduction of *D. polymorpha* and extirpation of Unionid mussels in many ecosystems (Ricciardi et al., 2003). They compete for food and habitat with native mussels, although it is believed that epizoic colonization (infestation) of the surface of Unionid mussel shells is the most direct and ecologically destructive characteristic of *D. polymorpha* (Hunter and Bailey, 1992; Mackie, 1993). Infestation densities of zebra mussels have been found to exceed 10,000/Freshwater mussel (Nalepa et al., 1993).

Fig. 4. Invasive zebra mussels, *Dreissena polymorpha,* infesting a native fatmucket, *Lampsilis siliquoidea*. Photo courtesy of Chris Barnhart (http://unionid.missouristate.edu).

5.5 Habitat destruction and alteration

Many researchers believe that habitat destruction and alteration are one of the greatest threats to freshwater ecosystems and mussel populations worldwide (Ricciardi and Rasmussen, 1999; Richter et al., 1997; Sala et al., 2000; Osterling et al. 2010). Habitat modification is a general term that encompasses many of the threats described earlier, such as sedimentation, flow alteration, substrate modification, and others, but also include activities such as gravel and sand mining, channelization for boat transportation, clearing of riparian vegetation, and bridge construction (Watters, 1999). Increasing amounts of sediments, either from land surface runoff or instream erosion, is one of the largest contributors to mussel habitat loss, as it makes existing habitat unsuitable for many mussel species (Brim Box and Mossa, 1999). Altered stream behavior caused by modified flows, poor riparian zone management, and runoff from impervious cover can also result in habitat loss through bed scouring, channel morphology changes, and altered sediment regimes in the system (Brierley and Fryirs, 2005).

Headcutting, channelization, and other modifications in river geomorphology are also major causes of habitat alteration in mussel species. Headcutting occurs when an alteration on the bottom of a stream causes a localized washout that progressively moves up the river channel, deepening and widening the channel and releasing large amounts of sediment into the water column. Not only does this process physically destroy mussel habitat, the release of sediment smothers previously suitable downstream habitat as well (Harfield, 1993). Many rivers and streams have been channelized to allow easier boat and barge traffic and for transport of felled logs downstream. Dredging stream channels deposits huge amounts of sediment on the stream bottom, smothering mussels already present and preventing recolonization of future generations. Dredging also drastically alters the natural flow regime and homogenizes habitat, the natural flow regime, and results in habitat homogenization (Watters, 1999). Instream gravel mining operations have been shown to modify the spacing and structure of pools and riffles, change species diversity and abundance of fishes and invertebrates, and alter ecosystem functioning in streams (Brown et al., 1998). These changes can strongly impact freshwater mussels, as most unionids have evolved to thrive in shallow riffle areas with stable, moderately coarse substrate, and are extremely intolerant to disturbance, especially in their larval stages (Brim Box and Mossa, 1999).

5.6 Climate change

There is now strong evidence that both global and regional climate change is occurring and will cause an increase in mean air temperature, more erratic precipitation patterns, and more severe floods and droughts. (Bates et al., 2008) These changing patterns pose serious threats to both terrestrial (Thomas et al., 2004) and freshwater (Sala et al., 2000) ecosystems. One group of researchers predicted that up to 75% of fish species could become extinct in rivers suffering from declining flows as a result of both climate change and human withdrawals (Xenopoulos et al., 2005). Most of the research done on the effects of climate change in freshwater systems has focused on fish and other vertebrates, with very little direct study of the effect on unionids. However, it is well known that temperature affects several aspects of mussel physiology and life history, including reproduction, growth, and recruitment of juveniles (Bauer, 1998; Kendall et al., 2010; Roberts and Barnhart, 1999). It is possible that some mussel species will be able to acclimate to a gradual increase in water temperature, but it is the predicted spikes in maximum temperature and prolonged duration of high temperatures that are likely to impact many mussel populations, especially in small streams where water temperature is more closely linked to air temperature (Hastie et al., 2003).

The change in precipitation patterns could also impact mussel populations through increased flooding and prolonged droughts. Although periodic, low-intensity flooding can have beneficial effects on mussel populations such as flushing fine sediments and pollutants out of substrates (Gordon et al., 1992), extreme storm events can dislodge mussels from the sediment and alter mussel bed habitat (Hastie et al., 2001). In a record multi-year drought in Georgia, Golladay et al. (2004) observed a greater than 50% loss in total mussel abundance in some reaches in the study area. As mussels are limited in their ability to move horizontally, they are unlikely to reach refuges in response to complete dewatering of their habitat. Even reduced flows can have negative effects on respiration, feeding, growth, and glochidial recruitment; and can increase predation by terrestrial consumers like raccoons (Golladay et al., 2004; Hastie et al., 2003).

The response by unionid mussels to climate change will vary depending on several factors. Geographic location will play an important role as climate change is expected to affect different parts of the world differently (Parry et al., 2007). Climate change, as with most types of ecological changes, will produce winners as well as losers (McKinney and Lockwood, 1999; Somero, 2010). Endemic species with restricted geographical ranges are expected to be especially hard hit (Malcolm et al., 2006), as are species that are already close to their upper thermal tolerance ranges (Spooner and Vaughn, 2008). The threat of climate change does not exist in isolation. It also interacts with other disturbances such as land use, direct human-caused flow alterations, and biotic exchange of non-native species; and the severity of these other threats along with geographic location will influence the effects caused by a changing climate (Sala et al., 2000).

5.7 The extinction debt

As serious as the current conservation status of many freshwater mussels are, there most likely exists a substantial extinction debt in many mussel populations (Haag, 2010). Freshwater mussels naturally exist in spatially "patchy" populations separated by areas occupied by no or only a few individuals. These patches remained connected, however, by glochidia transported by host fishes travelling throughout the matrix of mussel beds and unoccupied areas (Strayer, 2008a). Thus, population declines caused by stochastic events such as major floods or droughts could be restored through recruitment from neighboring populations. Many of the threats unionids are facing today, though, such as the building of dams, decline or extinction of host species, increased difficulty of host fish finding female mussels' "lures" or conglutinates due to decreased visibility, and lack of suitable habitat for juvenile mussels, limit reproductive success and gene flow between patches.

As pelagic spawners that release sperm into the water column, it has also been shown that reproductive success declines dramatically with decreasing mussel density, with almost no fertilization occurring at densities below 10 individuals/m^2 (Downing et al., 1993). This lack of reproductive connectivity creates a genetic bottleneck in the remaining populations. These life history characteristics, along with the well-documented decline in mussel diversity and abundance, point to significant future losses in even seemingly stable mussel populations unless action is taken to reduce the perturbations causing the initial decline and increase connectivity between populations (Haag, 2010).

6. Solutions to the decline of freshwater mussels

Because of the growing awareness of the importance of freshwater mussel diversity and freshwater ecosystems in general, there have been increasing efforts to restore and rehabilitate mussel populations and their habitats. Most strategies focus on reversing the root causes of the decline in unionid abundance and diversity listed in the preceding section, along with restoring and protecting existing mussel populations.

6.1 Reduction in commercial harvesting

Although the commercial harvest of freshwater mussels has greatly contributed to the historic decline of Unionids, it is not generally considered to be a major threat to them at present. There are several reasons for the reduction in commercial harvesting of freshwater mussels. The replacement of mussel shell with plastics in the 1940s and 50s in the button industry reduced demand for shells, and more recently the collapse of the Japanese oyster pearl fishery has reduced the demand for pearl nuclei in that industry (Neves, 1999). Enforced regulation on commercial harvesting, as well as low prices for mussel shell, have also provided a respite for mussel populations (Strayer et al., 2004).

6.2 Best management practices to reduce pollution

Although water pollution has significantly declined in many industrial countries thanks to national-level legislation such as the Clean Water Act in the United States and the Water Resources Act in the UK, it is still a major threat to freshwater ecosystems and unionid mussels in most parts of the world. Acute toxicity studies in freshwater mussels have been performed on only a small number of known organic and inorganic contaminants present in the surface water of North America, and sublethal toxicity studies are even more rare (Keller et al., 2007). More studies are needed on a broader array of substances to provide regulators with better information for setting acceptable pollution standards in surface waters where freshwater mussels are found.

Non-point source nutrient and sediment pollution from agriculture, timber extraction, and urban runoff is regularly cited as one of the most serious threats to freshwater ecosystems (Richter, 1997). Best management practices that control runoff into surface water have been shown in numerous studies to improve the physical and chemical quality of streams (Caruso, 2000; D'Arcy & Frost, 2001; Lowrance et al., 1997). One of the most effective ways controlling sediment and nutrient inputs into streams is an intact, functional riparian zone. Well-vegetated riparian zones slow and reduce surface run-off into streams, capture large amounts of sediment in the runoff, store excess nutrients for uptake into riparian vegetation, and stabilize stream banks which further reduces instream sedimentation (Allan, 2004).

6.3 Restoring natural and adequate stream flows

Reversing the trend of increasing human control of the flow of rivers and streams worldwide is not likely in the near future. As the human population grows over the foreseeable future, the global demand for domestic and irrigation water is projected to increase correspondingly (Robarts and Wetzel, 2000). Although the world's rivers have been fragmented and controlled by more than 1 million dams (Jackson et al., 2001), there are methods of operating these dams to minimize the negative effects they have on downstream ecosystems. In several case studies in the United States, water managers, conservation organizations, and scientists have attempted to regulate releases from dams to mimic the timing, duration, and magnitude of natural flood events, and to minimize the number of low flow days in the rivers downstream (Poff et al. 1997; Richter et al., 2003). In one study in Tennessee, recolonization of mussel populations occurred after hydroelectric dam managers altered their release schedule to ensure minimum flows (Layzer and Scott, 2006). There is also a growing movement for the complete removal of dams. As their ecological implications are being realized by scientists and the public, and as dam managers are facing higher operating costs in maintaining aging structures and complying with federal endangered species laws, dam removal is being seen as a viable option for river restoration in many circumstances (Hart et al., 2002; Pejchar and Warner, 2001).

When water levels drop, either through natural wet and dry cycles or through human withdrawals or regulation, the amount of physical habitat available to mussels and other benthic organisms is reduced. Many states and countries have passed legislation that requires minimum ecological flows in streams and rivers. There are over 200 methods for determining exactly how much water is needed for a particular stretch of river, all of which take into consideration the specific ecological function or species water managers are trying to preserve (Arthington et al., 2006). Most of these methods focus on fish or other vertebrate species, and often flows suitable for the preservation of these target species is not sufficient for freshwater mussels or other invertebrates (Gore et al., 2001, Layzer and Madison, 1995). Obviously, more study into the flow requirements of freshwater mussels along with a greater emphasis on this group by regulators is necessary if the hurdle of inadequate flows is to be overcome.

6.4 Control of non-native species

Controlling invasive, non-native organisms in freshwater ecosystems has met with limited success for most species, despite passage of laws such as the Non-indigenous Aquatic Nuisance Prevention and Control Act of 1990 in the United States. The zebra mussel is still expanding its range, although the rate of spread has slowed in recent years as the most easily colonized waterways have already been occupied (Johnson et al., 2006). The early spread of *D. polymorpha* was due to physical connectivity of waterways to infected areas, whereas current range expansion is due to overland human-facilitated transport by recreational boaters (Johnson et al., 2001). Thus, it seems, the future distribution of *D. polymorpha* will depend on human behavior, although their ultimate range will be limited to ecosystems with suitable pH, calcium concentrations, and temperature (Strayer, 2008b). Although various chemical, thermal, mechanical, and thermal treatment options have been somewhat successful in controlling *D. polymorpha* near shoreline structures and water intake valves, and consumption by natural predators can be high (Hamilton et al., 1994, Perry et al., 1997), the overall fecundity of the species makes eradication or control in most open-water areas unlikely (Strayer, 2008b).

6.5 Restoring habitat

Many of the solutions to physical habitat loss have already been addressed in the previous sections, such as restoration of riparian vegetation; the use and enforcement of best management practices in construction, agriculture, and forestry; dam removal; and restoration of natural flow regimes. These practices will reduce terrestrial inputs of substrate-smothering sediment, ensure that adequate amounts of water are present, and restore natural stream channel morphology more suitable for freshwater mussels. It is also possible to directly restore benthic habitat through riparian and instream construction projects designed to stabilize banks and stream channels and increase the habitat heterogeneity that supports high levels of benthic diversity. Several studies in Finland (Muotka et al., 2002), Japan (Nakano & Nakamura, 2006), and the United States (Miller et al., 2009) have found increased macroinvertebrate abundance and richness in streams following stream channel restoration projects, and while these studies did not look at freshwater mussels specifically, they provide a basis of reference for mussel-specific restoration techniques. Osterling et al. (2010) indicated that restoration activities to improve environmental conditions of mussels' habitats should focus on reducing fine material transport into streams, because sedimentation of inorganic and organic materials and high turbidity can impact mussel recruitment.

6.6 Minimizing the effects of global climate change

The ability of freshwater organisms to adapt to climate change is dependent on a particular species' ability to disperse and migrate to cooler environments in higher latitudes or elevations (Poff et al., 2002). As unionid mussels have limited dispersal and reproductive potentials under the best of circumstances, this puts this group at a higher risk than many other groups (Hastie et al., 2003). There are two main approaches to dealing with the threat posed by climate change: (1) to reduce further changes in climate and (2) to manage the consequences of current and predicted changes. To review the numerous methods being debated and currently attempted to control climate change is beyond the scope of this chapter; however it is important to note that a few of these methods (construction of dams for "clean" hydroelectric power, intensification of agriculture for biofuels) have the potential to further degrade freshwater ecosystems beyond their current state if not planned and managed correctly (Bates et al., 2008).

As far as managing the effects of climate change on freshwater ecosystems, there are two major aspects to this as well: (1) to reduce pollution, habitat loss, and other anthropogenic disturbances that are already placing stress on freshwater systems, and (2) to establish a network of protected areas based on species' current and projected ranges, and to manage the connecting matrix between them (Hannah et al., 2002; Heino et al., 2009; Poff et al., 2002). Ways of reducing anthropogenic stress on freshwater ecosystems include riparian zone management, reducing nutrient loading, habitat restoration, and minimizing human-driven water withdrawal (Poff et al., 2002), and have already been discussed in previous sections. The concept of freshwater protected areas and dispersal corridors between populations will be covered in the following section.

6.7 Protecting and restoring freshwater mussel populations

Protected areas have been a mainstay of terrestrial and marine conservation efforts for decades, yet have only recently been part of the discussion about conserving freshwater species and habitat (Abell et al., 2007). Freshwater protected areas (FPAs) have been used in the past mostly to protect fish species from overharvesting by providing areas closed to fishing for at least part of the time. FPAs have the potential to do more than just limit fish harvests, though. Effectively planned and executed protected areas can protect specific habitat types against degradation, ensure minimum surface and groundwater flows, protect riparian zones, and protect rare and endangered species (Saunders et al., 2002; Suski and Cooke, 2007).

One of the key aspects that have limited the effectiveness of FPAs against ecosystem degradation, especially in rivers and streams, is that many of the stressors affecting these systems come from diverse, non-point sources upstream from critical habitat and threatened populations. The success of localized protected areas or catchment management strategies can be limited due to the large scale connection of aquatic ecosystems with terrestrial activities, especially where streams with their longitudinal connectivity are concerned (Saunders et al., 2000). Therefore, many researchers have pointed out the need for catchment-scale protection for threatened freshwater ecosystems that truly limit the impacts to sensitive areas (Abell et al., 2007; Dudgeon et al., 2006; Heino et al., 2009). Although there has been little published data on freshwater mussels and protected areas, some researchers have noted the possibility of refuges for some species (Ricciardi et al., 1998; Saunders et al., 2002), and preservation and protection of critical mussel habitat has the potential to significantly aid in the recovery of unionids.

Naturally reproducing unionid populations can take decades to recover after severe and prolonged disturbances. As mentioned earlier, mussels are dependent on critical densities to facilitate successful reproduction (Downing et al., 1993), and many areas where unionids have been extirpated lack access to restocking populations (Strayer et al., 2004). In these situations, artificially stocking mussels can help restore populations and eventually enable them to become self-sustaining (Strayer et al., 2004). Mussel relocation and reintroduction have been met with varying levels of success, mostly due to lack of knowledge of specific habitat requirements and handling techniques (Cope and Waller, 1995). Many successful propagation techniques have also been developed over the last few years (Barnhart, 2006; Henley et al., 2001), and although field trials of lab-reared mussels are limited, artificial propagation techniques hold much promise to enhance unionid populations in the future, provided the degraded environmental conditions that caused the decline in the first place are corrected.

7. Conclusions

The loss of biodiversity across biomes and habitats has direct and profound implications for human populations around the world (Sala et al., 2000). The functioning of both terrestrial and aquatic ecosystems is dependent on the diversity of their constituent organisms (Covich et al., 2004; Kinzig et al., 2002; Loreau et al., 2001), and the dependence of humans on these ecosystem services makes protecting and restoring biodiversity a priority for both the present and future generations. Freshwater ecosystems have received less consideration from the public and researchers, despite the critical linkages between freshwater systems and human well-being (Aylward et al., 2005; Costanza et al., 1997; Jackson et al., 2001). It is clear that through our actions we are degrading and damaging our freshwater ecosystems beyond their abilities to recover (Allan and Flecker, 1993; Dudgeon et al., 2006; Richter et al., 1997; Strayer and Dudgeon, 2010), and continuing these unsustainable activities puts all the world's inhabitants at risk.

Freshwater unionid mussels are an often-overlooked part of freshwater biodiversity, and one that is the most threatened (Ricciardi and Rasmussen, 1999; Williams et al., 1993). Unionids are key components to their ecosystems, carrying out many important ecological functions (McCall et al., 1995; Strayer et al., 1999; Vaughn and Hakencamp, 2001) and influencing the diversity of benthic communities (Aldridge et al., 2006; Gutierrez et al., 2003; Spooner and Vaughn, 2006). Their unique reproduction strategy, feeding behaviors, specific habitat requirements, and valuable shell and pearls have put them at risk to human-driven disturbances, and have contributed to their worldwide decline in both abundance and richness (Bogan, 1993; Vaughn, 1997). The drivers of the decline in unionid biodiversity are the same as those of freshwater diversity in general: pollution, habitat destruction, overharvest, altered flows, invasion by non-native species, and climate change, but because of their lifestyles and high degree of endemism, they are being especially hard hit (Strayer et al., 2004).

The solutions to the decline in unionid biodiversity are simple, but not easy. Reducing pollution (Caruso, 2000; Lowrance et al., 1997), restricting harvesting (Strayer et al., 2004), ensuring ecologically sustainable flows (Arthington et al., 2006; Layzer and Scott, 2006), habitat protection and restoration (Miller et al., 2010; Muotka et al., 2002; Wilson et al., 2011), combating non-native invaders (Strayer, 2008b), mitigating and planning for the effects of climate change (Heino et al., 2009; Poff et al., 2002), creating connected freshwater protected areas (Heino et al., 2009; Saunders et al., 2002) and artificially enhancing wild populations (Cope and Waller, 1995; Strayer et al., 2004) are all necessary to restore freshwater ecosystems and the mussels that occupy them.

It is clear that any successful freshwater conservation plans must be large in scale and long-term in scope, and take into consideration the multiple chronic stressors that are causing the alarming decline in freshwater pearly mussels. It is equally clear that failure to take concrete steps to halt and reverse the trend of biodiversity loss in unionid mussels could result in the permanent loss of this unique and important group of animals.

8. References

Abell, R. (2001). Conservation Biology for the Biodiversity Crisis: a Freshwater Follow-up. *Conservation Biology*, Vol.16, No.5, (October 2002), pp. 1435-1437, ISSN 0888-8892

Abell, R.; Allan,J. & Lehner, B. (2007). Unlocking the potential of protected areas for freshwaters *Biological Conservation*, Vol.134, No.1, (January 2007), pp. 48-63, ISSN 0006-3207

Aldridge, D.; Fayle, T. & Jackson, N. (2007). Freshwater mussel abundance predicts biodiversity in UK lowland rivers. *Aquatic Conservation: Marine and Freshwater Ecosystems*, Vol.17, No.6, (September/October 2007), pp. 554–564, ISSN 1099-0755

Allan, J. (2004). Landscapes and Riverscapes: The Influence of Land Use on Stream Ecosystems. *Annual Review of Ecology, Evolution, and Systematics*, Vol.35, No.1, (2004), pp. 257-284, ISSN 1543-592X

Allan, J. & Flecker, A. (1993). Biodiversity conservation in running waters. *Bioscience*, Vol.43, No.1, (January 1993), pp. 32-43, ISSN 0006-3568

Allen, D. C. & Vaughn, C. C. (2011). Density-dependent biodiversity effects on physical habitat modification by freshwater bivalves. Ecology, Vol.92, No.5, (2011), pp. 1013-1019. ISSN *0012-9658*

Amyot J. & Downing J. (1998). Locomotion in Elliptio complanata (Mollusca:Unionidae): A reproductive function? *Freshwater Biology*, Vol.39, No.2, (March 1998), pp. 351–358, ISSN 0046-5070

Anthony, J. & Downing, J. (2001). Exploitation trajectory of a declining fauna: a century of freshwater mussel fisheries in North America. *Canadian Journal of Fisheries and Aquatic Sciences*, Vol.58, No.10, (October 2001), pp. 2071-2090, ISSN 0706-652X

Arbuckle K. & Downing J. (2008). Freshwater mussel abundance and species richness: GIS relationships with watershed land use and geology. *Canadian Journal of Fisheries and Aquatic Sciences*, Vol.59, No.2, (February 2002), pp. 310-316, ISSN 0706-652X

Arthington, A.; Bunn, S.; Poff, N. & Naiman, R. (2006). The challenge of providing environmental flow rules to sustain river ecosystems. *Ecological Applications,* Vol.16, No.4, (August 2006), pp. 1311–1318, ISSN 1051-0761

Aylward, B.; Bandyopadhyay, J.; Belausteguigotia, J.; Borkey, P.; Cassar, A.; Meadors, L.; Saade, L.; Siebentritt, M.; Stein, R.; Tognetti, S.; Tortajada, C.; Allan, T.; Bauer, C.; Bruch, C.; Guimaraes-Pereira, A.; Kendall, M.; Kiersch, B.; Landry, C.; Mestre Rodriguez, E.; Meinzen-Dick, R.; Moellendorf, S.; Pagiola, S.; Porras, I.; Ratner, B.; Shea, A.; Swallow, B.; Thomich, T. & Voutchkov, N. (2005). Freshwater ecosystem services, In: *Ecosystems and Human Well-Being: Policy Responses: Findings of the Responses Working Group (Millennium Ecosystem Assessment Series),* Millennium Ecosystem Assessment, pp. 213-255, Island Press, ISBN 1559632704, Washington D.C.

Balfour, D. & Smock, L. (1995). Distribution, age structure, and movements of the freshwater mussel Elliptio complanata (Mollusca: Unionidae) in a headwater stream. *Journal of Freshwater Ecology*, Vol.10, No.3, (September 1995), pp. 255–268, ISSN 0270-5060

Balian, E.; Segers, H.; Leveque, C. & Martens, K. (2008). The Freshwater Animal Diversity Assessment: an overview of the results. *Hydrobiologia*, Vol.595, Num.1, (November 2008), pp 627-637, ISSN 0018-8158

Barnhart, M. (2006). Buckets of muckets: a compact system for rearing juvenile freshwater mussels. *Aquaculture*, Vol.254, No.1-4, (April 2006), pp. 227–233, ISSN 0044-8486

Barnhart, M. (2008). Unio Gallery at Missouri State University, 05/27/11, Available from: http://unionid.missouristate.edu/Default.htm

Bates, B.; Kundzewicz, Z.;Wu, S. & Palutikof, J. (Eds.), (2008). *Climate Change and Water.* Technical Paper of the Intergovernmental Panel on Climate Change, IPCC Secretariat, ISBN: 978-92-9169-123-4, Geneva

Bauer, G. (1987).The parasitic stage of the freshwater pearl mussel (Margaritifera margaritifera L.). III. Host relationships. Archiv für Hydrobiologie 76: 413–423.

Bauer, G. (1992). Variation in the life span and size of the freshwater pearl mussel. Journal of Animal Ecology 61: 425–436.

Bauer, G. (1998). Allocation policy of female freshwater pearl mussels. *Oecologia,* Vol.117, Nos.1-2, (November 1998), pp. 90-94, ISSN 0029-8549

Bennett, E. M. Peterson, G. D. & Gordon, L. (2009). Understanding relationships among multiple ecosystem services. Ecology Letters 12: 1394-1404.

Bogan, A. (1993). Freshwater Bivalve Extinctions (Mollusca: Unionoida): A Search for Causes. *American Zoologist, Vol.33,* No.6, (December 1993), pp. 599-609, ISSN 0003-1569

Bogan, A. (2008). Global diversity of freshwater mussels (Mollusca, Bivalvia) in freshwater. *Hydrobiologia*, Vol.595, No.1, (November 2007), pp. 139-147, ISSN 0018-8158

Bogan, A. & Roe, K. (2008). Freshwater bivalve (Unioniformes) diversity, systematics, and evolution: status and future directions. *Journal of the North American Benthological Society*, Vol.27, No.2, (June 2008), pp. 349-369, ISSN 0887-3593

Brierley, G. & Fryirs, K. (2005). *Geomorphology and River Management: Applications of the River Styles Framework*, Blackwell Publishing, ISBN 1-4051-1516-5, Oxford, UK

Brim Box, J. & Mossa, J. (1999). Sediment, land use, and freshwater mussels: prospects and problems. *Journal of the North American Benthological Society,* Vol.18, No.1, (March 1999), pp. 99-117, ISSN 0887-3593

Brown, A.; Lyttle, M. & Brown, K. (1998). Impacts of Gravel Mining on Gravel Bed Streams. *Transactions of the American Fisheries Society,* Vol.127, No.6, (1998), pp. 979 - 994, ISSN 1548-8659

Burlakova, L.; Karatayev, A.; Karatayev, V.; May, M.; Bennett, D. & Cook, M. (2010). Endemic species: Contribution to community uniqueness, effect of habitat alteration, and conservation priorities. *Biological Conservation,* Vol.144, No.1, (January 2011), pp. 155-165, ISSN 0006-3207

Caruso, B. (2000). Comparative analysis of New Zealand and US approaches for agricultural nonpoint source pollution management. *Environmental Management,* Vol.25, No.1, (January 2000), pp. 9–22, ISSN *0364-152X*

Cope, W. & Waller, D. (1995). Evaluation of freshwater mussel relocation as a conservation and management strategy. Regulated Rivers: Research and Management, Vol.11, No.2, (October 1995), pp. 147–155, ISSN 1535-1467

Cope, W.; Bringolf, R.; Buchwalter, D.; Newton, T.; Ingersoll, C.; Wang, N.; Augspurger, T.; Dwyer, F.; Barnhart, M.; Neves, R. & Hammer, E. (2008). Differential exposure, duration, and sensitivity of unionoidean bivalve life stages to environmental contaminants. *Journal of the North American Benthological Society*, Vol.27, No.2, (June 2008), pp. 451–462, ISSN 0887-3593

Costanza, R.; d'Arge, R.; de Groot, R.; Farberk, S.; Grasso, M.; Hannon, B.; Limburg, K.; Naeem, S.; O'Neill, R.; Paruelo, J.; Raskin, R.; Suttonkk, P. & van den Belt, M. (1997). The value of the world's ecosystem services and natural capital. *Nature,* Vol.387, No.6630, (May 1997), pp. 253–260, ISSN 0028-0836

Covich, A.; Austen, M.; Barlocher, F.; Chauvet, E.; Cardinale, B.; Biles, C.; Inchausti, P.; Dangles, O.; Solan, M.; Gessner, M.; Statzner, B. & Moss, B. (2004). The role of biodiversity in the functioning of freshwater and marine benthic ecosystems. *BioScience,* Vol.54, No.4, (August 2004), pp. 767–775, ISSN 0006-3568

D'Arcy, B. & Frost, A. (2001). The role of best management practices in alleviating water quality problems associated with diffuse pollution. *Science of the Total Environment,* Vol.265, No.1, (January 2001), pp. 359–367, ISSN 0048-9697

Downing, J.; Rochon, Y.; Pérusse, M. & Harvey, H. (1993). Spatial aggregation, body size, and reproductive success in the freshwater mussel Elliptio complanata. *Journal of the North American Benthological Society*, Vol.12, No.2, (June 1993), pp. 148-156, ISSN 0887-3593

Downing J. A.; Van Meter P. & Woolnough D. A. (2010). Suspects and evidence: a review of the causes of extirpation and decline in freshwater mussels. *Animal biodiversity and conservation,* Vol.33, No.2, (2011), pp. 151-185, ISSN 1578-665X

Dudgeon D.; Arthington A.; Gessner M.; Kawabata, Z.; Knowler, D.; Leveque, C.; Naiman, R.; Prieur-Richard, A.; Soto, D.; Stiassny, M. & Sullivan, C. (2006). Freshwater biodiversity: importance, threats, status and conservation challenges. *Biological Reviews,* Vol.81, No.2, (May 2006), pp. 163–182, ISSN 1469-185X

Gaston, K. (1998). Species-range size distributions: products of speciation, extinction and transformations. *Philosophical Transactions of the Royal Society B,* Vol.353, No.1366, (February 1998), pp. 219-230. ISSN 0962-8436

Golladay, S.; Gagnon, P., Kearns, M., Battle, J. & Hicks, D. (2004). Response of freshwater mussel assemblages (Bivalvia: Unionidae) to a record drought in the Gulf Coastal Plain of southwestern Georgia. *Journal of the North American Benthological Society,* Vol.23, No.3, (September 2004), pp. 494–507, ISSN 0887-3593

Gordon, N.; McMahon, T. & Finlayson, B. (1992). *Stream hydrology: an introduction for ecologists.* John Wiley and Sons, ISBN 9780470843581, New York

Gore, J,; Layzer, J, & Mead, J. (2001). Macroinvertebrate instream flow studies after 20 years: a role in stream management and restoration. *Regulated Rivers: Research & Management,* Vol,17, No.4-5, (July-October 2001), pp. 527–542, ISSN 1099-1646

Grabarkiewicz, J. & Davis, W. (2008). *An Introduction to Freshwater Mussels as Biological Indicators (Including Accounts of Interior Basin, Cumberlandian and Atlantic Slope Species).* EPA-260-R-08-015. U.S. Environmental Protection Agency, Office of Environment. Washington, D.C.

Graf, D. & Cummings, K. (2007). Review of the systematics and global diversity of freshwater mussel species (Bivalvia: Unionoida). *Journal of Molluscan Studies,* Vol.73, No.4, (November 2007), pp. 291–314, ISSN 0260-1230

Gutiérrez, J.; Jones, C.; Strayer, D. & Iribarne, O. (2003). Mollusks as Ecosystem Engineers: The Role of Shell Production in Aquatic Habitats. *Oikos,* Vol.101, No.1, (April 2003), pp. 79-90, ISSN *0030-1299*

Haag, W.; Butler, R. & Hartfield, P. (1995). An extraordinary reproductive strategy in freshwater bivalves: Prey mimicry to facilitate larval dispersal. *Freshwater Biology,* Vol.43, No.3, (December 1995), pp. 471–476, ISSN 0046-5070

Haag W. & Warren M. Jr. (1999). Mantle displays of freshwater mussels elicit attacks from fish. *Freshwater Biology,* Vol.42, No.1, (August 1999), pp. 35–40, ISSN 0046-5070

Haag, W. (2010). Past and future patterns of freshwater mussel extinctions in North America during the
Holocene, In: *Holocene Extinctions,* S.T. Turvey (Ed.), 107–128, Oxford University Press, ISBN 0199535094, Oxford, UK

Hamilton, D.; Ankney, C. & Bailey, R. (1994). Predation of zebra mussels by diving ducks: an exclosure study. *Ecology,* Vol.75, No.2, (March 1994), pp. 521-531, ISSN *0012-9658*

Hannah, L.; Midgley, G.; Lovejoy, T; Bond, W.; Bush, M.; Lovett, J.; Scott, D. & Woodward, F. (2002). Conservation of biodiversity in a changing climate. *Conservation Biology,* Vol.16, No.1, (February 2002), pp. 264– 268, ISSN *0888-8892*

Hart, D.; Johnson, T.; Bushaw-Newton, K.; Horwitz, R.; Bednarek, A.; Charles, D.; Kreeger, D. & Velinsky, D. (2002). Dam Removal: Challenges and Opportunities for Ecological Research and River Restoration. *BioScience,* Vol.52, No.8, (August 2002), pp. 669-682, ISSN 0006-3568

Hartfield, P. (1993). Headcuts and their effect on fresh water mussels, In: *Conservation and Management of freshwater mussels.* K. S. Cummings, A. C. Buchanan & L. M. Koch, (Eds.), Upper Mississippi River Conservation Committee, Rock Island, Illinois, U.S.A.

Hastie, L.; Boon, P.; Young, M. & Way, S. (2001). The effects of a major flood on an endangered freshwater mussel population. *Biological Conservation,* Vol.98, No.1, (March 2001), pp. 107–115, ISSN 0006-3207

Hastie, L.; Cosgrove, P.; Ellis, N. & Gaywood, M. (2003). The Threat of Climate Change to Freshwater Pearl Mussel Populations. *AMBIO: A Journal of the Human Environment*, Vol.*32, No.*1, (2003), pp. 40-46, ISSN 0044-7447

Heino, J.; Virkkala, R. & Heikki Toivonen. H. (2002). Climate change and freshwater biodiversity: detected patterns, future trends and adaptations in northern regions. *Biological Reviews*, Vol.84, No.1, (February 2009), pages 39–54, 1469-185X

Heino, J.; Virkkala, R. & Toivonen, H. (2009). Climate change and freshwater biodiversity: detected patterns, future trends and adaptations in northern regions. *Biological Reviews*, Vol.84, No.1, (February 2009), pp. 39–54, ISSN 1469-185X

Henley, W.; Zimmerman, L.; Neves, R. & Kidd, M. (2001). Design and evaluation of recirculating water systems for maintenance and propagation of freshwater mussels. *North American Journal of Aquaculture*, Vol.63, No.2, (2001), pp. 144–155, ISSN 1548-8454

Hunter, R. & Bailey, J. (1992). Dreissena polymorpha (zebra mussel): Colonization of soft substrata and some effects on unionid bivalves. *The Nautilus*, Vol.106, No.2, (June 1992), pp. 60-67, ISSN 0028-1344

IUCN (International Union for Conservation of Nature and Natural Resources). (2007). *2007 IUCN Red List of threatened species*. 03.03.2011. Available from: www.iucnredlist.org

Jackson, R.; Carpenter, S.; Dahm, C.; McKnight, D.; Naiman, R.; Postel, S. & Running, S. (2001). Water in a changing world. *Ecological Applications*, Vol.11, No.4, (August 2001), pp. 1027-1045, ISSN 1051-0761

Jobling, S. & Tyler, C. (2003). Endocrine disruption, parasites and pollutants in wild freshwater fish. *Parasitology*, Vol.126, No.2, (February 2003), pp. 103–108, ISSN 0031-1820

Johnson, L.; Ricciardi, A. & Carlton, J. (2001). Overland dispersal of aquatic invasive species: a risk assessment of transient recreational boating. *Ecological Applications*, Vol.11, No.6, (December 2001), pp. 1789–1799, ISSN 1051-0761

Johnson, L.; Bossenbroek, J. & Kraft, C. (2006). Patterns and pathways in the post-establishment spread of nonindigenous aquatic species: the slowing invasion of North American inland lakes by the zebra mussel. *Biological Invasions*, Vol.8, No.3, (April 2006), pp. 475–489, ISSN 1387-3547

Keller, A.; Lydy, M. & Ruessler D. (2007). Unionid mussel sensitivity to environmental contaminants, In: *Freshwater bivalve ecotoxicology*, J. L. Farris and J. H. Van Hassel (Eds.), 151–167, CRC Press, Boca Raton, Florida, and SETAC Press, Pensacola, Florida, ISBN 142004284X

Kendall, N.; Rich, H.; Jensen, L. & Quinn, T. (2010). Climate effects on inter-annual variation in growth of the freshwater mussel (Anodonta beringiana) in an Alaskan lake. *Freshwater Biology*, Vol.55, No.11, (November 2010), pp. 2339–2346, ISSN 0046-5070

Kinzig, A.; Pacala, S. & Tilman, D. (eds.). (2002) *The Functional Consequences of Biodiversity: Empirical Progress and Theoretical Extensions*. Princeton University Press, ISBN 9780691088228, Princeton, NJ

Layzer, J.; Gordon, M. & Anderson, R. (1993). Mussels: the forgotten fauna of regulated rivers. A case study of the Caney Fork River. *Regulated Rivers: Research and Management*, Vol.8, No.1-3, (May 1993), pp. 63-71, ISSN 1099-1646

Layzer, J. & Madison, L. (1995). Microhabitat use by freshwater mussels and recommendations for determining their instream flow needs. *Regulated Rivers: Research & Management*, Vol.10, No.2-4, (August 1995), pp. 329–345, ISSN 1099-1646

Layzer, J. & Scott, E. Jr. (2006). Restoration and colonization of freshwater mussels and fish in a Southeastern United States tailwater. *River Research and Applications*, Vol.22, No.4, (May 2006), pp. 475–491, ISSN 1535-1467

Leach, J. (1993). Impacts of the Zebra Mussel (Dreissena polymorpha) on Water Quality and Fish Spawning Reefs in Western Lake Erie, In: *Zebra mussels: Biology, impacts, and control.* T. Nalepa & D. Schloesser, (Eds.), 381-397, Lewis/CRC Press, Inc., ISBN 0873716965, Boca Raton, Florida

Loreau, M.; Naeem, S.; Inchausti, P.; Bengtsson, J.; Grime, J.; Hector, A.; Hooper, D.; Huston, M.; Raffaelli, D.; Schmid, B.; Tilman, D. & Wardle, D. (2001). Biodiversity and ecosystem functioning: current knowledge and future challenges. *Science,* Vol.294, No.5543, (October 2001), pp. 804-808, ISSN 0036-8075

Matisoff, G.; Fisher, J. & Matis, S. (1985). Effects of benthic macroinvertebrates on the exchange of solutes between sediments and freshwater. *Hydrobiologia,* Vol.122, No.1, (March 1985), pp. 19-33, ISSN 0018-8158

Lowrance, R.; Vellidis, G.; Wauchope, R.; Gay, P. & Bosch, D. (1997). Herbicide transport in a managed riparian forest buffer system. *Transactions of the American Society of Agricultural and Biological Engineers,* Vol.40, No.4, (1997), pp. 1047-1057, ISSN 0001-2351

Malcolm, J.; Liu, C.; Neilson, R.; Hansen, L. & Hannah, L. (2006). Global Warming and Extinctions of Endemic Species from Biodiversity Hotspots. *Conservation biology,* Vol.20, No.2, (April 2006), pp. 538-548, ISSN *0888-8892*

Master, L.; Stein, B.; Kutner, L. & Hammerson, G. (2000). Vanishing assets: conservation status of U.S. species, In: *Precious heritage: the status of biodiversity in the United States,* B. A. Stein, L. S. Kutner, and J. S. Adams (Eds.). pp. 93–118. Oxford University Press, ISBN 0195125193, New York.

McCall P.; Tevesz M.; Wang X. & Jackson J. (1995). Particle mixing rates of freshwater bivalves: Anodonta grandis (Unionidae) and Sphaerium striatinium (Pisidiidae). *Journal of Great Lakes Research*, Vol.21, No.3, (1995), pp. 333-339, ISSN 0380-1330

McKinney, M. & Lockwood, J. (1999). Biotic homogenization: a few winners replacing many losers in the next mass extinction. *Trends in Ecology & Evolution*, Vol.14, No.11, (November 1999), pp. 450-453, ISSN 0169-5347

McMahon, R. & Bogan A. (2001). Mollusca: Bivalvia, In: *Ecology and classification of North American freshwater invertebrates, 2nd ed.,* Thorp, J.H. & Covich A.P. (Eds.), 331–429, Academic Press, ISBN 0126906475, San Diego, CA

Meyers, N.; Mittermeier, R.; Mittermeier, C.; de Fonesca, G. & Kent, J. (2000). Biodiversity hotspots for conservation priorities. *Nature*, Vol.403, No.6772, (February 2000), pp. 853–858, ISSN 0028-0836

Miller, S.; Budy, P. & Schmidt, J. (2010). Quantifying Macroinvertebrate Responses to In-Stream Habitat Restoration: Applications of Meta-Analysis to River Restoration. *Restoration Ecology*, Vol.18, No.1, (January 2010), pp. 8–19, ISSN 1526-100X

Muotka, T.; Paavola, R.; Haapala, A.; Novikmec, M. & Laasonen, P. (2002). Long-term recovery of stream habitat structure and benthic invertebrate communities from instream restoration. *Biological Conservation*, Vol.105, No.2, (June 2002), pp. 243-253, ISSN 0006-3207

Myers, N.; Mittermeier, R.; Mittermeier, C.; da Fonseca, G. & Kent, J. (2000). Biodiversity hotspots for conservation priorities. *Nature,* Vol. 403, (February 2000), pp. 853–858, ISSN 0028-0836

Nakano, D. & Nakamura, F. (2006). Responses of macroinvertebrate communities to river restoration in a channelized segment of the Shibetsu River, Northern Japan. *River Research and Applications,* Vol.22, No.6, (July 2006), pp. 681–689, ISSN 1535-1467

Nalepa, T. & Schloesser, D. (Eds.) (1993). *Zebra mussels: Biology, impacts, and control.* Lewis/ CRC Press, Inc., ISBN 0873716965, Boca Raton, Florida

Newton, T.; Woolnough, D. & Strayer, D. (2008). Using landscape ecology to understand and manage freshwater mussel populations. *Journal of the North American Benthological Society,* Vol.27, No.2, (June 2008), pp. 424-439, ISSN 0887-3593

Neves, R. (1999). Conservation and commerce: Management of freshwater mussel (Bivalvia: Unionoidea) resources in the United States. *Malacologia,* Vol.41, No.2, (December 1999), pp. 461–474, ISSN 0076-2997

Osterling, M. E.; Arvidsson, B. L.; Greenberg, L. A. (2010). Habitat degradation and the decline of the threatened mussel *Margaritifera margaritifera*: influence of turbidity and sedimentation on the mussel and its host. *Journal of Applied Ecology,* Vol.47, No.4, August 2010), pp. 759-768, ISSN 0021-8901

Parry, M.; Canziani, O.; Palutikof, J.; van der Linden, P. & Hanson, C. (Eds). (2007). *Contribution of Working Group II to the Fourth Assessment Report of the Intergovernmental Panel on Climate Change, 2007.* Cambridge University Press, ISBN 978 0521 88009-1, Cambridge, United Kingdom and New York, NY, USA.

Pejchar, L. & Warren, K. (2001). A River Might Run Through It Again: Criteria for Consideration of Dam Removal and Interim Lessons from California. *Environmental Management,* Vol.28, No.5, (November 2001), pp. 561–575, ISSN 0364-152X

Perry, W.; Lodge, D. & Lamberti, G. (1997). Impact of crayfish predation on exotic zebra mussels and native invertebrates in a lake-outlet stream. *Canadian Journal of Fisheries and Aquatic Sciences,* Vol.54, No.1, (January 1997), pp. 120-125, ISSN 0706-652X

Poff, N.; Allan, J.; Bain, M.; Karr, J.; Prestegaard, K.; Richter, B.; Sparks, R. & Stromberg, J. (1997). The natural flow regime. *BioScience,* Vol.47, No.11, (December 1997), pp. 769-784, ISSN 0006-3568

Poff, N.; Brinson, M. & Day, J. JR. (2002). *Aquatic ecosystems and global climate change. Technical Report,* Pew Center on Global Climate Change, Arlington, USA.

Poff N.; Olden, J.; Merritt, D. & Pepin, D. (2007). Homogenization of regional river dynamics by dams and global biodiversity implications. *Proceedings of the National Academies of Science,* Vol.104, No.14, (April 2007), pp. 5732–5737, ISSN 0027-8424

Postel, S. & Carpenter, S. (1997). Freshwater ecosystem services, In: *Nature's services: societal dependence on natural ecosystems,* G.C. Daily (Ed.), pp. 195-214, Island Press, ISBN 1559634766, Washington D.C.

Ricciardi, A.; Neves, R. & Rasmussen, J. (1998) Impending extinctions of North American freshwater mussels (Unionoida) following the zebra mussel (Dreissena polymorpha) invasion. *Journal of Animal Ecology,* Vol.67, No.4, (July 1998), pp. 613–619, ISSN 1365-2656

Ricciardi, A. & Rasmussen, J. (1999). Extinction rates of North American freshwater fauna. *Conservation Biology*, Vol.13, No.5, (October 1999), pp. 1220–1222, ISSN 0888-8892.

Ricciardi, A. (2003). Predicting the impacts of an introduced species from its invasion history: an empirical approach applied to zebra mussel invasions. *Freshwater Biology*, Vol.48, No.6, (June 2003), pp. 972–981, ISSN 0046-5070

Richter, B.; Braun, D.; Mendelson, M. & Master L. (1997). Threats to imperiled freshwater fauna. *Conservation Biology*, Vol.11, No. 5, (October 1997), pp. 1081-1093, ISSN 0888-8892

Richter, B.; Mathews, R.; Harrison, D. & Wigington, R. (2003). Ecologically sustainable water management: managing river flows for ecological integrity. *Ecological Applications*, Vol.13, No.1, (February 2003), pp. 206–224, ISSN 1051-0761

Robarts, R. & Wetzel, R. (2000). The looming global water crisis and the need for international education and cooperation. *Societas Internationalis Limnologiae News*, Vol.29, No.1, (January 2000), pp. 1-3

Roberts, A. & Barnhart, M. (1999). Effects of temperature, pH, and CO2 on transformation of the glochidia of Anodonta suborbiculata on fish hosts and in vitro. *Journal of the North American Benthological Society*, Vol.18, No.4, (December 1999), pp. 477-487, ISSN 0887-3593

Roditi, H.; Strayer D. & Findlay S. (1997). Characteristics of zebra mussel (Dreissena polymorpha) biodeposits in a tidal freshwater estuary. *Archiv für Hydrobiologie*, Vol.140, No.2, (1997), pp. 207-219, ISSN 0003-9136

Sala, O.; Chapin, F.; Armesto, J.; Berlow, R.; Bloomfield, J.; Dirzo, R.; Huber-Sanwald, E.; Huenneke, L.; Jackson, R.; Kinzig, A.; Leemans, R.; Lodge, D.; Mooney, H.; Oesterheld, M.; Poff, N.; Sykes, M.; Walker, B.; Walker, M. & Wall, D. (2000). Global biodiversity scenarios for the year 2100. *Science*, Vol.287, No.5459, (March 2000), pp. 1770–1774, ISSN 0036-8075

Saunders D.; Meeuwig J. & Vincent C. (2002). Freshwater protected areas: strategies for conservation. *Conservation Biology*, Vol.16, No.1, (February 2002), pp. 30–41, ISSN 1523-1739

Small, C. & Cohen, J. (1999). Continental Physiography, Climate and the Global Distribution of Human Population, pp. 965-971, *Proceedings of the International Symposium on Digital Earth*, Beijing, China, November 29-December 2, 1999

Smith, V.; Tilman, G. & Nekola, J. (1999). Eutrophication: impacts of excess nutrient inputs on freshwater, marine, and terrestrial ecosystems. *Environmental Pollution*, Vol.100, No.1-3, (1997), pp. 179-196, ISSN 0269-7491

Somero, G. (2010). The physiology of climate change: how potentials for acclimatization and genetic adaptation will determine 'winners' and 'losers'. *The Journal of Experimental Biology*, Vol.213, No.6, (March 2010), pp. 912-920, ISSN 0022-0949

Spooner, D. & Vaughn, C. (2006). Context-dependent effects of freshwater mussels on the benthic community. *Freshwater Biology*, Vol.51, No.6, (June 2006), pp. 1016–1024, ISSN 0046-5070

Spooner, D. & Vaughn, C. (2008). A trait-based approach to species' roles in stream ecosystems: climate change, community structure, and material cycling. *Oecologia*, Vol.158, No.2, (2008), pp. 307-317, ISSN 0029-8549

Strayer, D. (1983). The effects of surface geology and stream size on freshwater mussel (Bivalvia, Unionidae) distribution in southeastern Michigan, U.S.A. *Freshwater Biology*, Vol.13, No.3, (June 1983), pp. 253–264, ISSN 0046-5070

Strayer, D. (1993). Macrohabitats of freshwater mussels (Bivalvia: Unionacea) in streams of the northern Atlantic slope. *Journal of the North American Benthological Society*, Vol.12, No.3, (September 1993), pp. 236–246, ISSN 0887-3593

Strayer, D.; Caraco, N.; Cole, J.; Findlay, S. & Pace, M. (1999). Transformation of freshwater ecosystems by bivalves. A case study of zebra mussels in the Hudson River. *Bioscience*, Vol.49, No.1, (January 1999), pp. 19-27, ISSN 0006-3568

Strayer, D.; Downing, J.; Haag, W.; King, T.; Layzer, J.; Newton, T. & Nichols, S. (2004). Changing perspectives on pearly mussels, North America's most imperiled animals. *BioScience*, Vol.54, No.5, (May 2004), pp. 429–439, ISSN 0006-3568

Strayer, D. (2008a). *Freshwater mussel ecology: a multifactor approach to distribution and abundance*. University of California Press, ISBN 9780520255265, Berkeley, California

Strayer, D. (2008b). Twenty years of zebra mussels: lessons from the mollusk that made headlines. *Frontiers in Ecology and the Environment*, Vol.7, No.3, (April 2008), pp. 135–141, ISSN 1540-9295

Strayer, D. & Dudgeon, D. (2010). Freshwater biodiversity conservation: recent progress and future challenges. *Journal of the North American Benthological Society*, Vol.29, No.1, (March 2010), pp. 344-358, ISSN 0887-3593

Suski, S. & Cooke, S. (2007) Conservation of Aquatic Resources through the Use of Freshwater Protected Areas: Opportunities and Challenges. *Biodiversity and Conservation*, Vol.16, No.7, (June 2007), pp. 2015-2029, ISSN 0960-3115

Thomas, C.; Cameron, A.; Green, R.; Bakkenes, M.; Beaumont, L.; Collingham, Y.; Erasmus, B.; Ferreira de Siqueira, M.; Grainger, A.; Hannah, L.; Hughes, L.; Huntley, B.; van Jaarsveld, A.; Midgley, G.; Miles, L.; Ortega-Huerta, M.; Peterson, A.; Phillips, O. & Williams, S. (2004). Extinction risk from climate change. *Nature*, Vol.427, (January 2004), pp. 145-148, ISSN 0028-0836

Vaughn, C. (1997). Regional Patterns of Mussel Species Distributions in North American Rivers. *Ecography*, Vol.20, No.2, (April 1997), pp. 107-115, ISSN 0906-7590

Vaughn C. & Taylor C. (1999). Impoundments and the decline of freshwater mussels: a case study of an extinction gradient. *Conservation Biology*, Vol.13, No.4, (August 1999), pp. 912-920, ISSN 0888-8892

Vaughn, C. & Hakenkamp C. (2001). The functional role of burrowing bivalves in freshwater ecosystems. *Freshwater Biology*, Vol.46, No.11, (November 2001), pp. 1431–1446, ISSN 0046-5070

Vaughn, C.; Spooner, D. & Galbraith, H. (2007). Context-dependent species identity effects within a functional group of filter-feeding bivalves. *Ecology*, Vol.88, No.7, (July 2007), pp. 1654–1662, ISSN 0012-9658

Vaughn, C. (2010). Biodiversity Losses and Ecosystem Function in Freshwaters: Emerging Conclusions and Research Directions. *BioScience*, Vol.60, No.1, (January 2010), pp. 25-35, ISSN 0006-3568

Wachtler, K.; Dreher-Mansur, M. & Richter, T. (2001). Larval types and early postlarval biology in naiads (Unionoida). In: *Ecological Studies-Volume 45: Ecology and evolution of the fresh-water mussels Unionoida*, G. Bauer and K. Wachtler (Eds.), 93-125, Springer-Verlag, Berlin, Germany

Walsh, C.; Roy, A.; Feminella, J.; Cottingham, P.; Groffman, P. & Morgan II, R. (2005). The urban stream syndrome: current knowledge and the search for a cure. *Journal of the North American Benthological Society*, Vol.24, No.3, (September 2005), pp. 706-723, ISSN 0887-3593

Watters, G. (1999). Freshwater mussels and water quality: A review of the effects of hydrologic and instream habitat alterations. *Proceedings of the First Freshwater Mollusk Conservation Society Symposium*, Chatanooga, Tennessee, 1999

Welker, M. & Walz, N. (1998). Can Mussels Control the Plankton in Rivers? A Planktological Approach Applying a Lagrangian Sampling Strategy. *Limnology and Oceanography*, Vol.43, No.5, (July 1998), pp. 753-762, ISSN 0024-3590

Williams, J.; Warren Jr., M., Cummings, K.; Harris, J. & Neves, R. (1993). Conservation status of freshwater mussels of the United States and Canada. *Fisheries*, Vol.18, No.6, (September 1993), pp. 6–22, ISSN 1548-8446

Wilson, C. D.; Roberts, D. & Reid, N. (2011). Applying species distribution modelling to identify areas of high conservation value for endangered species: A case study using *Margaritifera margaritifera* (L.). Biological Conservation, Vol. 144, No.2, (February 2011), pp. 821-829, ISSN 0006-3207

Xenopoulos, M.; Lodge, D.; Alcamo, J.; Märker, M.; Schulze, K. & Van Vuuren, D. (2005). Scenarios of freshwater fish extinctions from climate change and water withdrawal. *Global Change Biology*, Vol.11, No.10, (October 2005), pp. 1557–1564, ISSN 1365-2486

Yeager, M.; Cherry, D. & Neves, R. (1994). Feeding and Burrowing Behaviors of Juvenile Rainbow Mussels, Villosa iris (Bivalvia:Unionidae). *Journal of the North American Benthological Society*, Vol.13, No.2, (June 1994), pp. 217-222, ISSN 0887-3593

7

Effects of Climate Change in Amphibians and Reptiles

Saúl López-Alcaide[1] and Rodrigo Macip-Ríos[2]
[1]*Instituto de Biología, Universidad Nacional Autónoma de México*
[2]*Instituto de Ciencias de Gobierno y Desarrollo Estratégico, Benemérita Universidad Autónoma de Puebla*
México

1. Introduction

During the past decade it has been documented that the average of earth temperature increased 6 °C in a period of 100 years. The higher amount of this phenomenon has been recorded between 1910 and 1945 and, from 1976 to present date (Jones, 1999; Kerr, 1995; Oechel et al., 1994; Thomason, 1995). From 1976 to present the temperature rising has been the faster recorded in the last 10,000 years (Jones et al., 2001; Taylor, 1999; Walther et al., 2002), and this caused the maximum daily temperature increase in the southern hemisphere (Easterling et al., 2000), as well as a significant temperature increase in the tropical forest areas (Barnett et al., 2005; Houghton et al., 2001; Santer et al., 2003; Stott, 2003). According to projections, the average temperature of earth may increase up to 5.8 °C (Intergovernmental Panel on Climate Change IPCC, 2007) at the end of current century, which actually represents and enormous threat for biodiversity (Mc Carty, 2001; Parmesan & Yohe, 2003).

Although the historical data describes a changing climate during the last 350 million years of amphibians and reptiles history (Duellman & Trueb, 1985), the abrupt rising of the temperature during the last century could have a great impact on ectotherm organisms, which depend of environmental temperature to achieve physiological operative body temperatures (Walter et al., 2002; Zachos et al., 2001). Thus, the accelerated grow of earth temperature could affect physiological, reproductive, ecological, behavioral, and distribution traits among amphibians and reptiles (Cleland et al., 2006; Dorcas et al., 2004; Pough, 2001; Gvozdik & Castilla, 2001). In this context, a review of the published studies is necessary to evaluate and summarize the evidence of climate change effects in amphibians and reptiles. This review should provide an overview that should be helpful for researches, students, and policy makers, in order to address how climate change affects amphibians and reptiles, and the possible responses of these organisms to climate change.

Due to the available published information are not equal for amphibians and reptiles, the present chapter are divided in two: amphibians and reptiles. In each chapter subdivision a physiological, reproductive, and distribution effects are issued as long as information was available. Additional information in amphibians such as synergic effects of environmental factors due climate change, and evolutionary adaptations are addressed. At the end of the chapter a conclusion section is added in order to summarize the most important trends addressed in this review.

2.1 Amphibians

Amphibians are very susceptible to environmental variation since they have permeable skin, eggs without shell (not amniotic eggs), and a complex life cycle that expose them to changes in both the aquatic and terrestrial environment (Blaustein, 1994; Blaustein & Wake, 1990, 1995; Vitt et al., 1990), also they are particularly sensitive to changes in temperature and humidity, as well as to exposure to large doses of UV radiation (Blaustein & Bancroft, 2007).
During the last twenty years a decline o of more 500 populations of frogs and salamanders has been documented (Stuart, 2004; Vial & Saylor, 1989). The reasons for this decline stills unclear because of the factor complexity and their interactions (Alford & Richards, 1999; Blaustein & Kiesecker, 2002; Kiesecker et al., 2001). However many cases of species decline share ecological traits, life histories or demographic traits such as: 1) high habitat specialization, 2) reduced population size, 3) long generation time, 4) fluctuating abundance, 5) low reproductive rate, and 6) complex life cycles. These characteristics suit species more vulnerable to threats (Reed & Shine, 2002; Williams & Hero, 1998). The traits alone may not cause the decline, but cause that organisms become more vulnerable after an initial perturbation like the rise of environmental temperature (Lips et al., 2003). Climate change may have adverse effects by itself on the survival, distribution, reproductive biology, ecology, physiological performance, and immune system of organisms when they are exposed to higher environmental temperatures or dryness (beyond their threshold of tolerance). Also, the effects of climate change could act in synergy with various biotic and abiotic agents like diseases and infections, intense UV radiation, habitat loss, exotic species (competitors and depredators), and chemical pollution (Young, 2001, as cited in Lips et al., 2003).

2.1.1 Physiological performance

There are basic physiological aspects of amphibians that are highly sensitive to temperature increase; first water balance due to their permeable skin and high evaporation rates (Shoemaker et al., 1992), second since amphibians are poiquiloterms[1], the thermoregulatory performance are related to the water balance, digestion, oxygen supply, vision, hearing, emergence from hibernation, development, metamorphosis, growth, and immune response.
Amphibians do not perform physiological process when they are near of their critical body temperatures, instead they are able for perform many functions in suboptimal temperatures and do not exhibit a unique thermal optimum. Therefore, temperature may strongly influence on species' geographic distribution, so on the amphibian's distribution limits are determined by the extreme temperatures, in consequence the species that with broad tolerance to thermal regimens should be able to expand their ranges and then colonize new habitats. Finally, the two processes that are more sensitive to changes in temperature in amphibians are reproduction and development (Berven et al, 1979; Rome et al., 1992). The hormonal regulation of reproduction are affected by temperature, then an increase in temperature could affected the reproductive cycles of amphibians due to changes in the concentration of the hypothalamic hormone (GnRH) which acts directly on the gonads (Herman, 1992; Jorgensen, 1992).
Considering the information presented above, there are four aspects of climate change that could strongly affect the physiological performance in amphibians: 1) temperature increase,

[1]For which the temperature determines the appropriate functioning of many processes that have different thermal sensitivities.

2) the increase of the dry season length, 3) decrease of soil moisture (due changes of precipitation and temperature rise), and 4) increase in rainfall variation. This would affect organisms at population and community levels. As an example Carey & Bryant (1995) found that the individual growth rate, reproductive effort, and life span could change, with a parallel change in the activity patterns, microhabitat use, and thermoregulation.

2.1.2 Reproductive biology and phenology

The increase of temperature and thermal variation are the basic signal for emergence and reproduction in anurans, particularly in species from temperate zone (Duellman & Trueb, 1986; Jorgensen, 1992). Climate change linked with other factors such as photoperiod (Pancharatna & Patil, 1997) may affect the breeding patterns of amphibians. Amphibians in Canada are affected by the precipitation decrease and increased temperatures during summer (Herman & Scott, 1992; Ovaska, 1997). In The United Kingdom there has been evidence of early oviposition in *Bufo calamita, Rana sculenta* and *Rana temporaria* between 1978 and 1994, which is correlated with the increase of spring temperatures during the last 20 years (Beebee, 1995; Forchhammer et al., 1998). Also, an earlier beginning in the oviposition period (2 to 13 days between 1846 and 1986) in *Rana temporaria* in Finland has been recorded and correlated to changes in air and water temperature (Terhivou, 1988). Historical and recent climate data from Ithaca, New York suggest that the temperature rising during the 20th century has changed the matting patterns of *Pseudacris crucifer, Lithobates sylvaticus, Lithobates catesbeianus,* and *Hyla versicolor*, by promoting the vocalization behavior for 10 to 13 days earlier than the expected from historical data (Gibs & Breisch, 2000).

In addition, it has been proposed that the change in precipitation patterns (Duellman & Trueb, 1985) can affect the reproductive phenology of amphibian species that breed in ponds. If ponds are filled latter in the season, then the short water permanence should lead to an increase in competition and a higher predation rate. Meanwhile organisms are concentrated in the remaining ponds and they are more vulnerable diseases outbreaks. Changes on amphibian's phenology could present complex effects over populations, changing the population structure, and then a rapid decline of sensitive populations (Donelly & Crump, 1998).

2.1.3 Distribution

It has been suggested that climate change is the cause of species migration to higher altitudes and latitudes, and then the subsequent shrinkage or loss of amphibian populations and distribution (Pounds et al., 1997). Exploring the relationship between current amphibian distribution and the possible effects of global warming on species distribution in order to build distribution models based on different algorithms such as GARP or MAXENT (Stockwell & Peters, 1999), neural networks, and generalized lineal models (Thuiller, 2003), It would possible to model for the basic conditions for the species to survive. General models of amphibians in Europe suggest that a large portion of species could expand its distribution along Europe if they are able to disperse unlimited, but if species are not capable to disperse, the distribution range of most of the species could be drastically reduced. This scenario could be the most likely because the current levels of habitat fragmentation and degradation, especially in the aquatic habitats. According to the projections of Araujo et al. (2006) the majority of the amphibian species from Europe could lost most of their distribution areas for 2050. This supports the hypothesis that climate change will cause the future decline of amphibian's populations. As an example, the pletodontid salamanders of the central highlands of Mexico *Pseudoricea leprosa* and *P. cefalica* could face important levels of habitat loss for the increased rise of temperature. Models predict that *P. leprosa* could possible loss the 75% of its distribution range for the year 2050 (Parra-Olea et al., 2005). These salamanders have limited dispersal behavior; so on they must face the loss in its distributional range in situ (Easterling et al., 2000; Mc Carty, 2001).

2.1.4 Synergy with biotic and abiotic factors (diseases, infections and physiological performance)

It is clear that the amphibians survival is linked to abiotic factors such water availability, the increase in temperature (Pouns et al., 1999; Pounds, 2001), and the increase in cloud cover (Pounds, 2006) could alter weather patterns and hydrology of the places when amphibians inhabit. The association of this factors with the exposure to higher doses of UV-B radiation due the thinning of the ozone layer (Kiesecker et al., 1995), and the interaction with biotic (diseases and infections), and physiological performance (immune system and tolerances thresholds) could be the main precursors of the events that cause widespread mortality in amphibians and the decline of their populations.

The emergence of disease and infections outbreaks due climate change is related with the decline of several amphibian species (Pounds, 2006). The link between climate change and the coming out of epidemics in amphibians and their subsequence decline has been attributed to the changes in environmental conditions that enhance outbreaks (Pounds & Crump, 1994; Pouns & Puschendorf, 2004). As temperature increases amphibians can exceed their physiological tolerance limits, therefore pathogens could reach their suitable thresholds, and outbreaks should be suited with faster dispersion at high temperatures (Epstein, 2001; Harvel et al., 2002; Rodo et al., 2002). A significant association between rises of local temperature (until thermal optimum levels) with occurrence of chytridiomycosis outbreaks in temperate zones has been described (Bosch et al., 2007). In the Peruvian Andes, the ponds originated by the water of the melting glacial retreat are colonized by three species of frogs *Pleurodema marmorata, Bufo spinulosus,* and *Telmatobius marmoratus,* and subsecuenty by the pathogen *Batrachochytridium dendrobatidis* (*Bd*), which possibly causes dead in the adults and declines in metamorphic juveniles and tadpoles. The *Bd* outbreak could be related with the rise of temperature in the ponds, with a subsequent achieve of the tolerable levels by *Bd* (Seimon et al., 2007). Therefore, the amphibian decline are consequence of a complex multifactorial processes that are related with host and parasites, since there is a synergic effects in life histories of hosts and parasites, because both are influenced by the same factors: humidity and temperature (Lips et al., 2008).

2.1.5 Thinning of the ozone layer and UV-B radiation

It has been suggested that climate change can work in synergy with the ozone layer reduction to increase the exposure of organisms to UV-B radiation (Blustein et al., 1994a, 1998; Schindler et al., 1996; Yan et al., 1996). Worrest & Kimeldorf (1976) found that high levels of UV-B radiation (290-315nm) cause developmental abnormalities and mortality in *Bufo boreas* tadpoles before metamorphosis. Therefore amphibians that breed in mountain aquatic habitats such as temporary ponds or lakes can be quite susceptible to increased levels of UV-B radiation. Because amphibian embryos develop in shallow lakes and mountain ponds are often exposed to direct sunlight, and therefore, considering the reduction of the ozone layer and higher doses of UV-B radiation (Stebbins, 1995). A deeper water column can reduce UV-B radiation, but in sites/localities with decreased rainfall per year, and the subsequent shallowness of ponds or streams an increase in the embryos exposure to UV-B radiation is expected. In the Pacific Northwest an outbreak of *Saprolegnia ferax* was described as the cause of extensive mortality of *Bufo boreas* embryos in shallow waters, also the exposition to higher levels of UV-B radiation causes that amphibian embryo more sensitive to infection by pathogens (Blaustein et al., 1994b; Kiesecker et al., 1995; Kiesecker et al., 2001; Laurance et al., 1996; Pounds et al., 1994; Pounds, 2001).

In spite of this, it has be found significant variation among species in the activity levels of photolyaze, an enzyme that repairs DNA for damage caused by UV-B radiation (Hays et al., 1996). The hypothesis of UV-B radiation sensitivity predicts: 1) Significant differences between species of amphibians in relation to the photolyaze repair activity in eggs, and differential success between clutches exposed to solar radiation, 2) a correlation of these differences to the extent of exposure of eggs to sunlight, and 3) high-repair activity in species that are not in decline compared with those declining. Blaustein et al. (1994b) reviewed 10 species of amphibians in Oregon. The enzyme activity was higher in species that are not declining and low in declining species. Field experiments showed that in *Hylla regilla* embryos, a species that is not in decline had high activity of photolyaze with higher clutch success, differing from two species in decline (*Rana cascadae* and *Bufo boreas*) with low levels of photolyaze activity (Blaustein et al., 1994b).

2.1.6 Other climatic variables and their influence on amphibian's declines

It has been suggested that the cause of exposures of amphibian embryos to UV-B is not the thinning of the ozone layer, but the unusual weather patterns such as high temperatures, which is linked with the assumption that the increasing frequency and intensity of El Niño Southwester Oscillation (ENSO) phenomenon as a result of climate change, inducing a high embryo mortality in the Pacific Northwest amphibians (Kiesecker & Blaustein 1997, 1999). In extremely dry and warm years a large proportion of the oviposition sites of *B. boreas* recorded very low levels of water depth, causing that aquatic habitats provide few protection against UV-B. In shallow water (20 cm.) filaments of *S. ferax* attack up to 80% of the embryos on average compared to 12% at 50 cm of water depth. Otherwise, the exposure of organisms to extreme temperatures may be an alternative explanation for the increased mortality of amphibian embryos in shallow waters (Pounds, 2001). The alteration of precipitation patterns may be related to the occurrence of ENSO, in this way, the increased frequency and magnitude of this phenomenon can increase the incidence and severity of *S. ferax* outbreaks (Kiesecker & Blaustein, 2001).

In addition to the evidence mentioned above there are some other hypothesis that predicts the highlands clime alteration by the rise of temperature in the sea surface due the evaporation increase sea surface in the tropics, this trend accelerates the atmospheric warming of the nearby highlands (Graham, 1995). One of these scenarios is "weather-related outbreaks" (Pounds & Crump, 1994), which predicts that the host parasite relationships are affected by climate, so amphibian mass extinctions in forests areas that apparently have no significant disruption could be evidence of how deep and unpredictable climate change result of warming sea surface alter the ecological interactions and generates potentially devastating effects. The case of the Harlequin Frog (*Atelopus varius*) could be considered under this situation. As climate increased since 1983 in the localities when this frog was distributed, the increased of vulnerability to this frogs to lethal parasites also increased, with the subsequent loss of several populations of this frog (Crump & Pouns, 1985, 1989). Finally it has been suggested that as the habitat dries up and frogs gather near the available waterfalls, the probability of being attacked and infected by parasites increase, resulting in high mortality (Pounds et al., 1999).

2.1.7 Increased cloud cover

The Chytridio fungus pathogen is distributed from the deserts, rain forests in the lowlands to the cold mountains enviroments (Ron, 2005). It is consider as a non-lethal saprophyte parasite (Daszak et al., 2003, Retallick et al., 2004), which grows in the amphibian skin (Berguer et al., 2004; Piotrwski et al., 2004). However this fungus has become more common recently, with an increase of lethal records in places with low climate temperatures (Daszak et al., 2005; La Marca et al., 2005). This pathogen has been associated with amphibian mortality in upland regions even during winter (Alexander & Eischeid, 2001; Berguer et al., 2004). The global warming has been considered as a key factor in the spread of the *Bd*, and then the decline of amphibian populations, because of the sequence of events that propitiate the parasite development (altering the local temperature though the effect of changing ambient humidity and light) causing favorable conditions for the fungus growth and propagation.

Linked with the increase of the fungal infection, it is important to understand how environmental changes affect the immune system of amphibians. Environmental temperatures have a large effect on amphibians' immune system, and this is an important factor that makes them more susceptible to the emergence of pathogens (Carey et. al., 1999; Maniero & Carey, 1997; Rojas et al., 2005). The effects of temperature on the amphibian susceptibility to the outbreaks caused by the *Bd* fungus, which causes high mortality at low temperatures is a basic research issue that should be addressed to understand interactions between amphibian decline and climate change (Berguer et al. 2004; Woodhams et al., 2003). Immunity in ectotherms organisms depends on temperature, which can be decisive for the dynamics of disease in amphibians, especially in temperate regions. Seasonal changes in temperature can cause short-term decrease in optimal levels of immunity, causing a drop in the production of immune system cells during these periods until the organisms achieve to adapt as predicted by the hypothesis of the delay and the hypothesis of seasonal acclimatization (Raffel et al., 2006).

As an example, the Red-Spotted Newt (*Notophthalmus viridescens viridescens*) shows considerable variation at their basal levels in different immune parameters (lymphocytes, neutrophils and eosinophils) when they are affected by significant thermal variation. These findings suggest that temperature variability causes an increase in the susceptibility of amphibians to infection, and this has important implications for the emergence of diseases. The effects of delay and acclimatization may also cause infections after the unusual climate change, or could turn the evolution of parasites life-history strategies to take advantage of periods where host are more susceptible. Finally the increased variability of weather conditions predicted by the climate change scenarios could lead to longer or more frequent periods of immune suppression in amphibians which may increase their decline (Hegerl et al., 2004; Schar et al., 2004).

2.1.8 Consequences of climate change on the ecology of amphibians

It has been suggested that the climate change and its interaction with several factors could easily cause the disappearance of rare, endemic, and isolated species compared with common and broad distributed species with numerous sub-populations which are less threatened (Davies et al., 2000). Endemic species use to have unique environmental specializations, and the possibility of losing them could be higher (Grimm, 1993). It is predicted that populations of amphibians at the edges of their distributional ranges may be particularly vulnerable to changes in local and global climate (Pounds and Crump, 1994; Wyman, 1990). Therefore anurans assemblages may be dominated by species with wide ranges of ecological and physiological tolerance. Also, if weather patterns changes, some regions may be affected by the decreasing in the abundance of terrestrial invertebrates (basic diet of amphibians; Blaustein & Kiesecker, 2002; Bradford, 2002; Kiesecker & Blaustein 1997a). Populations also could be affected by changes in births, mortality, emigration, and immigration rates. This can alter the operational sex ratio, age-population structure, and genetic variability. If the population size declines, the young frogs can grow more slowly than adults and reduce the energy allocated to reproduction (Stearns, 1992).

I has been speculated that terrestrial frogs of the dry zones of the world could be forced to use the fewer wet areas available, consequently this large aggregations would propitiate higher contact between individuals and thus more vulnerability to parasites and predators, as has been demonstrated for *Atelopous varius* in Costa Rica (Pounds & Crump, 1987). Species with continuous reproductive cycle may experience a reduction of the breeding season, restricting the oviposition to wet periods only. In environments with unpredictable rainfall patterns, the frogs may experience hard times locating the adequate ponds for their larvae, and therefore populations of these species could decline more rapidly (Alford & Richards, 1999; Berven & Grudzien 1990; Gulve, 1994; Sjogren, 1991).

2.1.9 Climate change and possible evolutionary adaptations of amphibians

To understand the relationships between the biotic and abiotic factors causing the amphibians decline must be considered five evolutionary principles: 1) the development is limited by historical constraints, 2) not all evolution is adaptive, 3) the adaptations are often linked through trade-offs, 4) the development could only alter existing variations, and 5) the evolution takes a lot of time! Therefore, in an evolutionary context although amphibians have been exposed to sunlight and UV-B radiation through its evolutionary history, by now they are facing an unprecedented situation that put them at risk (Cockell, 2001). For example behavior, morphology, and lifestyles that have enabled them to persist for millions of years, today could be highly risky because exposure to sunlight involves receive high doses of UV-B radiation that cause mutations and cell death, decrease growth rates, disable the immune system (Trevini, 1993). Thus in the context described above, the exposure to UV-B radiation can be especially harmful to those species in which natural selection shaped behavioral strategies necessarily exposed to relatively high doses of solar radiation (e.i. species they have to put their eggs in water). In addition, many other selection pressures (biotic and abiotic) are intense and relatively new (i.e. chemical pollution, disease, etc.). Although these might cause a relatively rapid change in some populations, while other threats such as the emergence of infectious diseases combined with environmental changes can be so intense that amphibians cannot adapt to them (Kiesecker & Blaustein, 1997b). Since not all evolution is adaptive, as showed the evidence that in several species of amphibians still lay their eggs in shallow water and communal masses, exposing them to potentially harmful agents such as UV-B, exposure to temperatures that exceed their tolerance limits and conditions (Romer, 1968). Such behavior probably has persisted over millions of years, but under current conditions these behaviors may cause the damage by a large number of elements such as those mentioned above. Amphibians exhibit maladaptive characteristics because the evolution takes time. Obviously, amphibians, and other organisms have defenses against the harmful factors. Exposure to sunlight over evolutionary time has undoubtedly resulted in mechanisms that help animals to withstand UV-B radiation (Cockel, 2001; Hoffer, 2000).

2.2 Reptiles

As we mention before the global modification of ecosystems has been induced the global warming and is identified as a significant and immediate threat that could radically affect the ability of species to survive. It is of great interest the ability of species to adjust to changes in the thermal environment, habitat structure and other fundamental niche axis. For terrestrial ectotherms, an increase in average temperature may affect their spatial distribution, physiological performance, reproductive biology and behavior (Dunham, 1993; Grant & Porter, 1992). As reptiles depend of external heat sources to regulate their body temperature climate is a key factor influencing the distribution and abundance of species (Pough, 2001; Zug, 1993).

In contrast to the work conducted with amphibians, which has been extensive research on biological and ecological consequences of climate change, reptiles provide a scenario with broad potential. Although there are studies that suggest interesting perspectives on the issue, and then are exposed works concerning on the effects of ecological, physiological, reproductive, behavioral and evolutionary change in reptiles.

2.2.1 Physiological performance in reptiles and climate change

There are proposals that combine the spatial and temporal variation with the physiological (speed and strength) and morphological traits (shape of limbs) ecologically relevant. For example, in the *Urosaurus ornatus* lizard has been measured the speed and endurance of various populations, as well as the shape and size of their limbs through the altitudinal range where lizards are distributed. This lizard exhibit significant variation in the shape of his limbs corresponding to available perch types for each population, speed, and endurance. Thus the possible change in habitat structure and thermal regimes could result in an alteration of development patterns of lizards and cause changes in body shape and size of adults and in the way they use their habitat, behavior strategies, and physiological performance. Under this scenario only some population would be at risk, even though their evolutionary responses are consider slower compared with the speed of environmental change (Miles, 1994). On the other hand, long-term experimental studies with young *Notechis scutatus* snakes in enclosures with cold (19-22°C), intermediate (19-26°C), and hot (19-37°C) thermal gradients, suggested that these snakes compensated restricted thermal opportunities, although behavioral plasticity depending on thermal environment experienced to birth, therefore these conditions influenced subsequent thermoregulatory strategies (Aubret & Shine, 2010).

2.2.2 Life histories and distribution

In relation with the impact of climate change on the life-history of reptiles, there have been some changes in traits of species with limited dispersal ability such as *Lacerta vivipara*, which lives in the isolated mountain peaks of the southern Pyrenees (Chamaillé-Jammes et al.,

2006). According to the author's records individual body size increases dramatically in four populations studied for 18 years. Body size increase in all age classes appears to be associated with an increase in temperature experienced by the offspring in their first month of life (August). The maximum daily temperature in this region during August raised 2.2 °C and lizard snout-vent length increased over 28%. As a result, body size of adult females increased dramatically, with the following increase of the litter size and the reproductive effort. One of the populations surveyed by a capture-recapture study suggested that adult survivorship was correlated with May temperature. All the fitness components investigated responded positively to increase in temperature, so it can be concluded *Lacerta vivipara* has obtained benefits of climate change. Instead, it is possible that climate change drastically alters the marshes, the main habitat for the *Lacerta vivipara,* due to temperature increase could cause more evaporation and reduces its moisture (IPCC, 2001), and the species may not be able to cope with changes in their habitat. (Chamaillé-Jammes et al., 2006). Araujo et al. (2006) also suggest that a continuous increase in temperature could cause a long-term contraction in suitable habitat for lizards and therefore increase the risk of local extinction.

2.2.3 Thermoregulatory behavior and global warming

Global warming and the potential reduction in areas with suitable characteristics for the distribution of the lizard *Heteronotia binoei* was calculated the climate component of their fundamental niche through physiological measurements (thermal requirements for egg development, thermal preferences, and thermal tolerances). The environmental data analyzed was high-resolution climate data from Australia (air temperature, cloud cover, wind speed, humidity, radiation, etc.), and biophysical models projecting over the Australian subcontinent to predict the effects of global warming. Estimates predict relatively little effect on the maintenance of metabolic costs, mainly due to the buffering effect of thermoregulatory behavior of lizards. The lizards could be able to regulate their body temperature (and their metabolic rates) moving between thermally suitable places at the surface (shuttling), as has been shown for the nocturnal ectotherms, and other diurnal lizards like *Psammodroums algirus* in the Mediterranean region of Spain (Diaz & Cabezas-Díaz, 2004; Kearney & Predavec, 2000). On the other hand, lizards also should evade high thermal environments hiding in shelters in the hottest hours of day (Kearney, 2002), and changing their daily and stationary activity periods (Bawuens et al., 1996), and their habitat selection (Stevenson, 1995).

2.2.4 Effects of climate change on reproductive biology of reptiles

Climate change is a threat to reptile populations due that in some lineages the temperature experienced by embryos during incubation determines the offspring sex ratio. Increases or decreases of temperature could turn to bias in the proportion of the clutch toward one sex. Nests exposed to an increase heat produced dramatic differences in sex ratios compared to those placed in shaded sites (Doody et al, 2004; Janzen & Morjan, 2001; St. Juliana et al., 2004). For this reason reptiles can useful as indicators of biological impact due global warming (Janzen & Paukstis, 1991; Mrosovky & Provancha, 1992). For example, related studies has been conducted in North America with the loggerhead turtle (*Caretta caretta*); results suggest that organisms could alter their nesting behavior as adaptive mechanism to the warming of historical nesting sites, that range from southeastern Florida to southeastern

Virginia, where the sand temperature is lower and produces a higher proportion of males compared with those in Florida (Heppell et al., 2003). According to the IPCC (2001), Florida could experience a significant increase in temperature, so there is a high possibility of bias in the sex ratios, and a complete feminization of the setting of these the vast majority of the United Estates populations (Shoop & Kenney, 1992). The results indicates that an increase of 2° C is sufficient to cause feminization of clutches, and an increase of 3° C should drive to lethal incubation temperatures. As an alternative to temperature increase, turtles could potentially alter their nesting specific environment, looking for areas covered by vegetation, with greater proximity to the sea or groundwater. If turtles alter their oviposition season just a few days may be adapted to 1° C warming and if they did around a week, they could avoid the most extreme scenario (3° C); and this strategy could be the most viable adaptive mechanism for marine turtles in response to climate change (Hawkes et al., 2007).

In addition, evidence from genetic and behavioral analysis of the Painted Turtle populations *Chrysemis picta* in southeastern United States indicates that this turtle may disappear if do not develops traits that determine whether a balanced sex ratio, which directly affects their population dynamics (Girondot et al., 2004). Studies like those of Girondot et al., (2004) shows that species with temperature sex determination could be very sensitive to even modest variations (≤ 1 °C) in their local thermal environment. A slight increase in temperature could produce a high bias towards production of females (39° C). Such bias towards females results in a highly unequal sex ratio among adults and therefore if there are no males, females may lay eggs unfertilized eggs, and annual cohorts of offspring could lost and then the probability that the population becomes extinct.

The analysis of seasonal temperature variation and nesting behavior in *Chrysemis picta* suggest that pre-oviposition could mitigate climate change impacts on local populations located in less boreal latitudes and allow the production of males (Hays et al., 2001). Such nesting behavior modification may reduce the impact of local climatic variation, but may be insufficient for the populations living further north, since the young individuals may be low ability to survive the warm summer temperatures. It is believed that the metapopulation structure of these turtles among in the Mississippi River basin could help to mitigate the bias in sex ratio of the population caused by climate change if there is enough variation in both the thermal structure of suitable nesting areas, and migration rate between populations (Janzen, 1994).

In contrast, among the rhyncocephalians, lizards, and snakes with temperature sex determination, the threatened by global warming is higher in the tuatara (*Sphenodon punctatus*); a long generation time (which indicating limited potential to respond to rapid climate change), and extreme low temperature variation toward sex determination, with less than 1 °C drive the difference for the production of males or females (Nelson et al., 2004). Climate projections predict a significant increase among 1.4-5.8 °C in a very short period of time over the next 100 years (IPCC, 2001). Under this scenario reptiles may have four options to endure global warming: 1) modify its geographical range, 2) develop a genetic sex determination, 3) change their nesting behavior or 4) simply disappear (Janzen & Paukstis, 1991; Morjan, 2003). The tuatara may successfully manipulate the sex ratio of offspring by selecting the nesting site accord to vegetation, but apparently deeper nests in warm years to avoid bias toward males, which supports the proposition that the most viable strategy to deal with the effects of climate change is search sites with vegetal cover to nest. In the case of an abrupt increase in temperature that places the animals in the limit of their physiological tolerance, the peripheral populations of the tuatara would become extinct (Nelson et al., 2004). Finally some studies provide evidence that supports the proposal that the populations of reptiles compensate for differences in climate primarily through behavioral strategies (Doody et al., 2006; Gvozdika, 2002; Hertz & Huey, 1981). For example, in the Australian water dragon *Physignathus lesueurii* the maternal nesting behavior can respond for adjust the sex ratio and maintain viable populations across environmental extreme conditions, this compensate for climate differences by discriminating between potential nesting sites. This trait may be the most important to helping the species with thermal sex determination to compensate global warming (Janzen, 1992, 1994b).

3. General conclusions

Global climate change has influenced many aspects of the biology and ecology of amphibians and reptiles, which in some cases was caused the decline of their populations or serious threats. However evidence suggest that the phenomenon itself does not directly affect the organisms, but acts in combination with biotic and abiotic factors increasing its effects, as we illustrated in the case of diseases and infections the drying aquatic habitats draying up, the invasion of competing species, and the diminishing of the immune system due thermal stress regarding to the reproductive biology of organisms suggests that climate change affects several aspects among the most visible traits: phenology, survivorship and fecundity. However, it remains unclear if global warming will alter population dynamics of all populations or some one would be balanced due areas with suitable conditions for distribution and survival of organisms, mainly in the case of amphibians, whose survival depends largely on the presence of moisture and healthy aquatic habitats. However, there are non or very few data and projections turtles and crocodiles, comparing with lizards, which has been suggested that around of 50% of the Mexican *Sceloporus* lizards would disappear for 2080, since if maximum environmental temperature continues rising constantly due a overcome of physiological threshold of tolerance and the reduction of their daily activity times, which would cause an energetic shortfall as a consequence of low food intake (Sinervo et al., 2010).

Some potential adaptive responses already has bee suggested in different traits (behavior, physiology and morphology) among species affected by climate change. To test the likelihood of change in this traits due climate change requires the use of tools such as statistical analysis that incorporate phylogenetic hypotheses for the organisms under study, also an accurate estimate of the trait change rate both amphibians and reptiles is needed to understand the speed of the extraordinary rising of global environmental temperatures and their effects in biodiversity.

It is certain that climate global change will affect amphibians and reptiles around the world due synergic effects with other abiotic and biotic conditions. Our efforts should be concentrate in save as many populations and species we can, but first them all to understand the synergic effects and implement strategies to buffer them in regions when populations and species would be in the highest risk. It is quite possible that we cannot do anything against global warming and climate change, but we still can decide based on scientific evidence what, when and how to do about it.

4. Acknowledgements

We wish to thank the Posgrado en Ciencias Biológicas at The Universidad Nacional Autónoma de México and Consejo Nacional de Ciencia y Tecnología for their support and fellowships during our graduate studies.

5. References

Alexander, M. A. & Eischeid, J. K. 2001. Climate variability in regions of amphibians declines. *Conservation Biology*, Vol. 15, No 4 (August, 2001), pp. 930- 942. ISSN: 0888-8892.

Alford, R. A., & Richards, S. J. 1999. A problem in applied ecology. *Annual Rewiew of Ecology and Systematics*, Vol. 30, (November, 1999), pp.133-165. ISSN: 0027-8424.

Araujo, M. B., Thuiller W. &. Pearson R.G. 2006. Climate warming and the decline of amphibians and reptiles in Europe. *Journal of Biogeography*, Vol. 33, No 10 (December, 2006), pp. 1712-1728. ISSN: 0305-0270.

Aubert F., & Shine, R. 2010. Thermal plasticity in young snakes: how will climate change affect the thermoregulatory tactics of ectotherms? *The Journal of experimental Biology*, Vol. 213, No 2 (January, 2010). ISSN: 1477-9145.

Barnett ,T. P.,. Adam J. C & Lettenmaier D. P. 2005. Potential impacts of a warming climate on water availability in snow-dominated regions. *Nature*, Vol. 438, No 7066 (November, 2005), pp. 303-309. ISSN: 0028-0836.

Beebee, J. C. 1995. Amphibian breeding and climate change. *Nature*, Vol. 374, No 6519 (March, 1995), pp.219-220. ISSN: 0028-0836.

Berger, L., Speare, R., Hines, H.B, Marantelli, G., Hyatt, A.D., McDonald, K.R., Skerrat, L.F., Olsen, V., Clarke, J. M., Gillespie, G., Mahony, M., Sheppard, N., Williams C. &. Tyler M. J. 2004. Effect of season and temperature on mortality in amphibians due to chytridiomicosis. *Australian Veterinary journal*, Vol. 82, No 7 (March, 2008), pp. 434-439. ISSN: 1751-0813.

Berven, K.A., Gill D.E. & Smith-Gill S.J. 1979. Counter-gradient selection in the green frog, *Rana calamitans*. *Evolution*, Vol. 33, No 2 (June, 1979), pp.609-623. ISSN: 0014-3820.

Berven, K.A. 1990. Factors affecting populations fluctuations in larval and adults stages of the wood frogs (*Rana sylvatica*). *Ecology*, Vol. 71, No 4 (August, 1990), pp. 1599.1608. ISSN: 0012-9658.

Blaustein, A. R, & D.B. Wake 1990. Declining amphibian populations: A global phenomenon? *Trens in Ecology and Evolution*, Vol.5, No 7(July, 1990), pp. 203-204. ISSN: 0169-5347.

Blaustein A.R. 1994. Chicken Little or Nero's fiddle? A perspective on declining amphibian populations. *Herpetologica*, Vol. 50, No 1(March, 1994), pp. 85- 97. ISSN: 0018-0831.

Blaustein A.R., Hoffman P.D., Hokit D.G., Kiesecker J.M., Walls, S.C. &. Hays J.B. 1994a. UV repair and resistance to solar UV-B in amphibian eggs: a link to population declines? *Proceedings of the National Academy of Sciences, USA*, Vol. 91, No 5 (March 1994), pp. 1791-1795. ISSN: 0027-8424.

Blaustein A.R., Hokit D.G., O' Hara R.K. & Holt, R.A. 1994b. Pathogenic fungus contributes to the amphibians looses in the Pacific Norwest. *Biological Conservation*, Vol.67, No 3, pp. 251-254. ISSN: 0006-3207

Blaustein, A. R, & Wake, D.B. 1995. The puzzle of declining amphibian populations. *Scientific American*, Vol. 272, No 4 (April, 1995), pp. 52-57. ISSN: 0036-8733.

Blaustein, A.R., Kiesecker, J.M., Chivers, D.P., Hokit, D.P., Marco, D.G., Belden, A., & Hatch, L.K. 1998. Effects of ultraviolet radiation on amphibians: field experiments. *American Zoologist*, Vol. 38, No 6 (December, 1998), pp. 799-812. ISSN: 0003-1569.

Blaustein, A. R., Belden L.K., Olson, D.H., Green D.M., Root, T.L. & Kiesecker, J.M. 2001. Amphibian breeding and climate change. *Conservation Biology*, Vol. 15, No 6 (December 2001), pp. 1804-1809. ISSN: 0888-8892.

Blaustein, A. R. & Kiesecker J.M. 2002. Complexity in conservations: Lessons from the global decline of amphibians populations. *Ecology Letters*, Vol. 5, No 4 (July, 2002), pp. 587-608. ISSN: 1461-0248.

Blaustein, A.R. & Belden, L.K. 2003. Amphibian defenses against ultraviolet-B radiation. *Evolution & development*, Vol.5, No 1 (January, 2003), pp.89-97. ISSN: 1525-142X.

Blaustien A.R. & Bancroft B.A. 2007. Amphibian populations declines: evolutionary considerations. *Biosciences*, Vol. 57, No 5 (May, 2007), pp. 437-444. ISSN: 0006-3568.

Bradford, A.F. 2002.Amphibian declines and environmental change in the eastern Mojave Desert. Conference Proceedings. Spring-fed Wetlands: Important Scientific and Cultural Resources of the intermountain region pp7.

Bosch J., Carrascal,L. M., Durán, Walker, L, S. & Fisher, M. C. 2007. Climate Change and Outbreaks of Amphibian Chytridiomycosis in a Montane Area of Central Spain; Is There a Link? Proceedings of The Royal Society Biological Sciences, Vol. 274, No. 1607 (Jan. 22, 2007), pp. 253-260. ISSN: 1471-2954.

Carey, C. & Bryant, C.J. 1995. Possible interrelations among environmental toxicants, amphibian development and decline of amphibian populations. Environmental Health Perspectives, Vol. 103, No 4 (May, 1995), pp13-17. ISSN: 0091-6765.

Carey, C., Cohen N., & Rollins-Smith L. 1999. Amphibian declines an immunological perspective. *Developmental and Comparative Immunology*, Vol. 23, No 6 (September, 1999), pp. 459-472. ISSN: 0145-305X.

Chamaillè-Jammes, S., Massot, M., Aragon P. & J., Clobert 2006. Global warning and positive fitness response in mountain populations of common lizards *Lacerta vivipara*. *Global Change Biology*, Vol. 12, No 2 (February, 2006), pp. 392-402. ISSN: 1354-1013.

Christy, J. R., Clarke, R. A., Gruza, G. V., Jouzel, J., Mann, M. E., Oerlemans, J., Salinger, M. J., Wang, S. W. 2001. Observed Climate Variability and Change. In: *Climate change 2001: the scientific basis. Contributions of working group I to the third assessment report of the Intergovernmental Panel on Climate Change.* Houghton, J. T., Ding, Y., Griggs, D. J., Noguer, M., Van der, P. Linden, J. & Xiaosu, D. Published for the Intergovernmental Panel on Climate Change, pp. 99-184, ISBN: 0521014956, Cambridge University Press.

Cleland, E.E., Chiariello, N.R., Loarie, S.R., Mooney, H.A. & Field, C.B., 2006. Diverse responses of phenology to global changes in a grassland ecosystem. *Proceedings of the National Academy of Sciences, USA*, Vol. 103, No 37 (September 2006), pp. 13740–13744. ISSN: 0027-8424.

Crump, M.L. & Pounds, J.A. 1985.Lethal Parasitism of an Aposematic Anuran (*Atelopus varius*) by *Notochaeta bufonivora* (Diptera: Sarcophagidae) *The Journal of Parasitology*, Vol. 71, No 5 (October, 1985), pp. 588-591. ISSN: 0022-3395.

Crump, M.L. & Pounds J.A. 1989. Temporal variation in the dispersion of a tropical anuran. *Copeia*, Vol. 1989, no 1(February, 1989), pp. 209-211. ISSN: 0045-8511.

Cubasch, U., Meehl, G.A., Boer, G. J., Stouffer, R. J., Dix, M., Noda, A., Senior, C. A., Raper, S., & Yap, K. S. Abe-Ouchi, A., Brinkop, S., Claussen, M., Collins, M., Evans, J., Fischer-Bruns, I., Flato, G., Fyfe, J. C., Ganopolski, A., Gregory, J. M., Hu, Z. Z., Joos, F., Knutson, T., Knutti, R., Landsea, C., Mearns, L., Milly, C. J., F.B. Mitchell, T. Nozawa, H. Paeth, J. Räisänen, R. Sausen, S. Smith, T. Stocker, A. Timmermann, U. Ulbrich,. Weaver, A., Wegner, J., Whetton, P., Wigley, T., Winton, M., Zwiers, F. 2001. Projections of future climate change. In: *Climate change 2001: the scientific basis. Contributions of working group I to the third assessment report of the Intergovernmental Panel on Climate Change.* Houghton, J. T., Ding, Y., Griggs, D. J., Noguer, M., Van der, P. Linden, J. & Xiaosu, D. Published for the Intergovernmental Panel on Climate Change, pp. 527-578, ISBN: 0521014956, Cambridge University Press.

Daszack, P., Cunningham A. A., Hyatt, A.D., Green, D. E. & Speare, R. 1999. Infectious diseases and amphibians populations decline. *Emerging Infectious Diseases*, Vol 5, No 6 (Nov-Dec 1999) pp. 735-748. ISSN: 1080-6059.

Daszack, P., Scott, D.E., Kirkpatrick, A.M., Faggioni, C., Gibbons, J.W. & Porter, D.2005. Amphibian population declines at Savannah River site are linked to climate not chytridiomicosis. *Ecology*, Vol. 83, No 12 (December, 2005), pp.3232-3237. ISSN: 0012-9658.

Davies, K. F., Margules, C. R. &. Lawrence, J. F. 2000. Which traits of species predict population declines in experimental forest fragments? *Ecology*, Vol 81, No 5 (May, 2000), pp. 1450–1461. ISSN: 0012-9658.

Díaz, J. A. & Cabezas-Díaz, S. 2004. Seasonal variation in the contribution of different behavioral mechanisms to lizard thermoregulation. *Functional Ecology*, Vol. 18, No 6 (December, 2004) pp. 867-875.

Donelly, M. A. and. Crump, M. L 1998. Potential effects of climate change on two neotropical amphibian assemblages. *Climatic Change*, Vol. 39, No 2-3, pp. 541-561. ISSN: 0165-0009.

Dorcas, M. E., Hopkins W.A. & Roe, J. H. 2004. Effects of body mass and temperature on standard metabolic rate in the eastern Diamondback rattlesnake (*Crotalus adamanteus*). *Copeia*, Vol. 2004, No 1 (Feb, 2004), pp. 145–151. ISSN: 0045-8511.

Doody, J.S., Georges A. and Young J.E. 2004. Determinants of reproductive success and offspring sex in a turtle with environmental sex determination. *Biological Journal of the Linnean Society*, Vol. 81, No 1 (January 2004), pp 1-16. ISSN: 0024-4066.

Doody, J.S., Guarino, E., Georges, A., Corey, B, Murray, G. & Ewert, M. 2006. Nest site choice compensates in a lizard with environmental sex determination. Evolutionary *Ecology*, Vol. 20, No 4, pp 307-330. ISSN: 0012-9658.

Duellman, W. E. & Trueb, L. 1986. Biology of amphibians. McGraw-Hill Book Company, ISBN: 080184780X, Toronto.

Dunham, A. E. 1993. Population responses to environmental change. In: *Biotic interactions and global change*. Kareiva, J.G., Kingsolver, J.G. & Huey, R.B., pp. 5-119, Sinauer Sunderland, ISBN: 0878934308, Massachusetts.

Easterling, D. R., Meehl, G. A, Parmesan, C., Changnon, S. A., Karl, T. R. & Mearns, L. O. 2000. Climate extremes: observations, modeling, and impacts. *Science* Vol. 289, No 5487 (September, 2000), pp. 2068-2074. ISSN: 1095-9203.

Epstein, P. R 2001. Climate change and emerging infectious diseases. *Microbes and Infection*, Vol. 3, No 9 (July, 2001), pp. 747--754. ISSN: 1286-4579.

Ewert, M.A., Lang, J. W. & Nelson, C.E. 2005. Geographic variation in the pattern of temperature-dependant sex determination in the American snapping turtle (*Chelydra serpentina*). *Journal of Experimental Zoology*, Vol. 265, No 1, pp. 82-95. ISSN: 1097-010X.

Forchahammer, M.C., E. Post and N.C. Stenseth 1998. Breeding phenology and climate. Nature Vol. 391, No 6662 (January, 1998), pp. 29-30. ISSN: 0028-0836.

Gibbs, J.P., & Briesch A.R. 2001. Climate warning and calling phenology of frogs near Ithaca, New York, 1900-1999. *Conservation Biology*, Vol. 15, No 4 (August, 2001): 1175-1178. ISSN: 0888-8892.

Girondot, M., Delmas, V., Rivalan, P., Courchamp F., Prevot-Julliard A. &. Godfrey, M. H 2004. Implications of temperature-dependent sex determination for population dynamics. In: *Temperature-Dependent sex determination in vertebrates*. Valenzuela N. & Lance V., pp. 148-155, Smithsonian Institute Press, ISBN 1-58834-203-4, Washington, DC.

Grimm, N.B. 1993. Implications of climate change on stream communities. In: *Biotic Interactions and Global Change*. Kareiva, J.G., Kingsolver, J.G. & Huey, R.B., pp. 293–314, Sinauer, Sunderland, ISBN: 0878934308, Massachusetts.

Gulve, P. S. 1994. Distribution and extinction patterns within a northern metapopulation of the pool frog, *Rana lessonae*. *Ecology*, Vol. 75, No 5 (July, 1994), pp. 1357–1367. ISSN: 0012-9658.

Gvozˇdik L,and A.M. Castilla 2001. A comparative study of preferred body temperatures and critical thermal tolerance limits among populations of *Zootoca vivipara* (Squamata: Lacertidae) along an altitudinal gradient. *Journal of Herpetology*, Vol. 35 No 3 (September, 2001), pp. 486–492. ISSN: 0022-1511.

Hays, J.B.,. Blaustein, A. R, Kiesecker, J. M., Hoffman, P. D., Pandelova, I., Coyle, D. & Richardson, T. 1996. Developmental responses of amphibians to solar and artificial UV-B resources: a comparative study. *Photochemistry and Photobiology*, Vol. 64, No 3 (September, 1996), pp. 449-455. ISSN: 1751-1097.

Harvell, C. D., Mitchell, C. E. & Ward J. R. 2002. Climate warming and disease risks for terrestrial and marine biota. *Science*, Vol. 296, No 5576 (June, 2002), pp. 158–62. ISSN: 0036-8075.

Hawkes, L. A., Broderick, A. C., Godfrey, M. H. & Godley, B. J. 2007. Investigating the potential impacts of climate change on a marine turtle population. *Global Change Biology*, Vol. 13, No 5 (May, 2007), pp. 923-932. ISSN: 1365-2486.

Hays, G. C., Ashworth, J. S. & Barnsley M. J. 2001. The importance of sand albedo for the thermal conditions on sea turtle nesting beaches. *Oikos*, Vol 93, No 1 (April, 2001), pp. 87–94. ISSN: 1600-0706.

Hegerl, G. C., Zwiers, F.W., Sttot, P.A., & Kharin, K.. K. 2004. Detectability of anthropogenic changes in annual temperature and precipitation extremes. *Journal of Climate*, Vol. 17, No 19 (October, 2004), pp. 3683-3700. ISSN: 0894-8755.

Heppell, S. S., Snover, M. L. & Crowder L. B., 2003. Sea turtle population ecology. In: *The Biology of Sea Turtles. Volume II, CRC Marine Biology Series*, Lutz, P. L., Musick, J. A. & Wyneken, J. pp. 275-306, CRC Press, Inc., Boca Raton, ISBN: ISBN: 0849311233, London, New York, Washington D.C.

Herman, C. A. 1992. Endocrinology, In: *Environmental physiology of the amphibians*. Feder M. E. & Burggren W. W., pp. 40-54, University Chicago Press, ISBN: 0226239446, Chicago.

Hertz, P.E. & Huey, R.B. 1981. Compensation for altitudinal changes in the thermal environments by some *Anolis* lizards on Hispaniola. *Ecology*, Vol. 62, No 3 (June, 1981), pp. 515-521. ISSN: 0012-9658.

Houghton, J. T., Ding, Y., Griggs, D. J., Noguer, M., Van der, P., Linden, J. & Xiaosu, D. 2001. Observed variability and change. In: Climate change 2001: the scientific basis. Contributions of working group I to the third assessment report of the Intergovernmental Panel on Climate Change. Published for the Intergovernmental Panel on Climate Change, pp. 99-184, ISBN: 0521014956, Cambridge University Press.

IPCC. 2007. *Summary for policy makers*. Cambridge University Press, ISBN: 92-9169-322-7 Cambridge, U.K.

Janzen, F. J. 1994. Climate-change and temperature-dependent sex determination in reptiles. *Proceedings of the National Academy of Sciences, USA*, Vol. 91, No 16 (August, 1994), pp. 7484- 7490. ISSN: 0027-8424.

Janzen, F. J. 1994b.Vegetational cover predicts the sex ratio of hatching turtles in natural nest. *Ecology*, Vol. 75, No 6 (September, 1994), pp. 1593-1599. ISSN: 0012-9658.

Janzen, F. J. & Paukstis, G. L. 1991. Environmental sex determination in reptiles: ecology, evolution and experimental design. *The Quarterly Review of Biology*, Vol. 66, No 2 (June, 1991), pp. 149-179.ISSN: 15397718.

Janzen, F. J. 1992. Heritable variation for sex ratio under environmental sex determination in the common snapping turtle (*Chelydra serpentina*). *Genetics*, Vol. 131, No 1 (May, 1992), pp. 155-161. ISSN: 1943-2361.

Janzen F. J. & Morjan, C. L. 2001. Repeatability of microenvironment-specific nesting behavior in a turtle with environmental sex determination. *Animal Behaviour*, Vol. 62, No 1 (July, 2001), pp. 73-82. ISSN: 0003-3472.

Jones P. D., Osborn, T. J. &. Briffa, K. R. 2001. The evolution of climate over the last millennium. *Science*, Vol. 292, No 5517 (April, 2001), pp. 662-667. ISSN: 1095-9203.

Jones, P. D., New, M., Parker, D. E., Martin, S. & Rigor, I. G. 1999: Surface air temperature and its changes over the past 150 years. *Reviews of Geophysics*, Vol. 37, No 2 (May, 1999), pp. 173-199. ISSN: 8755-1209.

Kerr, R.A. 1995. It's official: First glimmer of greenhouse warming seen. Science. Vol. 270, No 5242 (December, 1995), pp. 1565-1657. ISSN: 1095-9203.

Jorgensen, C. B. 1992. Growth and reproduction, In: *Environmental physiology of the amphibians*. Feder M. E. & Burggren W. W., pp 439-366, University Chicago Press, ISBN: 0226239446, Chicago.

Kearney, M., & M. Predavec. 2000. Do nocturnal ectotherms thermoregulate? A study of the temperate gecko Christinus marmoratus. *Ecology* Vol. 81, No 11 (November, 2000), pp. 2984–2996. ISSN: 0012-9658.

Kearney, M. 2002. Hot rocks and much-too-hot rocks: Seasonal patterns of retreat-site selection by a nocturnal ectotherm. *Journal of Thermal Biology*, Vol. 27 No 3, (June), pp. 205-218. ISSN: 0306-4565.

Kearney, M. & Porter, W. P. 2004. Mapping the fundamental niche: physiology, climate, and the distribution of a nocturnal lizard. *Ecology*, Vol. 85, No 11 (November, 2004), pp. 3119-3131. ISSN: 0012-9658

Kiesecker, J. M., & Blaustein, A. R. 1995. Synergism between UV-B radiation and a pathogen magnifies amphibian embryo mortality in nature. *Proceedings of Natural Academy of Sciences USA*, Vol. 92, No 24 (November, 1995), pp. 11049-11052. ISSN: 0027-8424.

Kiesecker, J. M., & A. R. Blaustein 1997a. Populations differences in responses of red-legged frogs (*Rana aurora*) to introduced bullfrog. *Ecology*, Vol. 78, No 6 (September, 1997), pp. 1752-1760. ISSN: 0012-9658

Kiesecker, J. M. & Blaustein, A. R. 1997b. Influences of egg laying behavior on pathogenic infection of amphibian eggs. *Conservation Biology*, Vol. 11, No 1 (February, 1997), pp. 214-220. ISSN: 0888-8892.

Kiesecker, J. M. & Blaustein, A .R. 1999. Pathogen reverses competition between larval amphibians. *Ecology*, Vol. 80, No 7 (October, 1999), pp. 2442-2448. ISSN: 0012-9658.

Kiesecker, J. M., Blaustein, A. R. & Belden, L. K. 2001. Complex causes of amphibian population declines. *Nature*, Vol. 410, No (April, 2001) 681-684. ISSN: 0028-0836.

La Marca, E., Lips, K. R., Lötters, S., Puschendorf, R., Ibañez, R., Ron, S., Rueda-Almonacid, J. V., Schulte, R., Marty, C., Castro, F., Manzanilla-Pupo, J., García-Perez, J. E., Bustamante, M. R., Coloma, L. A., Merino-Viteri, A., Toral, E., Bolaños, F., Chaves, G., Pounds, A. & Young, B. A. (2005). Catastrophic population declines and extinctions in neotropical harlequin frogs (*Bufonidae: Atelopus*). *Biotropica*, Vol. 37, No 2 (June, 2005), pp. 190–201. ISSN: 0006-3606.

Laurence, W.F. 1996. Catastrophic declines of Australian rainforest frogs: is unusual weather responsible? *Biological Conservation*, Vol. 77, No 2-3, pp. 203-212. ISSN: 0006-3207.

Lips, K. R., Reeve, J. & Witters, L. 2003. Ecological factors predicting amphibian population declines in Central America. *Conservation Biology*, Vol. 17, No 4 (August, 2003), pp. 1078-1088. ISSN: 0888-8892.

Lips, K R, Diffendorfer J, Mendelson, J R III & Sears, M W, 2008. Riding the Wave: Reconciling the Roles of Disease and Climate Change in Amphibian Declines. Plos Biology, Vol.6, No 3, (March, 2008), pp. 441-453. ISSN-1545-7885.

Maniero G. D. & Carey C. 1997. Changes in selected aspects of immune function in the leopard frog, *Rana pipiens*, associated with exposure to cold. *Journal of comparative physiology*. Vol. 167, No 4 (May, 1997), pp. 256-263. ISSN: 1432-1351.

McCarty J. P. 2001. Ecological Consequences of Recent Climate Change. *Conservation Biology*, Vol. 15, No 2 (April 2001), pp. 320-331. ISSN: 0888-8892.

Miles, D. B. 1994. Population differentiation in locomotory performance and the potential response of a terrestrial organism to global environmental change. *American Zoologist*, Vol. 34, No 3, pp. 422–436. ISSN: 0003-1569.

Mrosovky, N., & Provancha, J. 1992. Sex ratio of hatchling loggerhead sea turtles: data and estimates from a 5-year study. *Canadian Journal of Zoology*, Vol. 70, No 3 (March, 1992), pp. 530-538. ISSN: 1480-3283

Morjan C.L. 2003. How rapidly can maternal behavior affecting primary sex ratio evolve in a reptile with environmental sex determination? *American. Naturalist*, Vol. 162, No 2 (August, 2003), pp. 205–219. ISSN: 00030147.

Nelson N. J., Thompson, M. B., Pledger, S., Keall, S. N. & Daugherty, C. H. 2004. Do TSD, sex ratios, and nest characteristics influence the vulnerability of tuatara global warming? *International Congress Series*. Vol. 1275, (December, 2004), pp. 250-257. ISSN 0531-5131.

Oechel, W.C. & Vourlitis, G.L. 1994: The effects of climate change on land-atmosphere feedbacks in Arctic tundra regions. *Trends in Ecology and Evolution*, Vol. 9, No 9 (September, 1994), pp. 324-329. ISSN: 0169-5347.

Ovaska K., Davies, T.M. & Flamarique, T.M. 1997. Hatching success and larval survival of the frog *Hyla regilla* and *Rana aurora* under ambient and artificially enhanced solar ultraviolet radiation. *Canadian journal of Zoology*, Vol. 75, No 7 (July ,1997) pp. 1081-1088. ISSN: 1480-3283.

Pancharatna, K. & Patil, M. M. 1997. Role of temperature and photoperiod in the onset of sexual maturity in female frogs, *Rana cyanophylictis*. *Journal of herpetology*, Vol. 31, No 1 (March, 1997), pp. 111-114. 0022-1511.

Parra-Olea G., Martínez-Meyer, E. & Pérez-Ponce de León G. 2005. Forecasting climate change effects on salamander distribution in the highlands of central México. *Biotropica*, Vol. 37, No 2 (June, 2005), pp. 202-208. ISSN: 1744-7429

Parmesan, C. &. Yohe, G. 2003. A globally coherent fingerprint of climate change impacts across natural systems. *Nature*, Vol. 421, No 6918 (January, 2003) pp. 37-42. ISSN: 0028-0836.

Piotrowski, J. S., Annis, S. L., & Longcore, J. E. 2004. Physiology of Batrachochytrium dendrobatidis, a chytrid pathogen of amphibians. *Mycologia*, Vol. 96, No 1 (January-February, 2004), pp. 9-15. ISSN: 0027-5514.

Pough, H., Andrews, R. M., Cadle, J. E., Crump, M. L., Savitzky, A. H. & Wells, K. D. 2001. *Herpetology*, (Second Edition), Prentice Hall, ISBN: 9780131008496, New Jersey.

Pounds, J. A & Crump, M.L. 1987. Harlequin frogs along a tropical montane stream aggregation and the risk of predation by frog eating fly. *Biotropica*, Vol 19, No 4 (December, 1987), pp. 306-309. ISSN: 1744-7429.

Pounds, J. A & Crump, M.L 1994. Amphibian declines and climate disturbance: the case of the golden toad and the harlequin frog. *Conservation Biology*. Vol. 8, No 1(March, 1994), pp. 72-85. ISSN 0888-8892.

Pounds, J. A., Fogden, M.L.P. Savage, J. M., & Gorman G. C. 1997. Test of null models for amphibian declines on a tropical mountain. *Conservation Biology,* Vol. 11, No 6 (December, 1997), pp. 1307-1322. ISSN 0888-8892.

Pounds, J. A., Fogden, M. P. & Campbell, J. H 1999. Biological response to climate change on a tropical mountain. *Nature,* Vol. 398, No 6728 (February, 1999), pp. 611-615. ISSN: 0028-0836.

Pounds, J. A 2001. Climate and amphibian declines. *Nature,* Vol. 410, No 6829 (April 2001) 639-640. ISSN: 0028-0836.

Pounds, J. & Puschendorf, R. 2004. Clouded futures. *Nature* Vol. 427, No 6970 (January, 2004), pp. 107-109. ISSN: 0028-0836.

Pounds, J. A 2006. Widespread amphibian declines from epidemic disease driven for global warning. Nature, Vol. 439, No 7073 (January, 2006), pp.161-167. ISSN: 0028-0836.

Raffel T. R., Rohr, J. R J., Kiesecker, M. & Hudson P. J. 2006. Negative effects of changing temperature on amphibian immunity under field conditions. *Functional Ecology,* Vol. 20, No 5 (October, 2006), pp. 819-828. ISSN: 1365-2435.

Reed, R. N., & Shine R. 2002. Lying in wait for extinction? Ecological correlates of conservation status among Australian elapid snakes. *Conservation Biology,* Vol. 16, No 2 (April, 2002), pp. 451-461. ISSN 0888-8892.

Retallick, R.W., McCallum, R.H. & Speare, R. 2004. Endemic infection of the amphibian chytrid fungus in a frog community post-decline. *Plos Biology,* Vol. 2, No 11(November, 2004), pp. 1965-1971. ISSN-1545-7885

Rodo´, X., Pascual, M., Fuchs, G. & Faruque, A. S. G. 2002. ENSO and cholera: A nonstationary link related to climate change? *Proceedings of the National Academy of Sciences,* USA, Vol. 99, No 20 (October, 2002), pp. 12901--12906. ISSN: 0027-8424.

Rojas, S., Richards, K. Jankovich, J.K. & Davison, E.W. 2005. Influence of the temperature on Ranavirus infection in larval salamanders *Ambystoma tigrinum. Diseases of Aquatic Organism,* Vol. 63, No 2-3 (February, 2005), pp.95-100. ISSN 1616-1580.

Romer A. S. 1986. *The procession of Life,* World Publishing Company, ISBN: 9781146617550, New York.

Rome L. C., Stevens, E. D. & John-Adler, H. B. 1992. The influence of temperature and thermal acclimation on a physiological function. In: *Environmental physiology of the amphibians.* Feder M. E. & Burggren W. W., pp. 183-205, University of Chicago Press, ISBN 0226239446,Chicago.

Ron, S. R. 2005. Predicting the distribution of the amphibian pathogen *Batrachochytrium dendrobatidis* in the New World. *Biotropica,* Vol. 37, No 2 (June, 2005), pp. 209-221. ISSN: 1744-7429.

Santer B. D., Wigley, T. M. L., Mears, C., Wentz, F. J. S., Klein, A., Seidel, D. J., Taylor, K. E., Thorne, P. W., Wehner, M. F., Gleckler, P. J., Boyle, J. S., Collins, W. D., Dixon, K. W., Doutriaux, C., Free, M., Fu, Q.,. Hansen, J. E., Jones, G. S., Ruedy, R., Karl, T. R., Lanzante, J. R., Meehl, G. A., Ramaswamy, V., Russell, G., Schmidt, G. A. 2005. Amplification of surface temperature trends and variability in the tropical atmosphere. *Science,* Vol. 309, No 5740 (August, 2004), pp.1551 – 1556. ISSN: 0027-8424.

Schar, C., Vidale, P. L., Luthi, D., Frei, C., Haberli, C., Liniger, M. A. & Appenzeller, C. 2004. The role of increasing temperature variability in European summers heatweaves. *Nature*, Vol. 427, No 6972 (January, 2004), pp. 332-336. ISSN: 0028-0836.

Schindler, D.W., Curtis, P. J., Parker, R.B., & Stainton, M.P. 1996. Consequences of climate warning and lake acidification for UV-B penetration in North American boreal lakes. *Nature*, Vol. 379, 6567 (February, 1996), pp. 705-708. ISSN: 0028-0836.

Shoemaker, V. H., Hillyard, S. D., Jackson, D. C., McClanahan, L. L., Whiters, P.C. & Wygoda M. L. 1992. Exchange of water ions, and respiratory gases in terrestrial amphibians. In: *Environmental physiology of the amphibians*. Feder M. E. & Burggren W. W., pp. 125-150, University of Chicago Press, ISBN0226239446, Chicago.

Shoop, C. R. & Kenney R. D. 1992. Seasonal Distributions and Abundances of Loggerhead and Leatherback Sea Turtles in Waters of the Northeastern United States. Herpetological Monographs, Vol 6, pp. 43-67. ISSN: 1938-5137.

Seimon, T. A., Seimon, A., Daszak, P., Halloy, S. R., Schloegel, L. M., Aguilar, C. A., Sowell, P., Hyatt, A. D., Konecky, B. & E. Simmonds, J. 2007. Upward range extension of Andean anurans and chytridiomycosis to extreme elevations in response to tropical deglaciation. *Global Change Biology*, Vol. 13, No 1 (January, 2007), pp. 288–299. ISSN: 1365-2486.

Sjögren, P. 1991. Extintion and isolation gradients in metapopulations: the case of the pool frog (*Rana lessonae*). *Biological Journal of the Linnean Society*, Vol. 42, No 1-2 (January, 1991), pp. 135-147. ISSN: 0024-4066.

Sinervo, B., Méndez-de-la-Cruz, F., Miles, D. B., Heulin, B., Bastiaans, E., Villagrán-Santa Cruz, M., Lara-Resendiz, R., Martínez-Méndez, N., Calderón-Espinosa, M. L., Meza-Lázaro, R. N., Gadsden, H., Ávila, L. J., Morando, M., De la Riva, I. J., Sepúlveda, P. V., Rocha, C. F. D., Ibarguengoytia, N., Puntriano, C. A., Massot, M., Lepetz, V., Oksanen, T. A., Chapple, D. G., Bauer, A. M., Branch, W. R., Clobert, J. & Sites, J. W., Jr. Erosion of lizard diversity by climate change and altered thermal niches. *Science*, Vol. 328, No 5980 (Mayo), pp. 894-899. ISSN: 0027-8424.

Stearns, S. C. 1992, *The Evolution of Life Histories*, Oxford University Press. ISBN: 978-0-19-857741-6, New York.Stebbins R. C. & Cohen, N. W. 1995. *A natural history of amphibians*, Princeton University Press, ISBN: 0691032815, Princeton, New Jersey, USA.

St. Juliana J.R., Bowden, R.M. & Janzen, F. J. 2004. The impact of behavioral and physiological maternal effects on offspring sex ratio in the common snapping turtle, *Chelydra serpentina Behavioral Ecology and Sociobiology*, Vol. 53, No 3 (July, 2004), pp. 270-278. ISSN: 1432-0762.

Stockwell, D. & Peters, D. 1999. The GARP medelling system. Problems and solutions to automated spatial prediction. International Journal of Geographic Information. *International Journal of Geographical Information Science*, Vol. 13, No 2, pp. 143-158 ISSN: 1365-8824.

Stott P. A. 2003. Attribution of regional-scale temperature changes to anthropogenic and natural causes. *Geophysical Research Letters*, Vol. 30, No 14 (July, 2003), pp. 1728-1732. ISSN: 0094-8276

Stuart, S.N., Chanson, J.S., Cox, N.A., Young, B.E., Rodriguez, A.S.L., Fischman, D.L., and Waller, R.W. 2004. Stratus and trends of amphibians declines and extinctions worldwide. *Science*, Vol. 306, No 5702 (December, 2004), pp. 1783-1786. ISSN: 0027-8424.

Taylor, K. 1999. Rapid climate change. *American Scientist*, Vol. 87, No 4 (July-August, 1999), pp. 320-327. ISSN: 1545-2786.

Terhivou, J. 1998. Phenology of spawning of the common frog (*Rana temporaria*) in Finland from 1846 to 1986. Annales Zoologica Fennicci, Vol. 25, No 2, pp.165- 175. ISSN: 0003-455X.

Thomson, D. J. 1995. The seasons, global temperature, and precession. *Science*, Vol. 268, No 5207 (April, 1995) pp. 59-68. ISSN: 0027-8424.

Thuiller W. 2003. BIOMOND: optimising predictions of species distributions and projecting potential future shifts under global change. *Global Change Biology*, Vol. 9 No 10 (October, 2003), pp. 1353-1362. ISSN: 1365-2486.

Trevini, M. 1993, Influences of Ozone Depletion on Human and Animal Health. In: *UV-B Radiation and Ozone Depletion: Effects on Humans, Animals, Plants, Microorganisms, and Materials*. Trevini, M., pp. 95-124, ISBN: 9780873719117, Lewis Publishers, Boca Raton, Florida.

Vitt, L. J., Caldwell, J. P., Wilbur, H. M., & Smith, D. C 1990. Amphibians as harbingers of decay. *Bioscience*, Vol. 40, No 6 (June, 1990), pp. 418. ISSN: 0006-3568.

Walther, G., R., Post, E., Convey, P., Menzel, A., Parmesan, C., Beebee, T J. C., Fromentin, J. M., Hoegh-Guldberg, O. & Bairlein, F. 2002. Ecological responses to recent climate change. *Nature*, Vol. 416, No 6879 (March, 2002), pp. 398-395. ISSN: 0028-0836.

Williams, S. E. & Hero J. M. 1998. Rainforest frogs of the Australian wet tropics: guild classification and the ecological simililarity of declining species. *Proceedings of the Royal Society of London*, Vol. 265, No 1329 (April, 1998), pp. 597-602. ISSN: 0080-4630.

Wollmuth, L. P., Crawshaw, L.I., Forbes, R. B. & Grahn, D. A.1987. Temperature selection during development in a montane anuran species, *Rana cascadae*, *Physiological Zoology*, Vol. 60, No 4 (July -August, 1987), pp. 472–480. ISSN: 0031-935X.

Woodhams, D. C., Alford, R. A. & Marantelli, G. 2003. Emerging disease of amphibian cured by elevated body temperatures. *Diseases of Aquatic Organisms*, Vol. 55, No. 1 (June), pp. 65-67. ISSN 1616-1580.

Worrest R. C. & Kimeldorf D. J. 1976. Distortions in amphibian development induced by ultraviolet-B enhancement (290-315nm) of a simulated solar spectrum. *Photochemistry and Photobiology*, Vol. 24, No 4 (October, 1976), pp. 377-382. ISSN: 1751-1097.

Wyman R.L. 1990. What's happening to the amphibians? *Conservation Biology*. Vol. 4, No 4 (December, 1990), pp. 350-352. ISSN: 1526-4629. ISSN: 0028-0836.

Yan, N. D., Keller, W., Scully, N. M., Lean, D. R. S., & Dillon, P. J. 1996. Increased UV-B penetration in a lake owing to drought-induced acidification. *Nature*, Vol. 381 No. 6578 (May, 1996)141-143 0028-0836.

Young, B., Lips, K., Reaser, J., Ibáñez, R., Salas, A., Cedeño, R., Coloma, L., Ron, S., La Marca, E., Meyer, J., Muñoz, A, Bolaños, F., Chavez, G. & Romo, D. 2001.

Population declines and priorities for amphibian conservation in Latin America. *Conservation Biology*, Vol. 15, No 5 (October, 2001) pp. 1213–1223. ISSN: 1526-4629.

Zachos J., M. Pagani, L. Sloan, E. Thomas & K. Billups 2001. Trends, Rhythms, and Aberrations in Global Climate 65 Ma. to Present. *Science* Vol. 292; No 5517 (April, 2001), pp. 686-693. 0027-8424.

Zug, G. R. 1993. *Herpetology and introductory biology of amphibians and reptiles*. Academic Press, ISBN: 9780127826226, San Diego California.

8

Identification and Analysis of Burned Areas in Ecological Stations of Brazilian Cerrado

Claudionor Ribeiro da Silva, Rejane Tavares Botrel,
Jeová Carreiro Martins and Jailson Silva Machado
Federal University of Uberlândia / Campus Monte Carmelo-MG / Geography Institute
Federal University of Piauí / Campus Bom Jesus-PI / Department of Forest Engineering
Brazil

1. Introduction

The degradation of the soil and native ecosystems and the dispersion of exotic species are the largest threats to the biodiversity. Brazil is considered one of the most biodiversity countries in the world, where it is concentrated about 10% of whole terrestrial biota (Mittermeier et al., 1997). Studies have shown that the Brazilian diversity was already greater than is recorded nowadays. Even with the disappearance of species, the biodiversity of the Cerrado is still quite expressive and notable. Depending on the taxonomic group, 20 to 50% of species occurring in the Brazilian Cerrado.

The great diversity of species of animals and plants of the Cerrado is associated with spatial heterogeneity (the variation of ecosystems over space) of this biome, which allows the coexistence of different physiognomic forms in the same region. The variation of environments makes possible species of animals and plants present a strong association with local ecosystems, being closely tied to natural environments, such as the *Antilophia galeata* which are found only in gallery forests, and the *Mauritia flexuosa* that are closely associated with the paths. This relation fauna-flora-Cerrado illustrates the importance of maintaining of the natural vegetable covering of this biome, as a basic strategy for maintaining biological diversity expressive.

Recent studies indicate that a loss up to 25% of bird species associated with the gallery forest of the Cerrado may occur, just if there is the destruction of natural environments to the neighboring woods, even though it remains untouched (Machado, 2000; Machado et al ., 2004). Furthermore, excessive reduction of native areas will cause the extinction of species from fragments of small size (Hass, 2002).

The Cerrado is a biome originally covered by vegetation ranging from grassland (country) to *Cerradão* and has two well-defined seasons: dry winter and rainy summer. This biome shelters deciduous species and presents a nutrient-poor acidic soil with high iron and aluminum content. According to Eiten (1979), the Cerrado vegetation varies from sparse trees to dense forest vegetation. The variation in the Cerrado physiognomies has been attributed to the action of fire, soil factors (Eiten, 1972; Coutinho, 1978; Rizzini, 1979), topography and water (Furley & Ratter, 1988).

The Cerrado biome is the second largest biogeographic region of Brazil covering an area of 2,036,448.00 km^2 (Fig. 1), around 23% of the national territory (Ratter et al., 1997).

Approximately 40% of this biome has already been completely anthropized for agricultural and livestock activities. Such activities may cause serious damage to the Cerrado such as burning, tree felling for charcoal production. Hoffmann & Jackson (2000) affirmed that the conversion of the natural Savannah in pasture can reduce the precipitation in at least 10%, and to increase the medium temperature of the superficial air in 0.5ºC. Those practices usually occur in a disorderly and non-sustainable way (Castro, 1999; Saucer, 1999; Araujo, 2005).

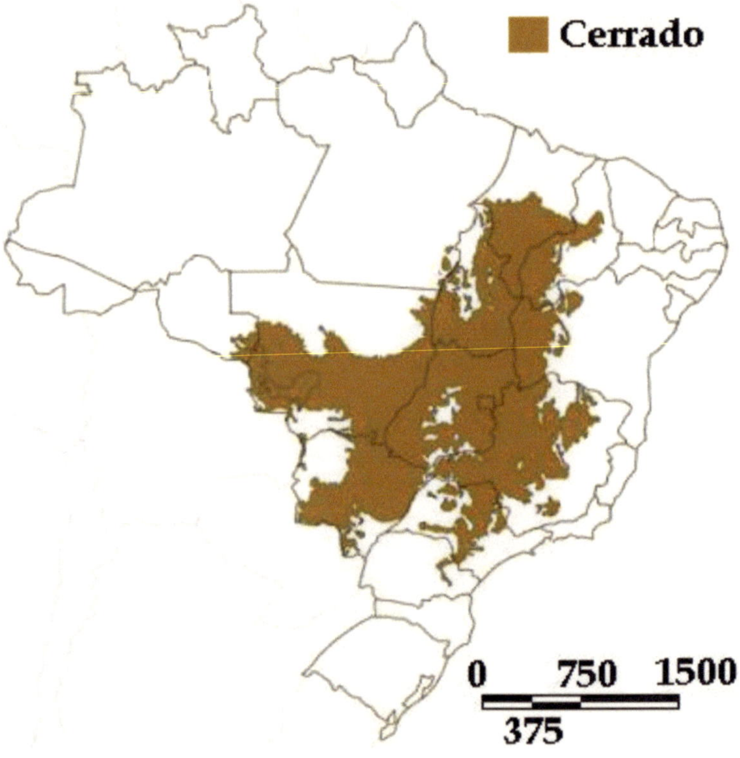

Fig. 1. Cerrado biome in Brazil (http://www.wwf.org.br/)

In Piauí state the Cerrado occupies approximately 115,000 km² (larger area of Northeast Brazil), presenting great potential for exploitation. Currently, the state has experienced a rapid occupation and consequently it is estimated that about 10% of this ecosystem have already been occupied by agricultural projects. From the 90's, this process was accelerated by the deployment of mechanized agriculture, especially in grain crops, including soybeans, corn, rice and beans (Araujo, 2005). Agricultural expansion occurs mainly in Southern and Southwestern of Piauí, because they are favoured by stable climate and topography, what consists of large plateaus on the tops of mountains (EMATER, 2009; Funaguas, 2009). The irrational use of this biome is followed by problems such as erosion and the consequent desertification. The last phenomenon has already been detected in this region.

Burning is the most harmful of all these factors. In drought periods, this phenomenon is a matter of constant concern due to use of passive fire protection products. There are two types of burning causes: natural and anthropic. In the first type, burnings usually occur in

areas of dry vegetation, leading to devastating fires; anthropogenic burnings, in turn, may occur in any type of vegetation (Eiten, 1972; Coutinho, 1978).
In Brazil, the main causes of burning in the Cerrado were due to improper use of fire equipment. Some pre-fire characteristics of ecological fires (Whelan, 1997) are not detected at the moment of the burning, as the firebreaks construction, the checking of climatic conditions and of the appropriate period, as well as the availability of fire control equipment. Additionally, criminal fires are frequent, which are caused by the action of arsonists, hunters, fishermen and balloons (Medeiros, 2002).
Great burnings may present serious risks for the conservation of the biodiversity and maintenance of ecological processes. These fires are particularly dangerous in small areas, which present rare species and/or species susceptible to extinction, which are usually very sensitive to the fire. This is the case of most Conservation Units (UC) in Brazil, including the Ecological Stations (ES) where many species are at risk of population decline because of these factors (Dias, 1998).
The geographical location of the fires within an ecological station can be used in strategies to fight fires in this area. A technique that has proven to be efficient is satellite monitoring (Cihlar et al., 1997; Carvalho, 2001; Remmel & Perera, 2001). The relatively low cost and the possibility of easily obtaining digital images have stimulated research on the theme. In digital, orbital or aerial, images it is possible to identify forest fires. Also, by temporary analysis, it is possible to characterize the previous types of vegetations that were destroyed by fires.
In this paper, a multi-temporary series of images of the Landsat 5 (American satellite) was used in the location and dimension of burned areas. Those images cover the area of the Uruçuí-Una Ecological Station (UUES) located in southern Piauí state / Brazil (approximate central coordinate: Latitude 8°52′ and Longitude 45°12′ South). The purpose of this study is to investigate the locations and frequencies of burnings, as well as the type of vegetation damaged by fires, besides analyzing if there is influence of the El-Niño and the La-Niña phenomena.

2. Ecological station

Currently, environmental problems have received special attention, and there has been much discussion on environmental preservation. The establishment of protected areas, called Conservation Units (CUs) emerges from this discussion. The purpose of these CUs is to protect the flora and fauna, by reducing the negative impacts of human activity on biodiversity. In Brazil, the creation of protected areas was legally established by the Brazilian Forest Code (Decree 23.793 - 1934). The first Ecological Park was created only in 1937 (Sick, 1997).
The Conservation Areas are divided into two major groups: Integral Protection Conservation Units (IPU) and Sustainable Use Conservation Units (SUU). The first group aims to balance nature conservation by promoting its sustainable use. The second goal is to preserve the nature, admitting only the indirect use of its natural resources, except in cases provided by Law. The two groups are subdivided into categories. The Ecological Stations belong to the second group.

2.1 Uruçui-Una ecological station

The Uruçuí-Una Ecological Station (UUES), with a total area exceeding 1,300 square kilometers, was established in 1981 with the purpose of protecting and preserving fragments of Cerrado ecosystems and promoting the development of scientific research.

The vegetation cover consists predominantly of UUES *cerrado stricto sensu*, but other types of vegetation, e.g. *campo cerrado* are also found. The *cerrado sensu stricto* in this area is composed of grasslands, with dense cover of grass and low trees.
The UUES located in the plateaux sub-region (scarps that resulted from the erosive action of waters) in the Cerrado of the Southwest of Piauí, within the boundaries of the cities of Baixa Grande do Ribeiro and Santa Filomena, about 730 km from Teresina, the State capital. Fig. 2 shows the location and boundaries of UUES.

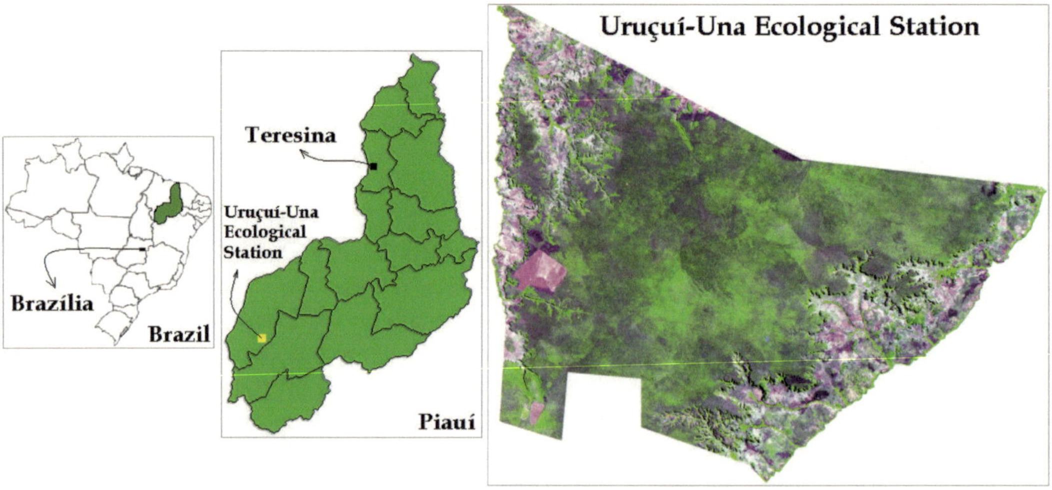

Fig. 2. Location of the Uruçuí-Una Ecological Station

The average altitude of the plateau in the UUES is 620 meters. With a typically tropical climate, this area has high average temperatures, varying between 24 and 26 °C, with absolute maximum annual reaching 40 °C. The relative humidity of air oscillates between 60 and 84% and the average annual precipitation levels are below 800mm. Regarding its physiography Piaui is a typical zone of transition, i. e., a mixture of biomes such as the *Caatinga*, the *Closed Deciduous Forest* and *Cerradão*. Finally, the aforementioned economies of the two municipal districts are predominantly agricultural, particularly grain production, such as soybean and corn (Medeiros & Cunha, 2006).

3. Forest fire

The fire is usually used to clean lands, e. g. the preparation for planting or pasture. According to Medeiros (2002), a forest fire is a fire that starts in several types of vegetation, running out of control, which can be intentionally started or have natural causes such as sunlight. Large forest fires can be considered a serious threat to biodiversity conservation and the maintenance of ecological processes. This threat is particularly serious for small areas, sections isolated by cities or agricultural monocultures and areas with rare and/or endangered species, because these ecosystems are very susceptible to fire.
The risk and intensity of damage are vary depending on the size of the area, age, intensity of the fire and time of year. Fires have been a matter of continuous concern in the dry season, when most of the damage to the ecosystem could be experienced in cities located in the referred area, which were covered by smoke and ash (Medeiros, 2002).

In Brazil, the causes of forest fires in the Cerrado were mostly associated to incorrect use of firefighting equipment: lack of fire lines, climate conditions and lack of fire control equipment. Also, illegal burning is one important cause of these fires. Recent studies have shown that 67% of the burned areas in Brazil (in 2000) were in Cerrado (Tansey et al. 2004).
According to Vicentini (1993) and Silva et al. (2009) the Cerrado has been increasingly occupied and converted into agricultural land. The author affirms that the Conservation Units located in this biome have been constantly impacted by the action of frequent forest fires. Intensive agriculture is one of the factors that contribute to the generation of Conservation Units of small areas, which presents one or more vulnerable characteristics due to the occurrence of fires, as previously mentioned.
Besides releasing carbon dioxide (CO2) into the atmosphere, burning can release other gases that cause global warming, and high frequency of fire affects the establishment of trees and shrubs (Hoffmann & Moreira, 2002, Krug et al., 2002).

4. Used data

The scanner Thematic Mapper (TM), coupled in Landsat 5, sense data in seven spectral bands simultaneously. Band 6 senses thermal infrared radiation and can only acquire night scenes. The resolution on the ground (spatial) in bands 1-5 and 7 are 30 square meters and in band 6 is 120 square meters.
Bands 1 (blue, 0.45-0.52μm), 2 (green, 0.52-0.60μm) and 3 (red, 0.63-0.69μm) are obtained in the waves lengths of the visible in the electromagnetic spectrum; while the bands 4-7 are collected in infrared region (4 - infrared near, 0.76-0.90μm; 5 - infrared medium, 1.55-1.75μm; 6 - infrared thermal, 10.4-12.5μm; and 7 - infrared medium, 2.08-2.35μm). Additional information on program Landsat can be seen in NASA (2011).
Only bands 3 and 4 were used in the experiments conducted in this paper. Band 2 was only used in the generation of the color composition (Fig. 3). An image was obtained in the quarter ASO (August, September and October) of every studied year, which corresponding to the drought periods. This study comprises the years 1985, 1987, 1989, 1996, 1998, 2000, 2007, 2008 and 2010, in agreement with the readiness of TM data and with the occurrence of the climatic phenomena El-Niño and La-Niña. Fig. 3 shows the UUES in 1989.

Fig. 3. Image of the UUES area in 1989

Imaging processing is performed using the software MatLabR2007a, Multispec3.2 and Envi4.2.

5. Method

For better understanding, the proposed method is divided into five distinct phases. The first stage concerns data collection and image pre-processing. In the second phase there is NDVI calculation and the isolation of band 4 (near infrared - IR). The creation of routines that detect fires using two test images and that is based on a threshold T is performed in the third phase. The information on the existing biodiversity in the UUES and the pre-processing and data collection steps occur in the fourth phase. Finally, the fifth phase concerns analysis of the results based on the quantification of burned areas and climatic phenomena.

5.1 Obtaining and pre-processing the data

All the nine images were geometrically corrected by means of image registration. The reference (georeferenced) image was also provided by INPE, at no cost, and the control points were collected in this image. Landsat TM images are first corrected by INPE, but with approximate coordinates, which simplifies the record of these scenes. For this reason, a first-degree polynomial model and the interpolator "nearest neighbor", which maintains the original Digital Values DV (or Gray Levels), were used. Since the digital values define the different features in digital images, depending on the intensity of the changes of the digital values, it could be difficult to carry out fire analyses.

5.2 Preparing the NDVI and IR images

The spectral signature of the "burned" feature varies according to the fire status at the moment of image collection. If there is a fire, the digital values in the green (G) and red (R) bands are medium or close; and in the infrared band (IR), the DV are low. Though, when there is no fire, the DV in the IR band is null, spectral characteristics of the ash and coal. In the G band, the DV are low, being however larger than the values found in the R band. In this context, the pixels of the "burned" class can be classified by vegetation indexes, which are based on the difference between the DV of the IR and R bands.

NDVI (Normalized Difference Vegetation Index) – index is selected because of its efficiency in the identification of burned areas. The method is widely recognized in the literature and uses simple calculations. This index is calculated by equation 1

$$NDVI = (IR-R)/(IR+R) \quad (1)$$

The isolation of the IR band (band 4 - TM) is a simple task, which is performed by ensuring the geometrical correction and cut out over the area of interest. There are no alterations in the original digital values in this band. Fig. 4 illustrates the image IR band and the resulting image of the NDVI calculation, both originated from the image shown in Fig. 3.

The simultaneous use of two images (NDVI and IR) is demonstrated in circles and rectangles illustrated in Fig. 4a and 4b. In the IR image (Fig. 4a) the burned area in the circle is not well identified as in the NDVI image. On the other hand, the cultivation area shown in the rectangle can be mistaken with the burned area in the NDVI image, whereas in the area in the IR image it is easily separated. Thus, the two images together provide a more accurate classification.

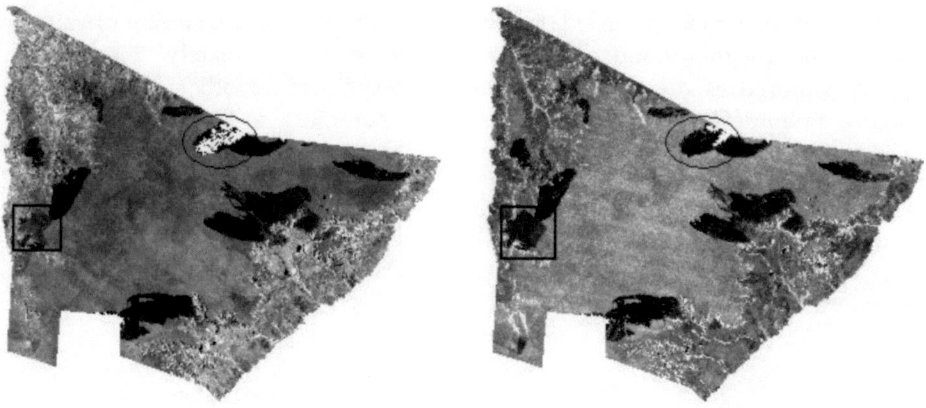

(a) (b)

Fig. 4. Near infrared band - IR (a) and NDVI image (b)

5.3 Identification of burned areas in digital images

The areas corresponding to the "burned" features are identified in the IR and NDVI images (Fig. 4a and 4b). Due to the spectral characteristics of this feature (burned), the digital values in the NDVI image are negative and close to value -1. So, a threshold T is empirically established to isolate the burned areas from the other features in the image. All digital values of the NDVI image smaller or equal to T are labeled as belonging to the feature “burned”, and the others (>T) as “unburned”. The same procedure is used in IR imaging, but the magnitude of new threshold is T1.

The resulting images of threshold T and T1 applications are submitted to convolution with the morphological operator “opening”, aiming to eliminate the noise in T and T1 operations. The opening of NDVI or IR images by operator B (structuring element) is obtained by the erosion of these images with B, followed again by dilation of the resulting image by B. The mathematical representation of the morphological operation with the NDVI image is given by equation 2.

$$NDVI \circ B = (NDVI \ominus B) \oplus B \qquad (2)$$

The structuring element B can be defined in many ways: linear, circular, rectangular or diagonal. The entire process involved in the identification of burned areas is accomplished by routines developed in MatLab2007a. Finally, the burned areas obtained from the NDVI and IR images are added (arithmetic operation), to fill up spaces that were not visible with the use of other techniques.

5.4 Classification of the Biome in UUES

The classification of vegetation types affected by forest fires within the study area (UUES) is done by *in loco* visits. The present study was conducted by a group of researchers and members of ICMBio (Chico Mendes Institute for Biodiversity Conservation) responsible for monitoring the UUES. The group crossed the UUES, passing through a rural road towards the North and the South, and used also a side access road.

Some central points of the burned areas detected in the digital images are used as reference in the collection of information concerning the type of vegetation in this area. In each point, the geodetic coordinates are recorded and the information of the vegetation physiognomy is collected in the neighborhood within a radius of approximately 200 meters. This information is stored as data points and then interpolated to other points on the area. Additionally, a photographic record is made at each point, e. g. in Fig. 5.

Fig. 5. Photographic record of local vegetation

Information on the vegetation is superimposed on burned areas, which are generated from the NDVI and IR, to measure and record biodiversity losses. The overlapping or crossing of this information is performed using routines developed in MatLab2007a.

5.5 Behavior of climatic phenomena

Finally, the results of all the experiments are assessed in relation to the climatic phenomena El-Niño and La-Niña, in order to determine whether or not these phenomena have influence on burnings. This analysis is done by checking the size of the burned area and the behavior of weather phenomena in the analyzed year. According to CPC (2010), the occurrence of the climatic phenomenon can be represented by indexes.

The quarterly values of the indices corresponding to the occurrence of El Niño and La Niña are graphically plotted (Fig. 6), in order to describe the behavior of these phenomena over time. The period represented on the graph varies between the years 1982 and 2010. Each year is punctuated by twelve quarters that overlap each other, for example, JFM (January, February and March) is the first quarter; the second is FMA (February, March and April);

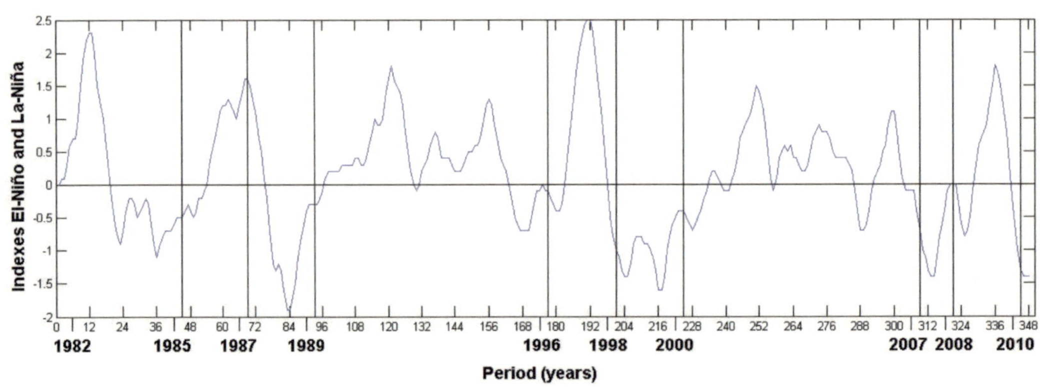

Fig. 6. El Niño and La Niña based in limit of ±0.5 °C for the Oceanic Niño Index

the third is MAM (March, April and May), and so on. This index is based on the limit of ±0.5 °C for the anomalous temperature at the sea surface (SST) in the Niño 3.4 region (5°N to 5°S latitude and 120°W 170°W longitude). Positive values of this index evince the occurrence of El Niño and negative values indicate the occurrence of La Niña.
In Fig. 6 the x-axis shows the year represented by the 12 quarters. Thus, year 1982 is represented by number 12, 1983 is represented by number 24, 1984 by 36, and so on. The assessed years in the present study are shown in vertical lines in Fig. 6.

6. Results and discussion

The constants used in this study are presented below. In the image registration procedure, the largest root mean square error (RMS) measured was 0.48 pixel. The values used for thresholds T and T1 in the identification of burned areas were 0.4 and 80, respectively. The structuring element B used in "opening" operation is the disc of radius 5. Essentially, opening removes small objects/noises (<5 pixels). The side effect is the elimination of the edges to round the objects. Although the burned areas may be partially (large areas) or totally (small areas) eliminated by the opening, this technique is very effective to eliminate noise. In Fig. 7 is shown an example in image IR-1989.

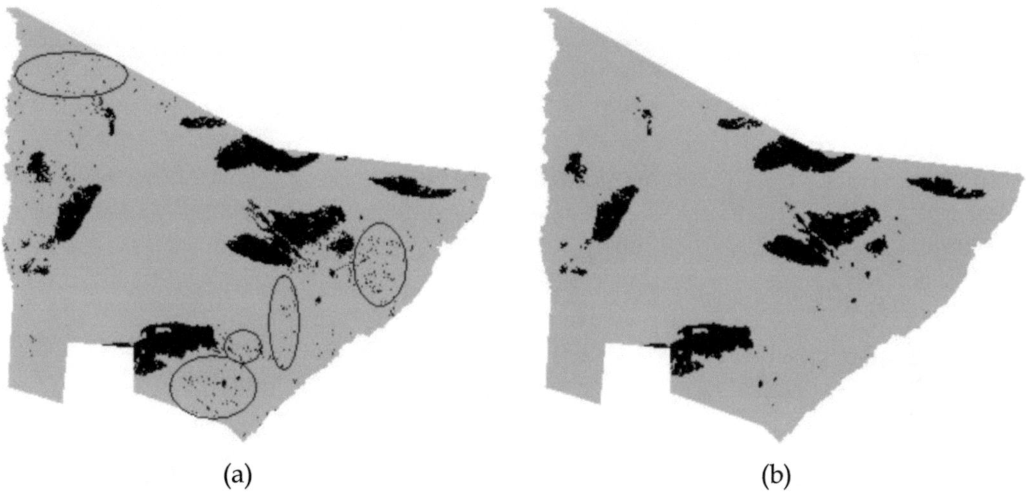

(a) (b)

Fig. 7. Points eliminated by the opening operation.

Several points detected as burning alone (Fig. 7a) were removed by opening operation (Fig. 7b). The circles exposed in Fig. 7a show concentration of these points (noise), what were eliminated with the application of that technique morphology.
The burned areas identified in the Landsat 5 TM images, concerning the nine years analyzed, can be seen in Fig. 8.
Visually, it can be inferred that in 1985 there were few fires, which were more frequent across the borders of the UUES area. One possible explanation is that some areas of the UUES are inhabited, and burning is used for the practice of subsistence farming. In the following years (1987, 1989 and 1996) an abrupt expansion of the burned areas was observed, although with a small decline between 1996 and 1998. According to members of ICMBio, this increase was due to the advent of intensive agriculture and the absence of fire-

fighting groups close to UUES in that period. In the 2000s, the fire brigades fought the fires in the area. Besides, climatic factors must be taken into consideration. However, they shall be discussed later.

For a quantitative analysis, all the areas were measured in square kilometers and represented graphically, as shown in Fig. 9.

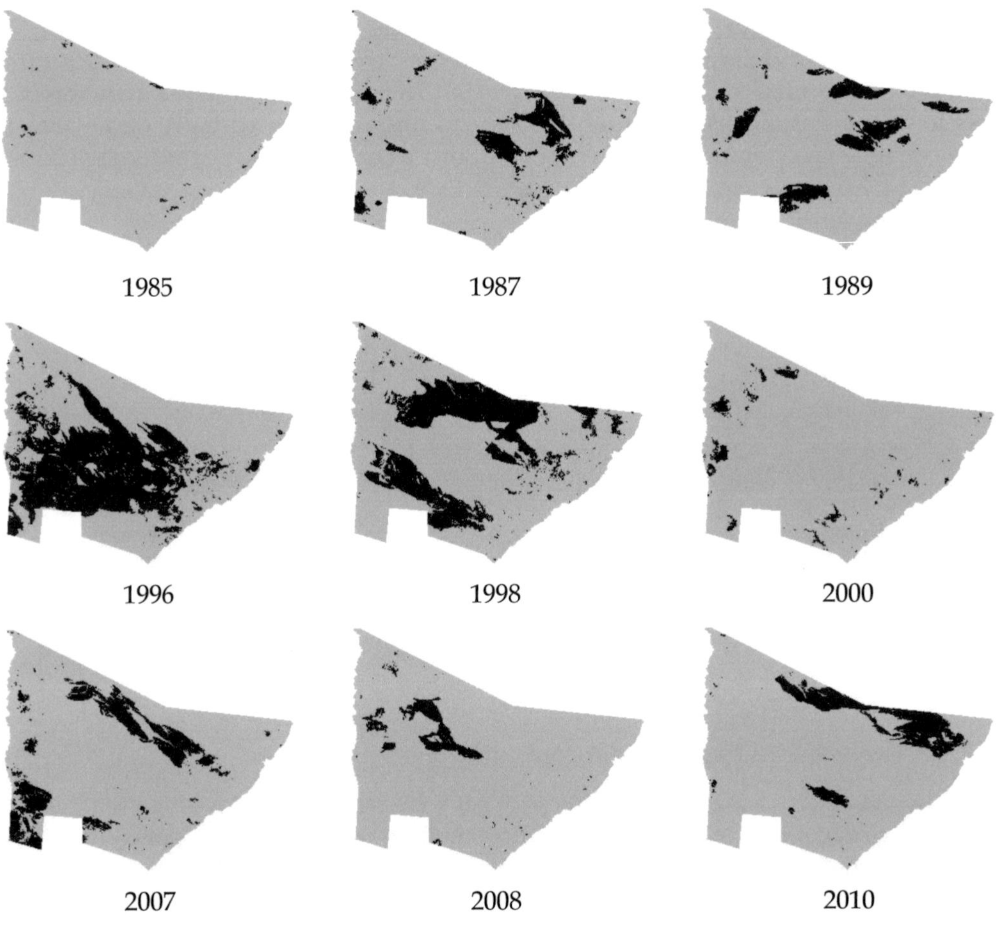

Fig. 8. Forest fires happened in the analyzed years

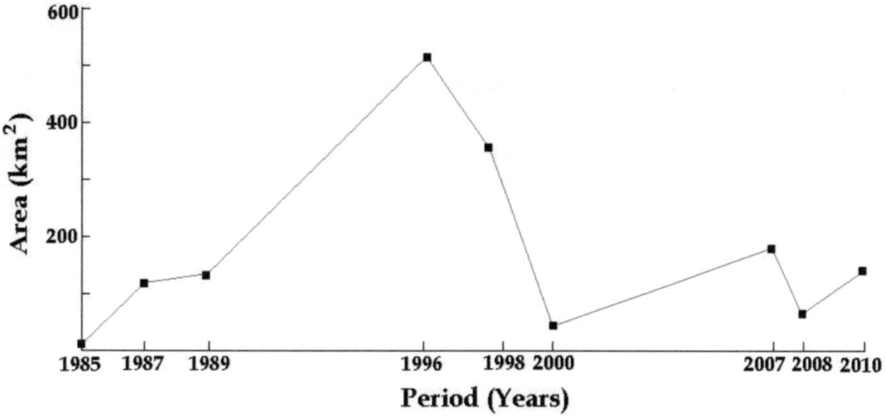

Fig. 9. Graphic representation of the measures of the burned areas

The peak that occurred in 1996 shows that 37.40% of the UUES area was reached by fire, which is more than the index measured (37.18%) for the sum of the burned areas in all the other years except for 1998 and 2007. The areas least affected by fires in 1985 and 2000 were 8.59 and 43.33 km^2, respectively. Fig. 10 shows that 67.25% (black area) of the UUES area was reached by fire in at least one of the years analyzed.

Fig. 10. Total area affected by fire in years analyzed.

In loco observation showed that most burned areas correspond to the physiognomies of *cerrado sensu stricto*, especially in upland areas (plateaux). Only one of the study areas was classified as *campo cerrado*. The areas located in the bottom (valley), which were also affected by fires corresponded to *cerrado stricto sensu anthropic*. According to Medeiros & Cunha

(2006), the dominant vegetation in the Ecological Station of Uruçui-Una is really the *cerrado strict sensu*.

It is known that depending on the intensity and frequency, forest fires may cause severe damage to the vegetation. However, during the *in loco* visit it was found that the vegetation in some areas was at an advanced stage of regeneration, indicating resilience. Fire is an important factor in maintaining biodiversity, since while some species are affected by fires, others may be benefited in the process of germination and dormancy break of seeds. Thus, the frequency of fires in UUES may be the cause of the predominance of the two existing physiognomic forms. If the germination of tree species is prevented from occurring because of periodic fires, there is a decrease in the density of trees in the region, which leads to a change in the *cerrado sensu amplo* forest landscape, with a larger concentration of trees.

Analysis of fire recurrence (Fig. 11) was carried out in areas adjacent to the checkpoint that characterized the physiognomic form *campo cerrado*. In this region, biodiversity loss is apparent (Fig. 5). Considering the combined damage of fauna and flora, this loss is still more serious.

The area of recurrence of fires presented in Fig. 11 measures 9.53 km^2. The measure of this area was determined in the intersection of the burned areas for the years 1978, 1989, 1996 and 1998. Since the burned areas coincided in only two years, a considerable variation occurs depending on the pairs of years analyzed. Table 1 shows this variation.

Fig. 11. Areas of recurrence of fires in period 1978-1998

All the indexes for the year 1985 are low due to the small extent and sparse spatial distribution of the burned areas. 1996, on the contrary, as the year with the largest area of intersection with another year, especially the 1996-1998 pair, which had an area of intersection of 135.25 km2. The intersection of burned areas in 1996 was due to the spatial extension and homogeneity of the area.

Analysis of the graphs of Figs. 6 and 9 shows that the La-Niña climate phenomenon has started two years before 1985. This phenomenon may also have contributed to the reduced number of fires that occurred in that year. The El-Niño climatic phenomenon occurred in the 1985-1987 period. Consequently, the graph of Fig. 9 shows an increase in the burned areas. Between 1987 and 1989 a transition from El-Niño to La-Niña was seen, with La Niña prevailing most of the time. The total amount of burned areas remained almost constant, with a slightly positive trend. Soon after, in a longer period (1989 to 1996) a predominance of El-Niño was observed, which corresponds to the period when the largest areas were destroyed by fires, being consistent with the heating caused by the phenomenon. Although smaller than the previous period, the size of the burned areas between 1996 and 1998 was relatively large, in agreement with the peak of El-Niño in Fig. 6. Cooling happened in the subsequent period (1998-2000) coinciding with the abrupt decrease in the total burned area. Despite a weak intensity the El-Niño phenomenon prevailed in the 2000–2007 interval. Hence, the curve of the area showed again a positive trend. In the last two intervals of time analyzed the consistency of reduction of the burned area in the period of La-Niña (2007-2008) was maintained and there was an increase in the El-Niño phenomenon (2008-2010).

Years	1987	1989	1996	1998	2000	2007	2008	2010
1985	0.93	1.33	3.55	1.82	0.35	0.89	0.22	1.57
1987		25.12	54.41	28.15	7.07	26.85	3.18	18.57
1989			66.10	78.02	4.27	31.68	4.75	24.79
1996				135.25	11.19	75.73	36.07	15.85
1998					6.40	77.25	20.06	57.14
2000						4.46	2.48	2.54
2007							0.15	21.92
2008								0.23

Table 1. Burned areas (km^2) coincide with each other

It is assumed that there is increased incidence of fires during the El-Niño phenomenon and that the opposite occurs during La-Niña. During the El-Niño periods there is an increase in temperature, which causes the vegetation to dry, favoring the fast propagation of the fire. On the other hand, during the La-Niña phenomenon, cooling is more frequent, which increases the intensity of rains, helping eliminate the fires. Thus, the two graphs (Fig. 6 and Fig. 9) show the existence of this correlation between the climatic phenomena and the intensity of burned areas in UUES.

7. Conclusion

The findings of the present study allow inferring the intensity, distribution form, location and cause of the fires that took place in the Uruçuí-Una Ecological Station (UUES), as well as the biodiversity loss provoked by these fires. Additionally, it is possible to infer the influence of the climatic phenomena El-Niño and La-Niña on the occurrence of fires.

In agreement with the results obtained, the Landsat 5 TM images were found to be useful in the analyses of fires in CUs, especially due to the collection frequency and quality spectral/spatial of the data.

The burned locations were correctly defined in the proposed method. They were located so much the burned areas individualized by year, as the total area reached by fire in the analyzed period. Additionally, all the areas were measured facilitating a more detailed analysis concerning the intensity of the fire.

The burning frequency was also accurately measured, indicating the recurrence of fire among pairs of years and in the most affected years. The study of the frequency of fires made it possible to identify the most critical areas, elaborate a strategy to support fire fighting, e. g., the creation of rural roads or accesses to these areas to facilitate the movement of the fire brigades.

The variation of the type of vegetation (biodiversity loss) was analyzed based on the occurrence of fire. It was noticed that in areas of high fire frequency the physiognomy of *cerrado sensu stricto* vegetation was changed into *campo cerrado* because of the damage caused to some species by the fires.

The products (burned map and descriptive information) generated in this research are important tools for governmental authorities in Brazilian Conservation Units (e. g. in ICMBio), particularly to promote innovation of public politics in the prevention and control management of fires within the ES (Ecological Stations). Besides, they provide insights for future research on environmental preservation and continuous monitoring of protected areas.

8. Acknowledgments

We thank INPE (National Institute for Space Research) for providing free access to Landsat 5 TM images.

At ICMBio (Chico Mendes Institute for Biodiversity Conservation) for providing transport and access to UUES facilities, which has made this research possible.

9. References

Araujo, A. A. (2005). Modernization of the Agricultural Frontier in Piauí Cerrado: the Case of Bom Jesus. 164p. *Dissertation* (Master in Development and Environment), Federal University of Piauí, Brazil.

Carvalho, L. M. T. (2001). Mapping and Monitoring Forest Remnants, a Multiscale Analysis of Space-Temporal data. 138p. *Thesis* (PhD in Remote Sensing), Wageningen University, Wageningen.

Castro, A. J. F. (1999). Cerrados do Brasil e do Nordeste: Caracterização, Área de Ocupação e Considerações sobre a sua Fitodiversidade. *Pesquisa Foco*, Vol.7, No.9, (December 1999), pp. 147-178, ISSN: 2176-0136.

Coutinho, L. M. (1978). O Conceito de Cerrado. *Revista Brasileira de Botânica*. Vol.1, No.1, (January 1978), pp. 17-23.

CPC - Climate Prediction Center. (December 2010). Available from http://www.cpc.noaa.gov/products/analysis_monitoring/ensostuff.

Eiten, G. (1972). The Cerrado Vegetation of Brazil. *The Botanical Review*. Vol.38, No.2, (June 1972), pp. 201-341.

Eiten, G. (1979). Formas Fisionômicas do Cerrado. *Revista Brasileira de Botânica*. Vol.2, No.2, (June 1979), pp. 139-148.

Furley, P.A & Ratter, J. A. (1988). Soil Resources and Plant Communities of the Central Brazilian Cerrado and their Development. *Journal of Biogeography*. Vol.15, No.1, (January 1988), pp. 97-108.

Hass, A. (2002). Efeitos da Criação da UHE Serra da Mesa (Goiás) sobre a Comunidade de Aves. *Thesis* (PhD in Ecology), University Campinas/SP, Brazil.

Hoffmann, W.A. & Jackson, R. B. (2000). Vegetation-climate feedbacks in the conversion of tropical savanna to grassland. *Journal of Climate*. Vol.13, No.10, (May 2000), pp. 1593-1602.

Hoffmann, W. A. & Moreira, A. G. (2002). The role of fire in population dynamics of woody plants. In: Oliveira, P. S. & Marquis, R. J. *The Cerrado of Brazil. Ecology and natural history of a neotropical savanna*. Columbia University Press, ISBN: 978-0-231-12043-2, Nova York, pp. 159-177.

Krug, T.; Figueiredo, H.; Sano, E.; Almeida, C.; Santos, J.; Miranda, H. S.; Sato, N. & Andrade, S. (2002). Emissões de gases de efeito estufa da queima de biomassa no Cerrado não-antrópico utilizando dados orbitais. Primeiro inventário brasileiro de emissões antrópicas de gases de efeito estufa relatórios de referência. *Ministério de Ciência e Tecnologia* (MCT), http://www.mct.gov.br/clima, Brasília.

Machado, R. B. (2000). A Fragmentação do Cerrado e Efeitos sobre a Avifauna na Região de Brasília-DF. 163p. *Thesis* (PhD in Ecology), University of Brasília, Brazil.

Machado, R. B.; Ramos Neto, M. B.; Pereira, P. G. P.; Caldas, E. F.; Gonçalves, D. A.; Santos, N.S.; Tabor, K. & Steininger, M. (2004). Estimativas de perda da área do Cerrado brasileiro. *Technical Report*. Conservation International, Brasília, DF. 23p.

Medeiros, F. C. & Cunha, A. M. C. (2006). Plano Operativo de Prevenção e Combate aos Incêndios Florestais da Estação Ecológica de Uruçui-Una/PI. Ministério do Meio Ambiente - *MMA*. 15p.

Medeiros, M. B. (2002). Manejo de Fogo em Unidades de Conservação do Cerrado. *Boletim do Herbário Ezechias Paulo Heringer*. Vol. 10, p. 75-88.

Mittermeier, R. A.; Gil, P. R. & Mittermeier, C. G. (1997). Megadiversidad - Los Países Biológicamente más Ricos del Mundo. *CEMEX*. Mexico, MX.

NASA - National Aeronautics and Space Administration. (March 2011). Available from http://landsat.gsfc.nasa. gov/about/tm.html.

Ratter, J. A.; Ribeiro, J.F. & Bridgewater, S. (1997). The Brazilian Cerrado Vegetation and Threats to its Biodiversity. *Annals of Botany*, Vol.80, No.3, (May 1997), pp. 223-230.

Remmel, T. K. & Perera, A. H. (2001). Fire Mapping in a Northern Boreal Forest: Assessing AVHRR/NDVI Methods of Change Detection. *Forest Ecology and Management*, Vol.152, No.3, (October 2001), pp. 119-129.

Sick, H. (1997). *Ornitologia brasileira*. Nova Fronteira, Rio de Janeiro, Brazil. 912p.

Silva, C. R.; Souza, K. B.; Aguiar, A. S.; Oliveira, O. A. & Silva, P. (2009). Análise Temporal da Variação do Dossel do Cerrado Piauiense. *Proceedings of II CONEFLOR 2009 2th Congresso Nordestino de Engenharia Florestal*, Patos, Paraíba, Brazil, November 09-13, 2009.

Tansey, K.; Grégoire, J. M.; Stroppiana, D.; Sousa, A.; Silva, J.; Pereira, J. M. C.; Boschetti, L.; Maggi, M.; Brivio, P. A.; Fraser, R.; Flasse, S.; Ershov, D.; Binaghi, E.; Graetz, D. & Peduzzi., P. (2004). Vegetation burning in the year 2000: global burned area estimates from SPOT VEGETATION data. *Journal of Geophysical Research*. Vol.109, No.7, (June 2004), pp. 1-22.

Vicentini, K. R. F. (1999). History of Fire in the Cerrado: A Pollen Analysis. 235p. *Thesis* (PhD in Ecology), Federal University of Brasília (UNB), Brasília/Brazil.

Whelan, R. J. (2001). *The Ecology of Fire.* Cambridge University Press, ISBN 9780521328722, Cambridge, England.

9

Provision of Natural Habitat for Biodiversity: Quantifying Recent Trends in New Zealand

Anne-Gaelle E. Ausseil[1], John R. Dymond[1] and Emily S. Weeks[2]
[1]*Landcare Research*
[2]*University of Waikato*
New Zealand

1. Introduction

1.1 Biodiversity and habitat provision in New Zealand

The Millennium Ecosystem Assessment (MEA) found that over the past 50 years, natural ecosystems have changed more rapidly and extensively than in any other period of human history (Millennium Ecosystem Assessment, 2005). In the 30 years after 1950, more land was converted to cropland than in the 150 years between 1700 and 1850, and now one quarter of the earth's surface is under cultivation. In the last decades of the twentieth century, approximately 20% of the world's coral reefs have disappeared and an additional 20% show serious degradation. Of the fourteen major biomes in the world, two have lost two thirds of their area to agriculture and four have lost one half of their area to agriculture. The distribution of species has become more homogeneous, primarily as a result of species introduction associated with increased travel and shipping. Over the past few hundred years, the species extinction rate has increased by a thousand times, with some 10–30% of mammal, bird, and amphibian species threatened with extinction. Genetic diversity has declined globally, particularly among cultivated species.

A framework of ecosystem services was developed to examine how these changes influence human well-being, including supporting, regulating, provisioning, and cultural services (Millennium Ecosystem Assessment, 2003). While overall there has been a net gain in human well-being and economic development, it has come at the cost of degradation to many ecosystem services and consequent diminished ecosystem benefits for future generations. Many ecosystem services are degrading because they are simply not considered in natural resource management decisions. Biodiversity plays a major role in human well-being and the provision of ecosystem services (Diaz et al., 2006). For example, natural ecosystems provide humans with clean air and water, play a major role in the decomposition of wastes and recycling of nutrients, maintain soil quality, aid pollination, regulate local climate and reduce flooding.

New Zealand has been identified as a biodiversity hotspot (Conservation International, 2010). Located in the Pacific Ocean, south east of Australia, New Zealand covers 270 thousand square kilometres on three main islands (North, South and Stewart Island). It has a wide variety of landscapes, with rugged mountains, rolling hills, and wide alluvial plains. Over 75 percent of New Zealand is above 200 meters in altitude, reaching a maximum of

3,700 meters on Mount Cook. Climate is highly variable and has played a key role in biodiversity distribution (Leathwick et al., 2003).

As New Zealand has been an isolated land for more than 80 million years, the level of endemism is very high, with more than 90% of insects, 85% of vascular plants, and a quarter of birds found only in New Zealand (Ministry for the Environment, 2007). One of the most notable characteristics of New Zealand's biodiversity is the absence of terrestrial mammals, apart from two bat species, and the dominance of slow-growing evergreen forest. New Zealand's indigenous biodiversity is not only unique within a global context - it is also of major cultural importance to the indigenous Maori people. Maori have traditionally relied on, and used, a range of ecosystem services including native flora and fauna for food, weaving, housing, and medicines.

The isolation of New Zealand has preserved its unique biodiversity, but also rendered the biodiversity vulnerable to later invasion. When Maori migrated from the Pacific Islands, circa 700 years ago, predation upon birds began and much lowland indigenous forest was cleared, especially in the South Island. Rats and dogs were also introduced. The birds, having evolved in an environment free of predators, were susceptible to disturbance and many began to decline to the edge of extinction. When Europeans arrived in the early 19th century, they extensively modified the landscape and natural habitats. Large tracts of land were cleared and converted into productive land for pastoral agriculture, cropping, horticulture, roads, and settlements. Only the steepest mountain land and hill country was left in indigenous forest and shrubland. Swamps were drained and tussock grasslands were burned. Not only was the natural habitat significantly altered, but a large range of exotic species were introduced, including deer, possums, stoats, ferrets, and weasels, causing a rapid decline in native birds and degrading native forest. Other introduced plants and animals have had significant effects in the tussock grasslands and alpine shrublands, most notably rabbits, deer, and pigs, and the spread of wilding pines, gorse, broom, and hieracium. Despite significant efforts to control weeds and pests and halt the loss of natural habitat, around 3,000 species are now considered threatened, including about 300 animals, and 900 vascular plants (Hitchmough et al., 2005).

The Economics of Ecosystems and Biodiversity study (TEEB) suggested that it is difficult to manage what is not measured (TEEB, 2010). To prevent further biodiversity loss, decision-makers need accurate information to assess and monitor biodiversity. However, biodiversity assessment is not a trivial task. As defined by the Convention on Biological Diversity (CBD), biodiversity encompasses "the variability among living organisms from all sources including, inter alia, terrestrial, marine and other aquatic ecosystems and the ecological complexes of which they are part; this includes diversity within species, between species and of ecosystems" (CBD, 1992). Conceptually, biodiversity is a nested hierarchy comprising genes, species, populations, and ecosystems. In order to assess status and trend, these multiple levels need to be assessed simultaneously. Noss (1990) suggested a conceptual framework with indicators providing measurable surrogates for the different levels of organisation. Loss of extent is one of the many indicators in this framework, and it has been widely used internationally in reporting to the CBD (Lee et al., 2005). It is relatively easy to report, and has been recognised as one of the main drivers for biodiversity loss (Department of Conservation [DOC] and Ministry for the Environment [MFE], 2000).

1.2 Previous assessments in natural habitat

Several national surveys of vegetation cover have been completed. The New Zealand Land Resource Inventory was derived by stereo photo-interpretation of aerial photographs combined with field work (Landcare Research, 2011). The survey scale was approximately 1:50,000 and had a nominal date of mid-1970. The legend included 42 vegetation classes, of which six were indigenous forests (coastal, kauri, podocarp-hardwood (lowland or mid-altitude), *nothofagus* (lowland or highland), and hardwood) and three were indigenous grass classes (snow tussock, red tussock, and short tussock). The Vegetative Cover of New Zealand was produced at the scale of 1:1,000,000 primarily from the NZLRI (Newsome, 1987). The small scale required mixed vegetation classes to be used, such as "grassland-forest" or "forest-scrub".

The Land Cover Database (LCDB) was derived by photo-interpretation of satellite imagery and has nominal dates of 1995–96 for LCDB1 and 2001–2002 for LCDB2 (Ministry for the Environment 2009). Indigenous classes included tussock grassland, manuka/kanuka, matagouri, broadleaved hardwoods, sub-alpine shrubland, and mangroves; however, different indigenous forest classes were not delineated and were lumped into one class of indigenous forest. Walker et al. (2006) used the LCDB to look at changes to natural habitat between 1995–96 and 2001–2002. They concluded that much of the highland natural habitats had been preserved since pre-Maori times, but also that much of the natural habitat of lowland ecosystems had been lost and continues to be lost. Limitations in the LCDB prevented reliable analysis of the changes in indigenous grassland, wetlands, and regeneration of shrublands to indigenous forest.

The recently completed Land Use Map (LUM) has extended the date range for indigenous forest to between 1990 and 2008 (Ministry for the Environment, 2010). LUM is primarily helping New Zealand meet its international reporting requirements under the Kyoto Protocol. It tracks and quantifies changes in New Zealand land use, particularly since 1990. For this purpose, it produced national coverages for 1990 and 2008 of five basic land cover classes (indigenous forest, exotic forest, woody-grassland, grassland, and other), from satellite imagery.

1.3 Proposed assessment of natural habitat provision

More recent work by Weeks et al. (in prep) has improved the accuracy and extended the analysis to between 1990 and 2008 on tussock grasslands. Ausseil et al. (2011) have improved the accuracy of wetland mapping and identified changes since pre-European time. These recent analyses, together with the LUM, permit a synthesis of information for assessing recent trends of natural habitat provision in New Zealand. This chapter presents this synthesis and describes a national measure of habitat provision for biodiversity. We look at New Zealand's natural habitat changes from pre-Maori to the present, and also at recent trends. We will focus this chapter on three natural ecosystems: indigenous forest, indigenous grasslands, and freshwater wetlands. The measure of habitat provision will combine information on current and historical extents with a condition index to quantify stress and disturbance.

2. Indigenous forests

Indigenous forests in New Zealand are generally divided into two main types. The first is dominated by beech trees (*Nothofagus*), and the second generally comprises an upper coniferous tier of trees with a sub-canopy of flowering trees and shrubs (the broadleaved species) (Wardle, 1991). However, these two types are not mutually exclusive and mixtures are common. Lowland podocarp-broadleaved forests are structured like forests of the tropics. Kahikatea (*Dacrycarpus dacrydioides*) and Kauri (*Agathis australis*) are the tallest trees in New Zealand, and can reach up to 50 metres in height. At maturity these trees tower above the broadleaved canopy with other emergent podocarps like rimu (*Dacrydium cupressinum*), totara (*Podocarpus totara*), matai (*Prumnopitys taxifolia*), and miro (*Prumnopitys ferruginea*), to give the forest a layered appearance. Below the upper canopy many shorter trees, shrubs, vines, tree- and ground-ferns compete for space, and below them, mosses. Beech forests tend to be associated with southern latitudes and higher elevations, such as in mountainous areas, and are generally sparser than the podocarp-broadleaf forests. Their understory may contain only young beech saplings, ferns, and mosses.

Indigenous forests provide unique habitat for a large range of plants, animals, algae, and fungi. Since the arrival of Maori, circa 700 years ago and the subsequent burning of large areas of forest, and then Europeans from ~1840, who cleared large areas for farming and settlement, the extent of indigenous forest has significantly declined and, in combination with many introduced pests, has placed enormous pressure on the survival of many species. MfE (1997) reported that 56 of the listed threatened plant species are from indigenous forest habitats. Also, many of the seriously threatened endemic birds are forest dwellers: wrybill, kiwi, fernbird, kokako, kakariki, saddleback, weka, yellowhead, kaka, and New Zealand falcon.

The extent of indigenous forest in 2008 can be mapped using a combination of LCDB2 and LUM. Theoretically, the LUM contains a recent extent of indigenous forest. However, because the class definitions are land-use rather than land-cover based (for Kyoto Protocol), the indigenous forest class is not the same as the standard definition in LCDB2 and contains much indigenous shrubland yet to reach the maturity of a forest. Hence the LUM should only be used to report on changes to forest if the LCDB definition of indigenous forest is to be used. We therefore combined all the changes "from" or "to" forest in the LUM with LCDB2 to produce a recent extent of indigenous forest.

Figure 1 compares the extent of indigenous forest and shrubland in 2008 with the estimated pre-Maori historic extent, derived by combining LCDB2 and a historic map of New Zealand (McGlone, 1988). In the North Island, the area of indigenous forest has reduced from 11.2 million hectares to 2.6 million hectares. Most remaining indigenous forest is in the hills and mountains. In contrast to indigenous forest, indigenous shrublands have now become extensive, comprising over 1.0 million hectares. These shrublands often comprise a wide variety of indigenous shrub species and could naturally regenerate to indigenous forest if left. In the South Island, the area of indigenous forest has reduced from 12.0 million hectares to 3.9 million hectares, and, similar to the North Island, the remaining forest is mainly in the hills and mountains. At 0.6 million hectares, the area of indigenous shrublands in the South Island is as large as in the North Island.

The loss of indigenous forest between 1990 and 2008 may be assessed directly from the LUM. In the North Island, 29 thousand hectares of indigenous forest have been lost, and in the South Island, 22 thousand hectares of indigenous forest have been lost. The spatial location of this loss is important as some types of forest are better represented than others. We follow the method of Walker et al. (2006) who considered the area of indigenous forest remaining in land environments. The land environments are defined by unique combinations of climate, topographic, and soil attributes, and are a surrogate for unique assemblages of ecosystems and habitats (Leathwick et al., 2003). Four levels of classification have been defined with 20 level I, 100 level II, 200 level III and 500 level IV environments.

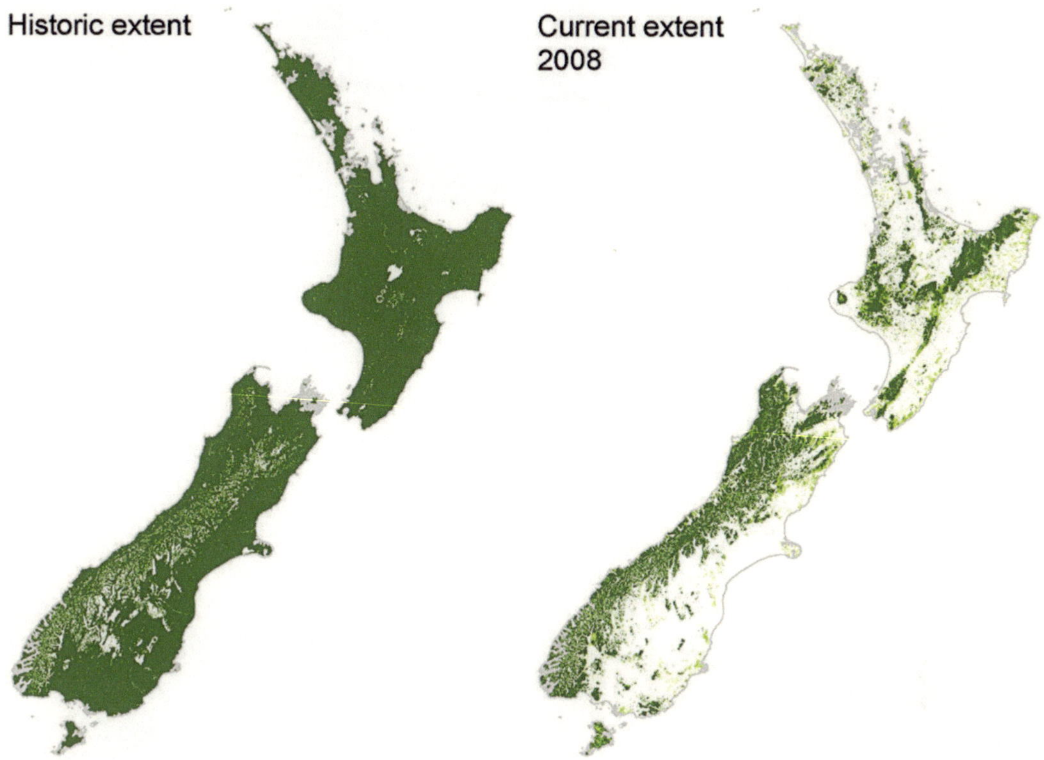

Fig. 1. Historic land cover (1000 AD) compared with recent land cover (2008). Dark green is indigenous forest, light green is indigenous shrubland.

Figure 2 shows the loss of indigenous forest in each of the Level II land environments over the last 18 years. Loss is still evident in many of the land environments. Indeed, nine land environments have lost more than 5% of their remaining indigenous forest. This could be critical, given that eight of those have less than 5% of the land environments remaining in indigenous forest.

3. Indigenous grasslands

Approximately one half of New Zealand's land area is made up of a variety of exotic and indigenous grassland ecosystems. Approximately one-fifth of these grasslands comprise modified indigenous short and tall-tussock communities, which are mostly located on the South Island. Unlike many other indigenous ecosystems in New Zealand, they have a unique, partially human-induced origin. Once largely distributed in areas of lowland montane forest and shrubland, large regions of grassland were created through burning by Maori, especially for moa hunting and for encouraging bracken fern (*Pteridium aquilinum*), an important food source (Stevens et al., 1988; Ewers et al., 2006). Lowland podocarp forests hosting such species as totara (*Podocarpus totora*) and matai (*Prumnoptiys taxifolia*) were replaced by a variety of fire adapted grassland species, in particular the short tussock species *Festuca novae-zelandia* and *Poa cita*. Some 200 years later these species were progressively replaced by taller large grain *Chionochloa* spps (McGlone, 2001).

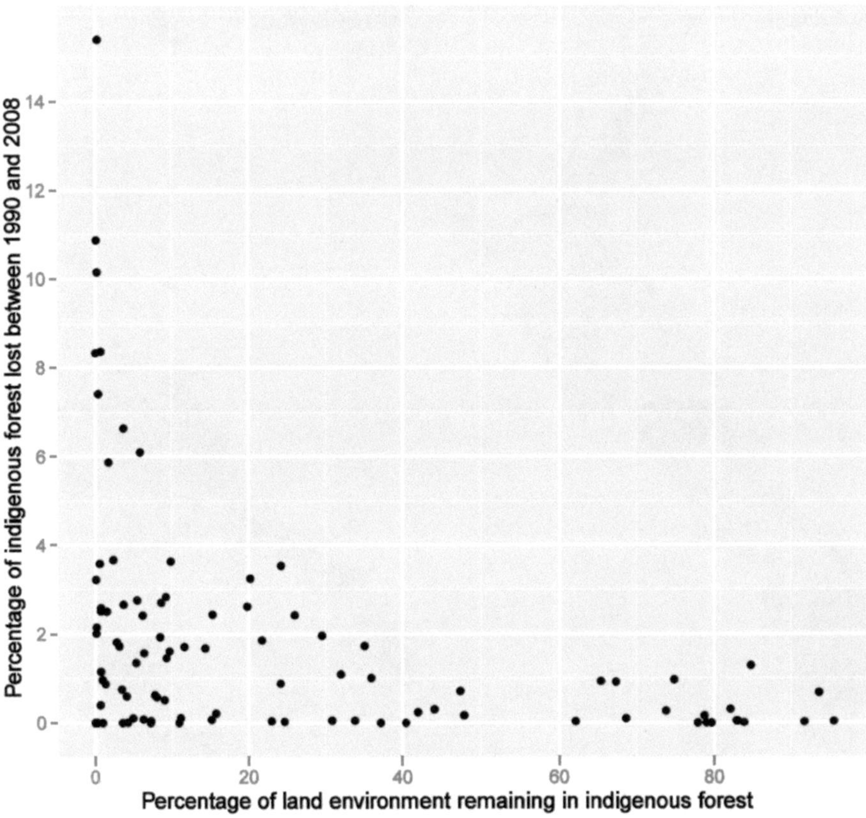

Fig. 2. Percentage of indigenous forest lost between 1990 and 2008.

New Zealand's tussock grasslands have undergone a variety of transformations. In the South Island, much of the high country (tussock grasslands) was acquired from the Maori between 1844 and 1864 (Brower, 2008). During this time, pastoral licenses were granted for 1 year in Canterbury and 14 years in Otago, and the tussock landscape rapidly began to change. Lease holders used fire both to ready land for grazing and to facilitate travel. The result was a huge reduction in area of lowland and montane red tussock grasslands, the elimination of snow tussock from lowland eastern parts, and the reduction of snow-tussock found near settled areas. By the 20th century there was substantial loss of native species through conversion to vigorous exotic grasses maintained by the widespread use of fertilizers and herbicides.

Today, New Zealand's indigenous grasslands are dominated by grass species (Poaceae family) characterised by tussock growth (elsewhere known as "bunch grasses") (Ashdown & Lucas, 1987; Levy, 1951; Mark, 1965; Mark, 1993). The plant communities, however, vary from highly modified to areas with no exotic species (predominantly at elevations above 700 meters (Walker et al., 2006; Cieraad, 2008). Though tussock species *Chionochloa, Poa,* and *Festuca* are the dominant species in the landscape, numerous woody species are also present. At higher and more exposed sites with shallow soils and less available moisture, shrubs including the species of *Brachyglottis*, *Coprosma*, *Dracophyllum*, *Carydium*, *Hebe*, *Podocarps* and other *Olearia* spp dominate; at lower altitudes native shrub species such as manuka

(*Leptospermum scoparium*) and kanuka (*L. ericoides*) are more common and through time have established themselves among the grasses (Newsome, 1987).

Though most New Zealand's indigenous grasslands have been modified to varying degrees by the indirect and direct effects of human activity, they continue to support a rich flora and fauna and are characterized by high species diversity (Dickinson et al., 1998; McGlone et al., 2001; Mark et al., 2009; Walker et al., 2008). However, recent changes in land-use activities have led to further fragmentation. An increasing area of indigenous grasslands (in the South Island), formerly used for extensive grazing, is being converted to intensive agriculture and areas once covered by indigenous grassland species are being progressively replaced with exotic pasture, forestry plantations, and perennial crops.

Mark and McLennan (2005) assessed the loss of New Zealand's indigenous grasslands since European settlement, comparing the Pre-European extent of five major tussock grassland types with their current extent (using LCDB1). They estimated that in 1840, 31% of New Zealand was covered by tussock grasslands dominated by endemic tussock grass species. In 2002, however, just 44% of this area of indigenous grasslands remained, of which most was in the interior areas of the South Island. Of this, approximately 28% was protected with a bias towards the high-alpine areas. Remaining subalpine grassland communities (i.e. short tussock grasslands) still persisted, but were severely degraded and/or modified and under protected. Figure 3 illustrates the change in extent from pre-human to pre-European to current times.

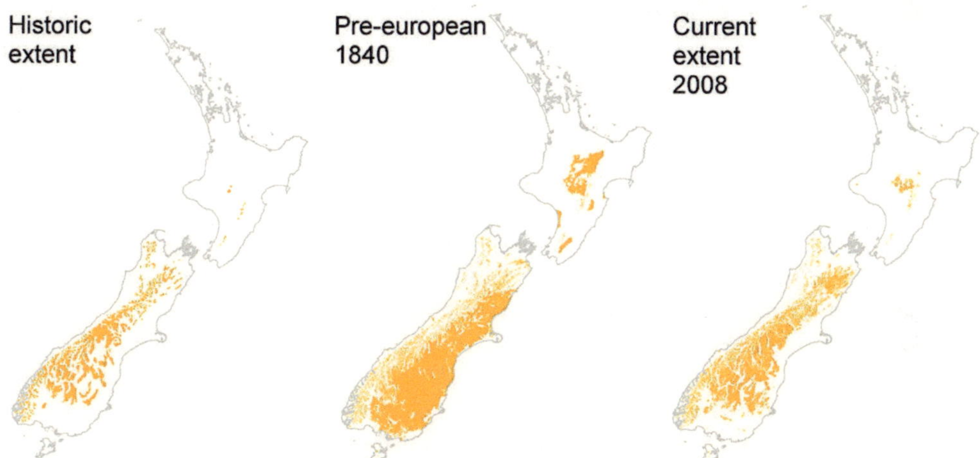

Fig. 3. Changes in the extent of New Zealand's indigenous grasslands since the arrival of humans.

Recent trends in land-use change suggest a movement towards increased production per hectare of land. Weeks et al. (in prep) estimated the current (2008) extent of indigenous grasslands and compared it with grassland in 1990. In 1990, 44% of New Zealand's indigenous grassland remained, by 2008 this was reduced to 43%. During this time there was an accelerated loss from 3,470 ha per year between 1990 and 2001 to 4,730 ha per year between 2001 and 2008. The majority of this change took place at lower altitudes (in short tussock grasslands) and on private or recently free-hold land. Most of the land-use change has been incremental and occurred at the paddock scale (less than 5 hectares).

Continued impacts and reduced indigenous biodiversity are expected over the next century. In grazed areas, plant community composition should continue to alter gradually depending on stocking rates and variability in climate and disturbance regimes. As for areas that are completely converted to new land cover types, changes should be much more immediate. These conversions are likely to have significant impacts on the ecosystem structure and provision of ecosystem services.

4. Freshwater wetlands

Wetlands are defined as permanently or intermittently wet areas, shallow water and land water margins that support a natural ecosystem of plants and animals that are adapted to wet conditions. They support a wide range of plants and animals. In New Zealand, wetland plants include 47 species of rush and 72 species of native sedge (Johnson & Brooke, 1998). Many of these plants have very specific environmental needs, with a number of plants species adapted to wet and oxygen deprived conditions. Wetlands support a high proportion of native birds, with 30% of native birds compared with less than 7% worldwide (Te Ara - the Encyclopedia of New Zealand, 2009). For instance, the australasian bittern (*Botaurus poiciloptilus*), brown teal (*Anas chlorotis*), fernbird (*Bowdleria punctata*), marsh crake (*Porzana pusilla)*, and white heron (*Egretta alba*) rely on New Zealand's remnant wetlands. Migratory species also depend on chains of suitable wetlands. Wetlands are also an essential habitat for native fish, with eight of 27 native fish species found in wetlands (McDowall, 1975). Among those are shortfin eel (*anguilla australis*) and inanga (*galaxias maculatus*), the major species in the whitebait catch, and species from the Galaxiid family like the giant kokopu (*galaxias argenteus*), which is usually found in swamps (Sorrell & Gerbeaux, 2004). Apart from provision of habitat for biodiversity, wetlands offer other valuable ecosystem services such as flood protection, nutrient retention for water quality, recreational services (Mitsch & Gosselink, 2000), and important cultural services for Māori, including food harvesting and weaving materials. The importance of wetland ecosystems is recognised internationally, and New Zealand is a signatory to the Ramsar Convention on Wetlands of International Significance. Six sites are currently designated as Wetlands of International Importance, with a total area of 55 thousand hectares.

In less than two centuries, New Zealand wetlands have been severely reduced in extent, particularly with the conversion to pastoral agriculture from the mid 19th century. The loss is attributed to human activities through fires, deforestation, draining wetlands, and ploughing (Sorrell et al., 2004; McGlone, 2009). Further degradation of the habitat has occurred since the introduction of livestock with consequent increases in nutrient flows, changing the fragile equilibrium in the wetlands and altering species composition (Sorrell & Gerbeaux, 2004). The loss of local fauna and flora has also been dramatic. Fifteen wetland birds species have become extinct (with 8 out of 15 being waterfowl species) (Williams, 2004), and ten species are on the list of threatened bird species (Miskelly et al., 2008). Among the plants, 52 wetland taxa species have been classified as threatened (de Lange et al., 2004). The decline in many native freshwater fish is also attributed to the loss and degradation of wetlands (Sorrell & Gerbeaux, 2004).

Ausseil et al. (2011) estimated that the pre-human extent of wetlands was about 2.4 million ha, that is, about 10% of the New Zealand mainland. The latest extent (mapped in 2003) was estimated at 250,000 ha or 10% of the original coverage.

Figure 4 compares the current extent of freshwater wetlands with its historic extent. The greatest losses occurred in the North Island where only 5% of historic wetlands remain compared with 16% in the South Island. The South and Stewart Islands contain 75% of all remaining wetland area, with the highest proportions persisting on the West Coast of the South Island and on Stewart Island. The remaining wetland sites are highly fragmented. Most sites (74%) are less than 10 ha in size, accounting for only 6% of national wetland area. Only 77 wetland sites are over 500 ha, accounting for over half of the national wetland area.

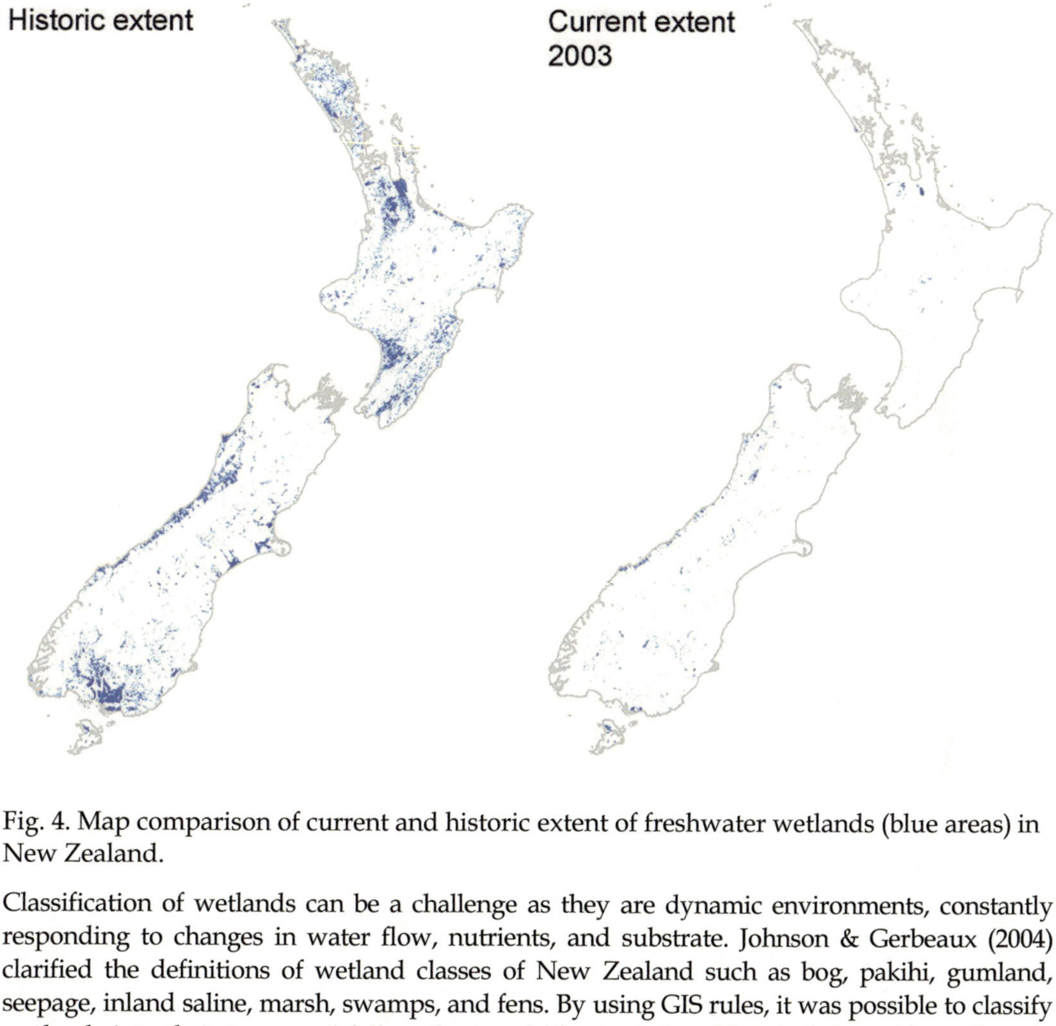

Fig. 4. Map comparison of current and historic extent of freshwater wetlands (blue areas) in New Zealand.

Classification of wetlands can be a challenge as they are dynamic environments, constantly responding to changes in water flow, nutrients, and substrate. Johnson & Gerbeaux (2004) clarified the definitions of wetland classes of New Zealand such as bog, pakihi, gumland, seepage, inland saline, marsh, swamps, and fens. By using GIS rules, it was possible to classify wetlands into their types and follow the trend of extent since historical times (Ausseil et al., 2011). Swamps and pakihi/gumland are the most common wetland types found in New Zealand. However, swamps have undergone the most extensive loss since European settlement, with only 6% of their original extent remaining (Figure 5). This is due to swamps sitting mainly in the lowland areas where conversion to productive land has been occurring.
Unlike indigenous forest and indigenous grasslands, there is no national study describing recent loss over the last ten to twenty years for wetlands in New Zealand. However, some

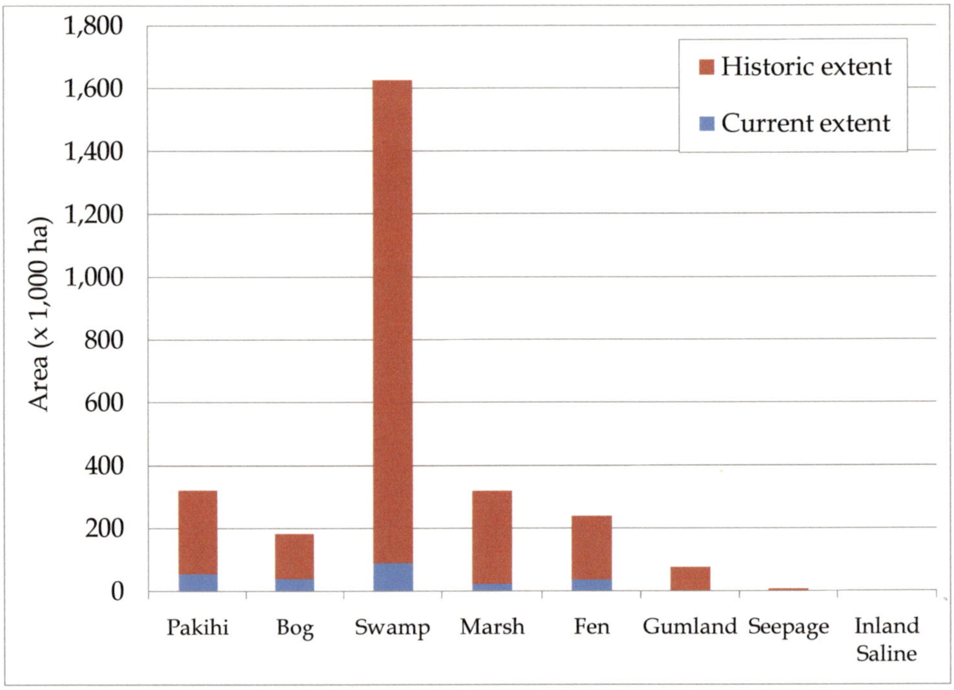

Fig. 5. Current and historic extent of wetland per class.

regional analyses suggest that wetland extent continues to decline, although at a slower rate, as land drainage and agricultural development continue (Grove, 2010; Newsome & Heke, 2010). Wetland mapping is a challenging task as wetlands are sometimes too small in area to be identified using common satellite resolution. Their extent can vary seasonally (e.g., dryness, wetness) and therefore can change markedly at the time of imagery acquisition. While satellite images are useful for providing information at national scale, automatic classification is not possible as vegetation types in wetlands are so variable, making them difficult to characterise through spectral signature. Thus wetlands have been mapped on a manual or semi-automated basis (Ausseil et al., 2007), and this requires a significant amount of effort for all of New Zealand.

5. Measure of natural habitat provision

Measures of habitat provision need to account for different types of habitat and their associated biodiversity. Dymond et al. (2008) showed how proportions of unique habitat remaining may be combined to give a national measure of habitat provision. The habitat measure is based on the contribution it makes to the New Zealand Government goal of maintaining and restoring a full range of remaining natural habitats to a healthy and functioning state. For measuring indigenous forest and grasslands, the historical unique habitats come from Land Environments New Zealand (LENZ) (Leathwick et al., 2003). Wetlands are at the interface of terrestrial and freshwater habitats, and therefore another habitat framework representing both aquatic and terrestrial biota (Leathwick et al., 2007) is used. As such, the measure of habitat provision for wetlands is applied separately from the indigenous forests and grasslands measure.

5.1 Indigenous forest and grasslands

We used LENZ at level II (suitable for national to regional scale) and the most recent land cover (2008) to characterise historic and present habitats. The measure of habitat provision for a land environment is defined as:

$$H_i = P_i \left(\frac{a_i}{A_i} \right)^{0.5} \tag{1}$$

where
a_i is the area of natural habitat remaining in land environment i,
A_i the area of land environment i, and
P_i is the biodiversity value of the ith land environment when fully natural.
The 0.5 power index is used to produce a function monotonically increasing from zero to one with a decreasing derivative in order to represent the higher biodiversity value of rare habitat. In the absence of comprehensive and detailed biodiversity information, Dymond et al. (2008) suggested using species-area relationships (Connor & McCoy, 2001) to estimate P_i as the land environment area to the power of 0.4. The varying condition, or degree of naturalness, of individual sites also needs to be taken into account in the habitat measure:

$$H_i = P_i \left(\frac{\sum_{j=1}^{n} c_{ij} b_{ij}}{A_i} \right)^{0.5} \tag{2}$$

where
c_{ij} is the condition,
b_{ij} is the area of of the jth habitat site in the ith land environment, and
n is the number of habitats in the ith land environment.
The condition of indigenous forest, subalpine shrublands, alpine habitats, and tussock grasslands above the treeline, are assumed to have a condition of 1.0. Tussock grasslands below the treeline have a condition of 0.8 and indigenous shrublands have a condition of 0.5. All other landcovers have a condition of 0.0.
Figure 6 shows the input layers (current land cover and land environments at level II) and the resulting habitat provision map. This map shows the contribution per hectare to the habitat measure (i.e. each pixel represents $\frac{c_{ij} b_{ij}}{\sum_{j=1}^{n} c_{ij} b_{ij}} H_i$).

5.2 Freshwater wetlands

Wetlands are at the interface between water and terrestrial dry environments. They have been categorised with freshwater environments in the past, and as such require a different definition of biogeographic units than the terrestrial environments. We replaced land environments data with biogeographic units defined by climatic and river basin characteristics (Leathwick et al., 2007). This framework was used to define priority conservation for rivers (Chadderton et al. 2004) and wetlands (Ausseil et al., 2011). A condition index for wetlands, similar to c_i in equation (1), was calculated for all current wetland sites in New Zealand (Ausseil et al., 2011). This condition index reflects the major

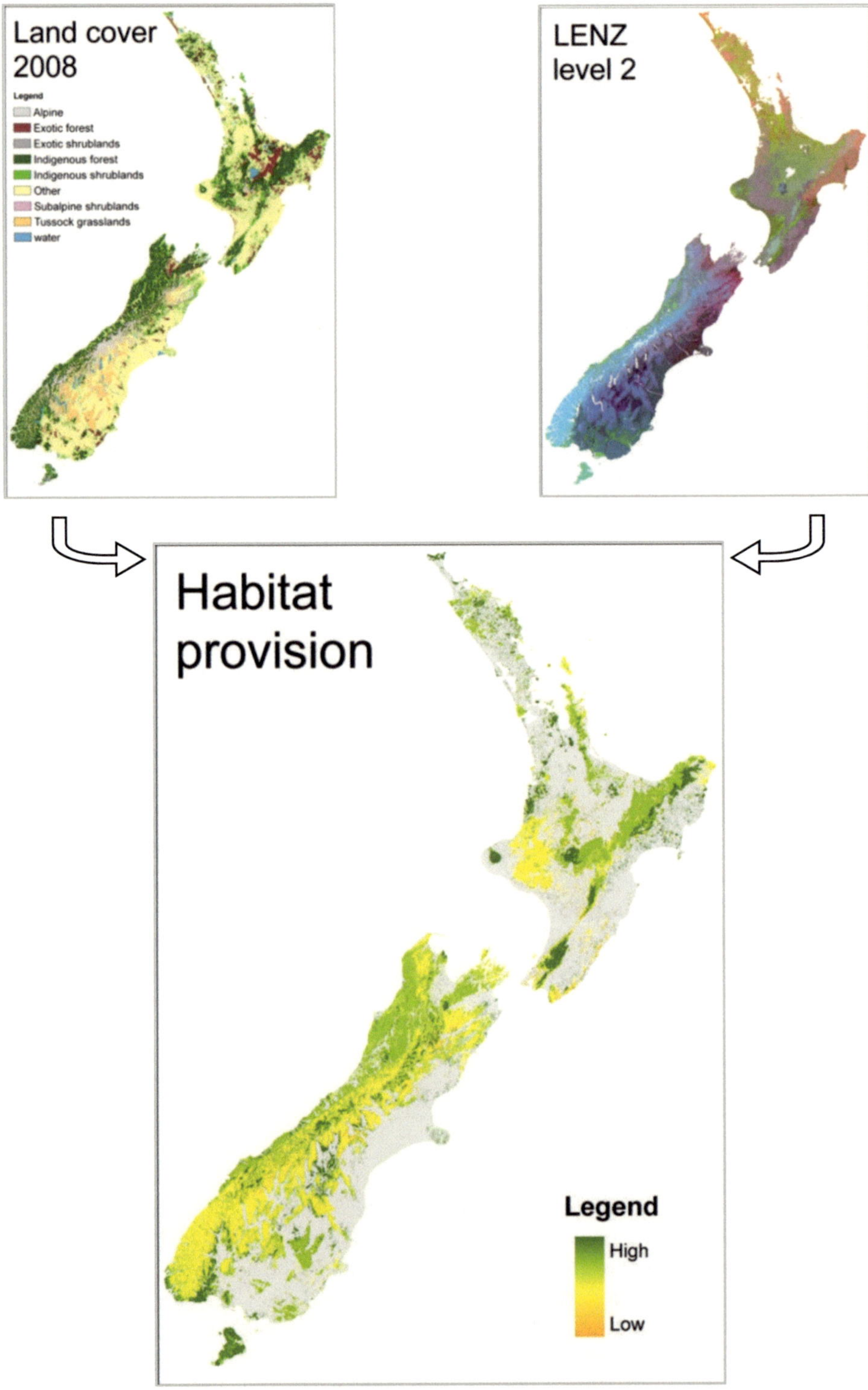

Fig. 6. Habitat provision per hectare from forests and grasslands.

anthropogenic pressures on wetlands, including nutrient leaching, introduced species, imperviousness, loss of naturalness, woody weeds, and drainage pressure.

The measure of habitat provision for wetlands in a biogeographic unit now needs to account for different wetland classes, so is defined as

$$W_i = \sum_{k=1}^{m} P_{ik} \left(\frac{\sum_{j=1}^{n} c_{ijk} b_{ijk}}{A_{ik}} \right)^{0.5} \quad (3)$$

where

c_{ijk} is the condition index of wetland site j in class k in biogeographic unit i,

b_{ijk} is the area of wetland site j in class k in biogeographic unit i,

A_{ik} is the historic area of class k in biogeographic unit i,

m is the number of wetland classes, and

n is the number of class k wetland sites in biogeographic unit i.

Wetland habitats are defined at the class level (m=8 classes) using the wetland classification of Johnson & Gerbeaux (2004). As with P_i in equation (2), P_{ik} is defined as the historical area per wetland class per biogeographic unit to the power of 0.4. The sum $\sum_{j=1}^{n} c_{ijk} b_{ijk}$ reflects the total area of class k wetlands in biogeographic unit i, weighted by the condition index c_{ijk} for each wetland site. If all the wetlands were in pristine condition, the sum would equal the total areal extent in that class.

Figure 7 shows wetland habitat provision for each biogeographic unit. The colours represent the value W_i from equation (3).

6. Discussion

Though there are still large areas of natural habitat remaining in New Zealand, there continues to be ongoing loss. Prior to the settlement of humans, there were 23 million hectares of indigenous forest. Today, only 6.5 million hectares of indigenous forest are remaining. While the total area remaining is large, little of that is in lowland forest ecosystems, and over the last 20 years more lowland ecosystems have been lost. Despite continuing losses in lowland ecosystems, the net area of indigenous forest may well be increasing due to regeneration of indigenous shrublands in marginal hill country. Indigenous grasslands have a similar pattern of change. Over the last 170 years, 4.7 million hectares of indigenous grasslands have been lost. Though the total area of remaining grasslands is large, little of that is in lowland ecosystems, and over the last 20 years more lowland ecosystems have been lost. Wetlands are the most severely impacted ecosystems. Of the 2.4 million hectares of wetlands existing pre-Maori, only 250 thousand hectares are remaining - that is, only 10% of what was there originally. Again, lowland wetlands are mostly affected, with a higher proportion of swamps lost. Recent trend analyses shown in this chapter reveal that loss is still continuing, and is a precursor to negative impacts on provision of ecosystem services and subsequent human well-being.

The habitat provision map for indigenous forest and grasslands show large spatial variability. High values are usually associated with rarer habitats in good condition, but also with habitats in very small land environments. For wetlands, the habitat provision map is

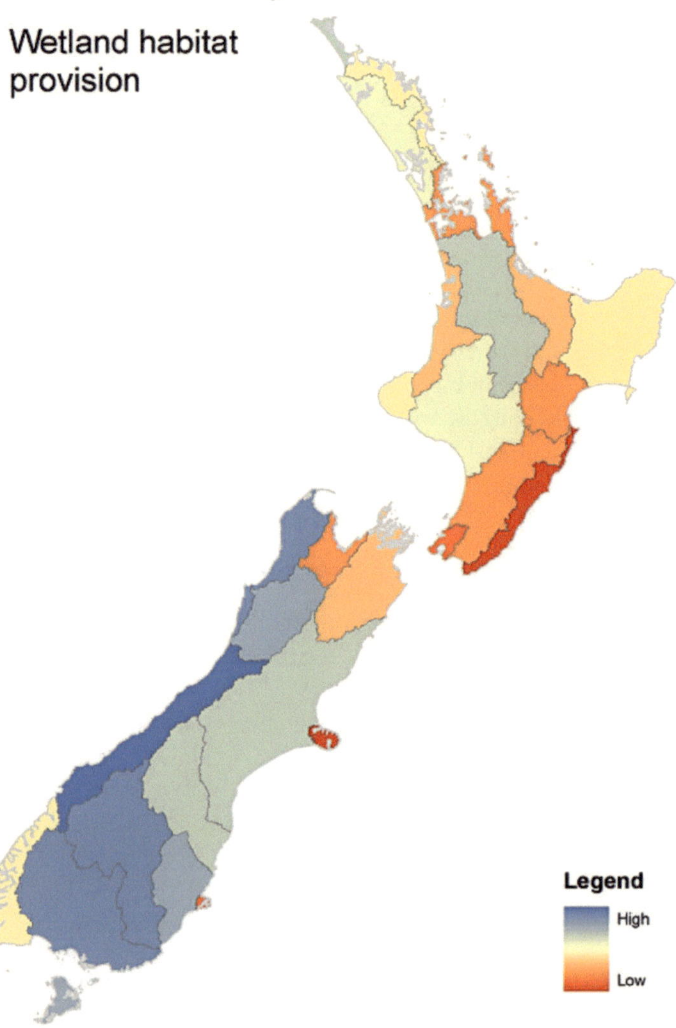

Fig. 7. Habitat provision for freshwater wetlands in each biogeographic unit.

shown at the biogeographic unit level, mainly because wetland boundaries are difficult to depict at the scale shown here. The contribution to the national habitat measure comes mostly from biogeographic units with minimal conversion to productive land. Low values represent units where wetland areas have depleted or where wetlands have been degraded. This information can be used by decision-makers to prioritise the allocation of conservation funds. For example, the maps can be intersected with legally protected areas, like those from Walker et al. (2008), which target areas under private ownership with high natural values. Several legislative tools can be used to protect remnant habitats, including the establishment of conservation covenants like the Queen Elizabeth the Second National Trust (QEII), Nga Whenua Rahui, and the National Heritage Fund.

QEII's goal is to help New Zealand farmers protect open space on private land for the benefit and enjoyment of the present and future generations of New Zealanders. The covenant is registered against the title of the land in perpetuity and there are obligations to manage the land in accordance with the covenant document. Over 70,000 ha are now protected by QEII covenants (Ministry for the Environment, 2007). Nga Whenua Rahui is a contestable fund to negotiate the voluntary protection of native forest on Maori-owned land. Legal protection is offered through covenants, setting aside areas as Maori reservations or through management agreements. About 150,000 ha of native ecosystems are now protected under this fund. The Nature Heritage Fund (NHF) is a third contestable fund for voluntary protection of nature on private land. Its aim is to add to public conservation land those ecosystems important for indigenous biodiversity that are not represented within the existing protected area network. Since 1990, the fund has protected over 100,000 hectares of indigenous ecosystems through direct land purchases, covenants on private land or fencing.

The information on habitat provision could feed into the Department of Conservation (DOC) management system. DOC is responsible for managing biodiversity on the conservation estate, and is developing the Natural Heritage Management System (NHMS). DOC's statement of intent is to legally protect the best possible examples of each native ecosystem type, by fencing, reinstating water levels, replanting, controlling pest animals and weeds, and reintroducing native species to restore and maintain natural ecosystems. The framework proposed in this chapter is envisaged to help achieve this goal through accurate information on habitat extent and ecosystem loss, and provides a measure for comparing habitats within and across land environments where species level assessments may not be possible.

Continuing loss of natural habitat may be due to a lack of market prices for associated ecosystem services (TEEB, 2010). Monetisation of habitat provision could partly redress this. In New Zealand, Patterson & Cole (1999) estimated natural forests and wetlands to both have total economic values of approximately 6 billion dollars in 1994. From this, it is possible to convert the units of habitat measure to economic value in dollars per year - Dymond et al. (2007) estimated this as 60 units to one (NZ) dollar per year (assuming areas are in hectares). This monetisation would permit the comparison of changes in habitat alongside changes in other ecosystem services in the same units. This reduces the complexity of results when analysing impacts of different land-use decisions. Using dollars also provides context for stakeholders unfamiliar with biodiversity impacts associated with habitat loss. The negative side of monetisation is that some stakeholders may be encouraged to make trade-offs on the basis of the monetisation alone, not realising the assumptions and limitations involved, or being aware of environmental bottom-lines. Indeed, the risk of valuation is to get the figure very wrong. There are numerous valuation methods, often based on subjective, hypothetical, and questionable assumptions, which can all give vastly different values (Spangenberg & Settele, 2010). Altogether, monetisation, although easy to comprehend, can be misleading and should be used with caution. It should be used in close consultation with decision-makers, so that they are fully aware of the pitfalls and assumptions behind the valuation, to avoid misallocation of resources.

The measure of habitat provision is a landscape approach which makes several assumptions. First, it uses particular GIS databases, each of which has a certain level of sensitivity and accuracy. Land environments has been tailored to forest ecosystems, and does not encompass the full breadth of other ecosystem types. Biogeographic units were used for wetlands, assuming that freshwater species would be concentrated within defined hydrological boundaries. Second, it assumes that landscape morphology and pattern can be used as a surrogate for species. Though this overcomes the issues surrounding availability of data, application is limited at the various levels and components of biodiversity. In other words, provisions can not be assessed at multiple scales (i.e. habit, community and/or species). Third, the condition of indigenous forest and grassland assumes that all sites are characterised by one condition, though condition could vary within large sites. For wetlands, the condition does vary per site, but it is based on landscape indicators. It is appropriate for a rapid assessment of sites, and can help for prioritising field visits (Ausseil et al., 2007), but does not necessarily reflect the true condition in the field.

The loss of indigenous forest is well characterised by the habitat provision analysis, but the gain of indigenous forest from regenerating indigenous shrublands is not. This is because both the LCDB and the LUM datasets focus on mapping change primarily between woody and herbaceous vegetation, and the subtle changes in the spectral signature of regenerating indigenous forest and mature forest are not accurately characterised or determined, making it difficult to decide whether indigenous vegetation is mature enough to be classified as forest. This is important because there are large areas of indigenous shrublands in New Zealand, approximately 1.6 million hectares. Much of these shrublands are currently regenerating to forest and could make a significant contribution to the areal extent of indigenous forest if this trend continues. If we assume a conservative time of 100 years to reach forest maturity and a uniform distribution of shrubland age, then we would expect about 1% of the shrubland area to change to indigenous forest each year - this amounts to 16 thousand hectares per year. Over 18 years this would equate to approximately 300 thousand hectares, which is six times the estimated current loss of indigenous forest. This fills an important information gap in our understanding of the changing areal extent of indigenous forest and indicates the importance of using objective mapping techniques to monitor change.

Conservation management in New Zealand is becoming increasingly strategic, systematic, and reliant on accurate information on which to plan and prioritise goals and actions. A range of sophisticated tools and approaches have been developed to support these efforts in the past ten years. These include measuring Conservation Achievement (Stephens et al., 2002), the Land Environments of New Zealand (Leathwick et al., 2003), and measuring provision of natural habitat (Dymond et al., 2008). In addition, these efforts have spawned considerable activity for acquiring underlying data, such as biodiversity value (Cieraad, 2008), land cover (the LCDB3 project), and threats to biodiversity (Overton et al., 2003; Walker et al., 2006). However, a national coordinated approach to conservation management taking into account species distributions is required. Overton et al. (2010) are developing a tool called Vital Sites to assess ecological integrity. This incorporates current and natural distributions of native species based on a modeling approach, pressures (e.g., pests or habitat loss) on biodiversity, and the effects of management on relieving pressures. It operates at two levels (species and landscape) and assessments of significance and priorities can be made at each separate level or by combining the two levels. This research tool will provide another step to helping achieve goals towards identifying the most vulnerable ecosystems in New Zealand requiring urgent protection and management.

7. Acknowledgment

This work was supported by the New Zealand Foundation for Research, Science, and Technology through Contract C09X0912 "An ecosystem services approach to optimise natural resource use for multiple outcomes" to Landcare Research. The authors would like to thank Fiona Carswell and Garth Harmsworth from Landcare Research for their valuable comments on an earlier draft and Anne Austin for internal editing.

8. References

Ashdown, M and Lucas, D (1987). *Tussock grasslands landscape values and vulnerability.* New Zealand Environmental Council, ISBN 0477058272, Wellington.

Ausseil, A-GE, Chadderton, WL, Gerbeaux, P, Theo Stephens, RT & Leathwick, JR (2011). Applying systematic conservation planning principles to palustrine and inland saline wetlands of New Zealand. *Freshwater Biology* Vol. 56, No 1, pp. 142-161.

Ausseil, A-GE, Dymond, JR & Shepherd, JD (2007). Rapid mapping and prioritisation of wetland sites in the Manawatu-Wanganui region, New Zealand. *Environmental Management,* Vol. 39, No 3, pp. 316-325.

Brower, A (2008). *Who owns the high country? The controversial story of tenure review in New Zealand.* Craig Potton Publishers, ISBN 9781877333781, Christchurch.

CBD (1992). *Convention on Biological Diversity.* Rio de Janeiro, UNCED.

Chadderton, L, Brown, D & Stephens, T (2004). *Identifying freshwater ecosystems of national importance for biodiversity: Criteria methods, and candidate list of nationally important rivers.* Department of Conservation Discussion Document, Wellington.

Cieraad, E (2008). *How much indigenous biodiversity remains in land under indigenous cover?* Unpublished contract report LC0708/145, Landcare Research Ltd, Lincoln.

Connor, EF & McCoy, ED (2001). Species-area relationship. In: *Encyclopaedia of Biodiversity.* Academic Press, San Diego. Vol 5, pp 397-441.

Conservation International. (2010). *Biodiversity hotspots.* Retrieved 21 December, 2010, available from http://www.conservation.org/Documents/cihotspotmap.pdf.

de Lange, PJ, Norton, DA, Courtney, S, Heenan, PB, Courtney, S, Molloy, BPJ, Ogle, CC, Rance, BD, Johnson, PN & Hitchmough, RA (2004). Threatened and uncommon plants of New Zealand. *New Zealand Journal of Botany,* Vol 42, pp. 45-76.

Diaz, S, Fargione, J, Chapin, FS, III & Tilman, D 2006. Biodiversity loss threatens human well-being. *PLoS Biol,* Vol 4, No 8, e277. doi:10.1371/journal.pbio.0040277.

Dickinson, KJM, Mark, AF, Barratt, BIP & Patrick, BH (1998). Rapid ecological survey, inventory and implementation: A case study from Waikaia ecological region, New Zealand. *Journal of The Royal Society of New Zealand,* Vol 28, pp. 83-156.

Department of Conservation & Ministry for the Environment (2000). *The New Zealand biodiversity strategy.* from http://biodiversity.govt.nz/picture/doing/nzbs/contents.html.

Dymond, JR, Ausseil, A-GE, Shepherd, JD & Janssen, H (2007). A landscape approach for assessing the biodiversity value of indigenous forest remnants: Case study of the Manawatu/Wanganui region of New Zealand. *Ecological Economics,* Vol 64, No 1, pp. 82-91.

Dymond, JR, Ausseil, A-GE & Overton, JM (2008). A landscape approach for estimating the conservation value of sites and site-based projects, with examples from New Zealand. *Ecological Economics,* Vol 66, No 2-3, pp. 275-281.

Ewers, RM, Kliskey, AD, Walker, S, Rutledge, D, Harding, JS & Didham, RK (2006). Past and future trajectories of forest loss in New Zealand. *Biological Conservation,* Vol 133, No 3, pp. 312-325.

Grove, P (2010). Monitoring extent of Canterbury freshwater wetlands by remote sensing, 1990–2008. *Proceedings of New Zealand Ecological Society conference*, University of Otago, Dunedin, November 2010.

Hitchmough, RA, Bull, LJ & Cromarty, P (2005). *New Zealand Threat Classification system lists.* Department of Conservation, Wellington.

Johnson, P & Gerbeaux, P (2004). *Wetland types in New Zealand*. Department of Conservation, Wellington.

Johnson, PN & Brooke, PA (1998). *Wetland plants in New Zealand*. DSIR, Wellington.

Landcare Research (2011). *New Zealand Land Resource Inventory*. Retrieved 20/01/2011, from http://lris.scinfo.org.nz.

Leathwick, JR, Collier, K & Chadderton, L (2007). *Identifying freshwater ecosystems with nationally important natural heritage values: Development of a biogeographic framework.* Science for Conservation report no. 274. Department of Conservation, Wellington.

Leathwick, JR, Wilson, G, Rutledge, D, Wardle, P, Morgan, F, Johnston, K, McLeod, M & Kirkpatrick, R (2003). *Land environments of New Zealand*. David Bateman Ltd, Auckland.

Lee, W, McGlone, M & Wright, E (2005). *Biodiversity Inventory and Monitoring: A review of national and international systems and a proposed framework for future biodiversity monitoring by the Department of Conservation*. Landcare Research Contract Report LC0405/122, Lincoln.

Levy, EB (1951). *Grasslands of New Zealand*. R.W. Stiles&Co, Wellington.

Mark, AF (1965). The environmental and growth rate of narrow-leaved snow tussock, *Chionochloa rigida*, in Otago. *New Zealand Journal of Botany,* Vol 43, No 1, pp. 73-103.

Mark, AF (1993). Ecological degradation. *New Zealand journal of Ecology,* Vol 17, No 1, pp. 1-4.

Mark, AF & McLennan, B (2005). The conservation status of New Zealand's indigenous grasslands. *New Zealand Journal of Botany,* Vol 43, No 1, pp. 245-270.

Mark, AF, Michel, P, Dickinson, KJM & McLennan, B (2009). The conservation (protected area) status of New Zealand's indigenous grasslands: An update. *New Zealand Journal of Botany,* Vol 47, pp. 53-60.

McDowall, RM (1975). *Reclamation and swamp drainage - their impact on fish and fisheries.* Fisheries Research Division Information Leaflet No. 6. New Zealand Ministry of Agriculture and Fisheries, Wellington.

McGlone, MS (1988). Glacial and Holocene vegetation history: New Zealand. In: B. Huntley and T. Webb (eds), *Vegetation History*. Kluwer Academic Publishers, Dordrecht. Vol 7, pp. 557-599.

McGlone, MS (2001). The origin of the indigenous grasslands of southeastern South Island in relation to pre-human woody ecosystems. *New Zealand Journal of Ecology,* Vol 25, No 1, pp. 1-15.

McGlone, MS (2009). Postglacial history of New Zealand wetlands and implications for their conservation. *New Zealand journal of Ecology,* Vol 33, No 1, pp. 1-23.

McGlone, MS, Duncan, RP & Heenan, PB (2001). Endemism, species selection and the origin and distribution of the vascular plant flora of New Zealand. *Journal of Biogeography,* Vol. 28, pp. 199-216.

Millennium Ecosystem Assessment (2003). *Ecosystems and human well-being: A framework for assessment.* Island Press, Washington D.C., USA.

Millennium Ecosystem Assessment (2005). *Millenium ecosystem assessment synthesis report.* Island Press, Washington D.C., USA.

Ministry for the Environment (1997). *The state of New Zealand's environment.* Ministry for the Environment, Wellington, New Zealand.

Ministry for the Environment (2007). *Environment New Zealand 2007.* Report ME847. Retrieved from http://www.mfe.govt.nz/publications/ser/enz07-dec07/index.html. Wellington, New Zealand.

Ministry for the Environment. (2009). *The New Zealand Land Cover Database.* Retrieved from http://www.mfe.govt.nz/issues/land/land-cover-dbase/.

Ministry for the Environment. (2010). *Measuring carbon emissions from land-use change and forestry: The New Zealand Land-Use and Carbon analysis system.* Retrieved from www.mfe.govt.nz/issues/climate/lucas/.

Miskelly, CM, Dowding, JE, Elliott, GP, Hitchmough, RA, Powlesland, RG, Robertson, HA, Sagar, PM, Scofield, RP & Taylor, GA (2008). Conservation status of New Zealand birds, 2008. *Notornis,* Vol 55, pp. 117-135.

Mitsch, WJ & Gosselink, JG (2000). *Wetlands* (3rd edition). John Wiley & Sons Inc., New York/Chichester.

Newsome, PJF (1987). *The vegetative cover of New Zealand.* Water and Soil Miscellaneous Publication No112. Ministry of Works and Development, Wellington.

Newsome, P & Heke, H (2010). *Mapping wetlands of the Taranaki region 2001-2007.* Unpublished contract report Landcare Research Ltd, Palmerston North.

Noss, RF 1990. Indicators for monitoring biodiversity: A hierarchical approach. *Conservation Biology,* Vol. 4, No 4, pp. 355-364.

Overton, JM, Kean, J, Price, R, Williams, PA, Barringer, JRF, Barron, M, Cooke, A, Martin, O & Bellingham, PJ (2003). *PestSpread Version 1.0: A prototype model to predict the spatial spread of pests.* Landcare Research unpublished contract report LC0405/048. Department of Conservation, Wellington.

Overton, JM, Price, R, Stephens, RTT, Cook, S, Earl, R, Wright, E & Walker, S (2010). *Conservation planning and reporting using the Vital Sites model.* Unpublished contract report LC0910/064. Landcare Research Ltd, Hamilton.

Patterson, M & Cole, A (1999). *Assessing the value of New Zealand's biodiversity.* Occasional paper no1. Massey University, Palmerston North.

Sorrell, B & Gerbeaux, P (2004). Wetland Ecosystems. In: *Freshwaters of New Zealand,* J. S. Harding, P. Mosley, C. Pearson & B. Sorrell (eds), chapter 28, New Zealand Hydrological Society and New Zealand Limnological Society, Christchurch.

Sorrell, B, Reeves, PN & Clarkson, BR (2004). Wetland Management and restoration. In: *Freshwaters of New Zealand,* J. S. Harding, P. Mosley, C. Pearson & B. Sorrell (eds), chapter 40, New Zealand Hydrological Society and New Zealand Limnological Society, Christchurch.

Spangenberg, JH & Settele, J (2010). Precisely incorrect? Monetising the value of ecosystem services. *Ecological Complexity,* Vol. 7, No 3, pp. 327-337.

Stephens, RTT, Brown, DJ & Thornley, NJ (2002). *Measuring conservation achievement: Concepts and their application over the Twizel Area.* Science for Conservation 200. Department of Conservation, Wellington.

Stevens, G, McGlone, M & McCulloch, B (1988). *Prehistoric New Zealand.* Auckland, Heinemann Reed.

Te Ara - the Encyclopedia of New Zealand (2009). *Wetland birds.* Retrieved 1 March 2009, from http://www.TeAra.govt.nz/en/wetland-birds/.

TEEB (2010). The economics of ecosystems and biodiversity. Mainstreaming the Economics of Nature: A synthesis of the approach, conclusions and recommendations of TEEB.

Walker, S, Price, R & Rutledge, D (2008). *New Zealand's remaining indigenous cover: Recent changes and biodiversity protection needs.* Department of Conservation, Wellington.

Walker, S, Price, R, Rutledge, D, Stephens, RTT & Lee, WG (2006). Recent loss of indigenous cover in New Zealand. *New Zealand journal of Ecology,* Vol. 30, No 2, pp. 169-177.

Wardle, P (1991). *Vegetation of New Zealand,* Cambridge University Press, Cambridge, United Kingdom.

Weeks, ES, Dymond, JR, Shepherd, JD & Walker, S (in prep.) Quantifying land-use change in New Zealand's indigenous grasslands.

Williams, M (2004). Bird communities of lakes and wetlands. In: *Freshwaters of New Zealand,* J. S. Harding, P. Mosley, C. Pearson & B. Sorrell (eds), chapter 26, New Zealand Hydrological Society and New Zealand Limnological Society, Christchurch.

10

Limited Bio-Diversity and Other Defects of the Immune System in the Inhabitants of the Islands of St Kilda, Scotland

Peter Stride
University of Queensland School of Medicine
Australia

1. Introduction

Intra-species variations in Homo sapiens can contribute to health and resistance to infection, or alternatively to death and disease. The small isolated population of the St Kilda archipelago in the Scottish Hebrides suffered severely from many infectious diseases in the seventeenth to nineteenth centuries, with greater morbidity and mortality than the inhabitants of similar Scottish communities on other remote islands.

Fig. 1. Map of Scotland

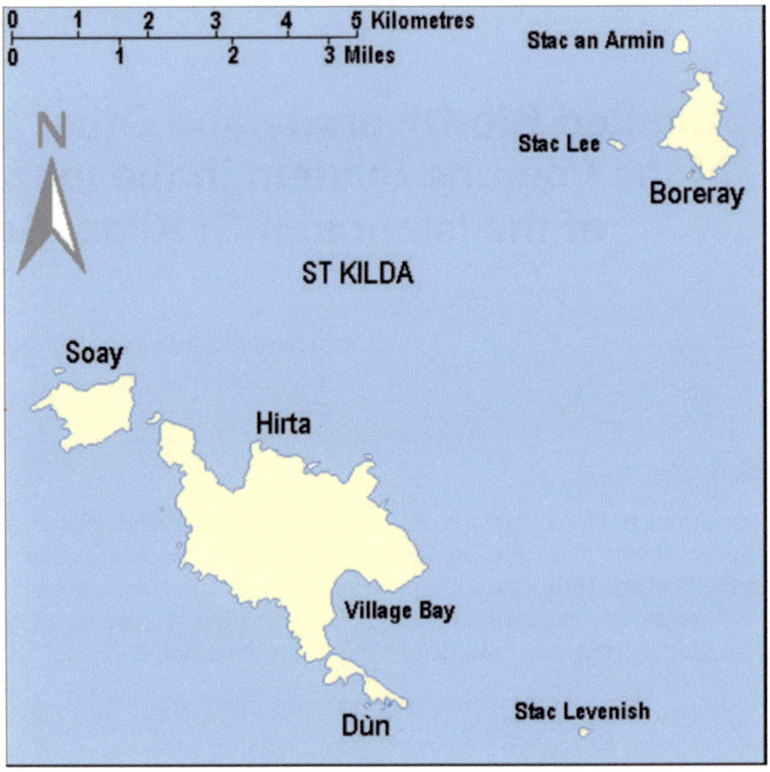

Fig. 2. Map of St Kilda

Fig. 3. Picture of St Kilda

These infections were one of the factors leading to the failure of the island society, culminating in its final evacuation in 1930. Limited population genetic biodiversity and other factors predisposing to infectious diseases including low herd immunity and isolation, the absence of any 'healthcare professionals', malnutrition, social conditions, climatic factors and dioxin toxicity are discussed.

St Kilda, located at 57° 48′ N, 8° 34′ W, is an inaccessible and isolated archipelago in the Scottish Hebrides, previously inhabited for approximately 2,000 years by a small struggling community until evacuated in 1930 as a non-sustainable society. Today it has been re-occupied by the National Trust of Scotland, and the Ministry of Defence.

The 430 metre cliffs of St Kilda can be seen on clear days from the nearest islands in the Outer Hebrides, Harris and North Uist, which have themselves been occupied continually for six thousand years, according to archaeological evidence. Simple stone tools found on the main St Kilda island of Hirte, suggest that travellers visited St Kilda some 4,000 to 5,000 years ago. [1, 2, 3] The other smaller islands composing the St Kilda archipelago are Boreray, Soay and Dun.

The date of the first permanent settlement on St Kilda is not clear, but evidence suggests at least transient occupation from prehistoric times, harvesting the abundant stocks of fish and the sea birds, growing crops and keeping animals. A souterrain, an Iron Age earth house store about 2,000 years old, with a central long passage, and shorter passages or cells branching off, was discovered in 1844. A possible Bronze Age burial structure was excavated in 1995.

Fig. 4. Souterrain

Fig. 5. Village bay

The presence of three early chapels and two incised stone crosses of early Christian style were recorded by the Rev. Kenneth Macaulay. [4, 5] The continued use of Norse place names, such as Oiseval - the east hill - and Ruaival - the red hill, and archaeological finds of Norse brooches and steatite, are strong evidence of continual occupation by the Norsemen and their descendents until the 20th century. However, prior to the steamship era, the inclement climate and the small exposed rocky harbour restricted access by visiting sailing vessels.

2. Genetic origins

The inhabitants of St Kilda were predominantly of Celtic origin. The Vikings from Scandinavia occupied the Scottish islands until the 13th century. Control of the Hebrides was largely transferred from the Scandinavians to the Scots after the battle of Largs in 1263, but many settlers of Viking origin remained in the Hebrides. Studies of mitochondrial and Y chromosome DNA reveal that 30% of Orcadians and Shetland Islanders have Norse maternal and paternal ancestors, but in the Hebrides the male DNA remains around 30%, but Norse mitochondrial DNA falls to 8%, indicating that the Scandinavians took their women to the Northern Isles of Orkney and Shetland, but predominantly took local Celtic females as their partners in the Hebrides. [6] The degree of consent cannot be determined. There have been no specific studies of the DNA of the current single survivor and the descendents of the evacuees from St Kilda, but the persistence of Norse names suggests at least some perhaps only male genetic biodiversity on St Kilda.

The St Kilda population up to the epidemic of 1727 were predominantly from two families of Morrisons and McDonalds generating inbred families with limited genetic diversity. St Kilda, as a 'virgin soil'environment, also had a non-immune adult population, with increased morbidity and mortality from most infections. Subsequent to that episode in 1727 which left one adult survivor on Hirte, the island was repopulated from neighbouring islands, though the inhabitants numbering over a hundred were essentially derived from only five resident families, the Gillies, MacQueens, MacDonalds, MacKinnons and Fergusons, for the remaining 200 years.[7] Close consanguineous marriages were carefully avoided, and an external review of insanity caused by close intermarriages found no evidence of this problem. Consanguinity may, however, have been closer than suspected following the 'religious' leadership of a self-appointed predatory character known as Roderick for six years at the end of the 17th century. Seduction formed part of his 'instruction' of women attending counselling before marriage or childbirth.[8]

Comparisons with pre-Columbian North America indigenous population who had not encountered European viruses are interesting and relevant. [7] Their genetic biodiversity of histocompatibility leukocyte antigens (HLA), the genetic key to immunological defence against viruses, was 64 times less than that of the Europeans. The indigenous North American population declined from perhaps 100 million to a few million in 300 years. Smallpox is incriminated, without indisputable evidence, as causing the death of 90% of non-immune indigenous Americans in the sixteenth and seventeenth centuries.

Crosby defines the term 'virgin soil' epidemics as '*those in which the populations at risk have had no previous contact with the diseases that strike them and are therefore immunologically almost defenceless*'. John Morgan, a Manchester physician, used the term when writing about his visit to St Kilda in 1860: '*May we not explain the accumulated fatality in all these cases by supposing that in the same manner as the different cereals flourish best when planted in virgin soil, or at longer intervals of time, so it is with infectious disease? The more distant their visitation, the richer the pabulum supplied for the epidemic.*'[7]

3. Animal evolution[2]

The islands of St Kilda were sufficiently isolated for animals to evolve different characteristics over a few centuries. The common house mouse probably introduced by the Vikings evolved into the now extinct, but larger subspecies of the St Kilda House Mouse (Mus musculus muralis), which was found only on St Kilda. The mouse was dependent on human habitation and died out after the evacuation in 1930.

Another local genetically different animal is the Soay sheep (*Ovis aries*). The name Soay derives from the Viking name of island of sheep. This breed is thought to be the descendents of the earliest domesticated sheep of Northern Europe, and is physically similar to the other wild ancestors of domestic sheep like the horned urial sheep of Central Asia and the Mediterranean mouflon. Soay sheep are hardy and extremely agile animals well adapted for survival on the high cliffs of St Kilda, though they are smaller than modern domesticated sheep. Unattended, their numbers tend to build up to peak, followed by a crash, perhaps due to over-grazing or parasite infection. In the autumn of 1960 for example, 1344 sheep were counted, yet by the following spring 820 had died.

They have a reduced genetic mechanism to select their coat colour compared to other sheep. Two genetic loci determine the colour which is limited to black, brown or, less commonly, white. The population of Soay sheep are a fascinating subject for researching evolution and

Fig. 6. Soay sheep

Fig. 7. St Kilda wren

population dynamics, as the numbers are unmanaged, closed to migration in or out, and without competitors or predators.

The third different local sub-species is the St Kilda Wren, *Troglodytes troglodytes hirtensis*, which differs from the common mainland wren by its larger size, its long strong bill and its colour which is more pale grey and less rufous.

Even the humans were rumoured to have evolved differently. The men over several centuries were hunter-gatherers of the birds and birds' eggs, which became their staple diet, from the sheer cliffs of St Kilda. They climbed barefoot and were said to have developed the 'St Kilda toe', an elongated big toe to give them more traction on the rocks. While this story is not substantiated, some tourists in 2008 were astonished by the size of the big toe of a man whose mother was a native of St Kilda.

4. Plant diversity[2]

St Kilda partially escaped the intense glaciations of the Great Ice Age, hence pre-glacial plants are found in its peaty soil. Two hundred varieties of lichen and a hundred and thirty different flowering plants have been discovered on the island. Some of these are extremely rare or ancient, such that a study of these plants helps to explain the distribution and origin of plants in the United Kingdom.

Fig. 8. St Kilda mouse

Fig. 9. Deserted house

5. Diet, vitamins and infections[9]

Malnutrition was clearly another factor contributing to the immune problems of St Kildans. The church ministers, who were usually the only literate members of the society, gave broad details of the islanders' food sources and diet, although none gave a detailed year-round description of their daily consumption of vegetables or other food, so the average daily intake of vitamins is uncertain. There are no laboratory studies that assessed nutrient levels in the islanders. Staff-Surgeon Scott, who visited St Kilda in 1887, detected rheumatism, dyspepsia, anaemia, childhood palpitations and incipient scurvy. The islanders' diet of flesh and eggs, with a lack of fruit and vegetables, was deficient in vitamin C and probably in the B group vitamins.

In 1912, after a severe winter Dougal MacLean, the island missionary, reported that the population had survived on tea, bread and butter for months. The islanders devoted time mainly to catching seabirds and collecting their eggs, secondly to attending to cattle and sheep and lastly to tending the limited arable land. Sheep and a few cattle were slaughtered for food only on special occasions. The St Kildans did not rotate crops nor leave areas fallow for a season. The arable land was a maximum of eighty acres and naturally poor, with thin stony topsoil and poor drainage. They destroyed large areas of potential farm land by stripping turf near the village for fires, rather than collecting peat from further away. Crops were mainly barley, oats and potatoes, but also sea-plants, dulse, (an edible seaweed), silverweed roots, dock, sorrel, scurvy-grass, rhubarb, turnips and cabbages. Peas and beans flowered without produce. Fruits were clearly a rarity.

Fig. 10. Cultivated land

Fig. 11. St Kilda cliffs, a food source

The major source of food was the gannet and fulmar, caught on the cliffs and consumed either as eggs or as young birds, both fresh and cured. Boiling was the usual method of cooking all meals, which would have further reduced the vitamin C content.
Fishing was important, but at times was impossible due to the unpredictable weather and heavy seas. Much of the St Kildans' food produce, including most of their agricultural crops, was paid as rent and taken to Harris or consumed by visitors, leaving a nutritionally limited diet of seabirds, eggs, fish and the less nutritious vegetables for the islanders.
Studies of Ascorbic acid supplements have shown some benefit in the treatment of respiratory infections, particularly in patients with more severe illness and pre-existing low vitamin C levels. Tetanus was another serious problem on St Kilda, and vitamin C deficiency may have contributed to the high death rate. A study has shown that the addition of 1,000 mg ascorbic acid to the standard therapy of anti-tetanus serum, sedatives and antibiotics reduced the mortality in children aged one to 12 years. [9]

6. Infectious diseases on St Kilda

The island population suffered from many diseases caused by micro-organisms including leprosy and tuberculosis, the more common conditions are discussed.

7. Neonatal tetanus[10]

The best known health problem of these islanders was neonatal tetanus, causing the desperately distressing and tragic death of up-to two thirds of the babies born on St Kilda for at least a hundred and fifty years, between 1750 and 1900.
A visit to the island for three weeks was made by Martin Martin MD in 1695. He was an astute observer of all facets of life, yet noted:

'They are not infested with several diseases which are so predominant in the other parts of the world'.

The first mention of *trismus nascentium* or *tetanus neonatorum* was made by the Rev. Kenneth Macaulay, minister of Ardnamurchan, after a visit to the island of St Kilda in 1758, 61 years after Martin's visit. His 278 page book contains less than one page, quoted in full below, about these tragic neonatal deaths for which he had no explanation.

'The St Kilda infants are peculiarly subject to an extraordinary kind of sickness; on the fourth, fifth or sixth night after their birth, many of them give up suckling; on the seventh day their gums are so clenched together that it is impossible to get anything down their throats; soon after this symptom appears, they are seized with convulsive fits, and after struggling against excessive torments, till their little strength is exhausted die generally on the eighth day. I have seen two of them expire after such agonies. It is surprising that Martin, who was himself bred to physic, and a person of unbounded curiosity, should have passed over in silence a circumstance so very striking, supposing that this very uncommon distemper had got any footing on Hirta in his time.'

It seems highly improbable that Martin could have overlooked this classical description of neonatal tetanus. Possibly some important sterile technique was not passed on from one village midwife to her successor early in the 18th century, unfortunately the delivery methods were never documented or observed by a doctor or indeed any other male. Handover remains an imperfect process in current medical circles. The deaths continued and were documented by Dr John Morgan MA, MB, MRCP, on his visit to St Kilda in 1860 from the parochial Island Register kept by the resident missionary, the Rev. Neil Mackenzie.

He noted thirty-three of the recorded 64 island deaths between July 1830 and September 1840 were attributed to the 'eight-day sickness'. Morgan spoke to the midwife, Betty Scott, who had herself lost 12 out of 14 of her own children to this condition about the clinical features. Scott excluded any obvious congenital problem stating:-

'At the time of birth, there was no appreciable physical inferiority on the part of those infants who were so prematurely and suddenly selected as a prey. They were all proper bairns, and so continued till about the fifth or sixth day. The mother's eye might then not infrequently observe on the part of her child a strange indisposition to take the breast.'

Date	Live births	No of Neonatal deaths in 28 days	mortality rate %
1830-39	61	35	57%
1840-49	5	-	-
1850-59	11	5	45%
1860-69	29	20	69%
1870-79	28	14	50%
1880-89	27	14	52%
1890-99	25	6	24%
1900-09	15	2	13%
1910-19	17	1	6%
1920-29	7	0	0%

Table 1. Figures from the Island and District Registers (incomplete 1840-56)

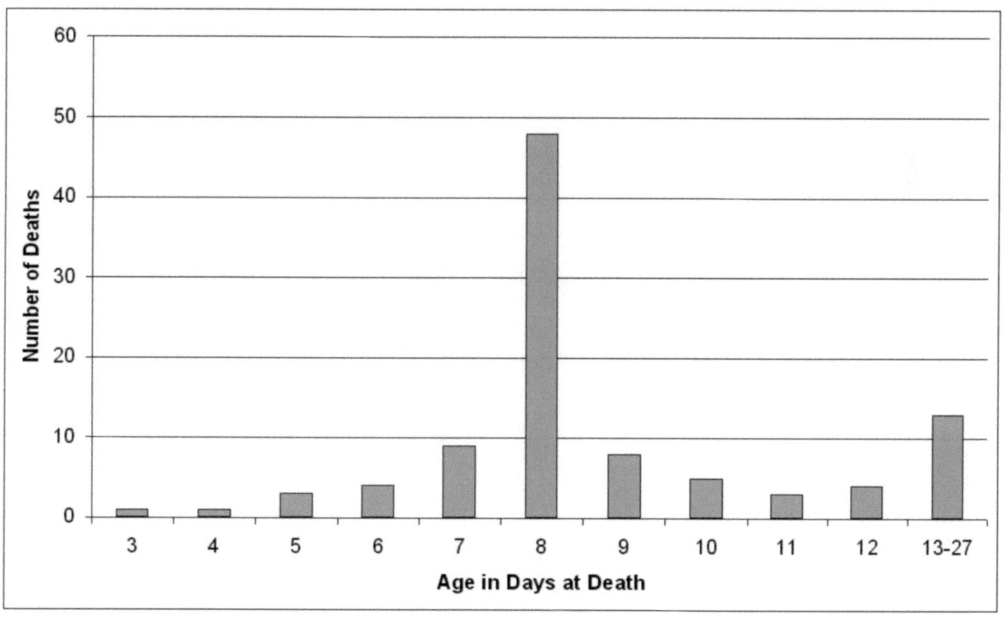

Graph 1. day of death

The problem persisted late in the 19th century, with frequent deaths, personal and family tragedies, and a failure to replenish the society with children who would ultimately sustain the workforce. George Murray, the island's schoolteacher, wrote in his diary on 12th December 1886:

'Last night at 10.30, after six days' intense suffering, the child departed this life. Every one expressed great wonder how it lived so long after being seized with the illness, as they generally succumb at the end of a week after they are born. This one was 13 days except one and a half hours. It had a frequent cry since it was born; but the first signs of its being dangerously ill was at the end of a week, when it ceased to suck the breast, but still sucked the bottle. The following day, the jaws fell (thuit na gilan), when all hope of its recovery were given up. From that time until its death it occasionally took a little milk in a spoon or out of a bottle. The last two days a little wine in water was given once or twice. It often yawned and sometimes looked hard at you. It was pitiful to see the poor little things in the pangs of death. May God prepare us all for the same end.'

When the child was buried, Murray wrote, showing significant insight into the cause of neonatal tetanus in the last sentence:

'In the grave which was opened, I saw the coffins of its two little brothers that died in the same way. The one coffin was quite whole, there being only about 16 months since it was interred. On looking through the churchyard, I felt sad at the sight of so many infant graves. One man, not yet 50 years old, I should say, pointed to the place to me where he buried nine children. He is left with four of a family. Another buried no less than a dozen infants and is left with two now grown up. Sad to think of the like. Bad treatment at birth must have been the cause of so many dying.'

Fig. 12. Graveyard

The solution was finally discovered in 1890 by the Rev. Angus Fiddes, a Free Church clergyman and scientist, who lived and worked on St Kilda. He visited the leading obstetricians in Glasgow, identified the problem as neonatal tetanus, and learned the latest delivery techniques, which when correctly applied prevented any further deaths, though some of the island women were initially opposed to new ideas and to a new midwife from outside.

The precipitating cause remained unknown, and for some 80 years, the unsubstantiated view of Dr George Gibson FRCPEdin, that anointing the umbilical cord with oil from the fulmar gull caused the infection, held sway. Clearly some unhygienic midwifery practice was the cause, and recent bacteriological studies revealing the gull oil to be sterile, yet the soil, inside and outside the house and storage cleits, to be heavily infected with tetanus, suggests cutting the umbilical cord with an unsterilized blade, such as the dirty rusting lancet still used on St Kilda for bloodletting in the early 20th century, a much more probable cause.

Biodiversity appears to play no part in the neonatal tetanus tragedy; however St Kildans were also susceptible to common viral contagious diseases which had severe effects.

Fig. 13. Fulmar gull

8. The Smallpox epidemic in St Kilda[7]

Smallpox outbreaks were common on the Scottish Islands in the eighteenth and nineteenth centuries. All the islands had a similar hostile climate, and varying degrees of isolation and malnutrition. Ten outbreaks occurred on the Shetland Islands between 1700 and 1830. The 1740 diary of Thomas Gifford of Busta, from Greig's *Annals of a Shetland parish,*[7] recorded that two of his daughters became unwell and bed-bound, developed a rash five days later and died a further eight and nine days respectively after that. Some of Gifford's 11 other children developed a rash but all survived, thus illustrating the typical features of smallpox:

- Severe symptoms preceding the infectious rash, with isolation preceding infectivity.
- Deterioration and death of only a few of those infected with smallpox.
- Greater vulnerability in children than adults
- Relatively slow spread to household contacts
- Most of those infected survive mild but obvious disease

In contrast to such episodes, in 1727 an outbreak of possible smallpox on St Kilda killed nearly the entire population. Ninety-four deaths in a population of 132 were recorded, leaving four adult and twenty-six children as the survivors.

Neil MacKenzie, minister on St Kilda from 1829 to 1843, writing a hundred years later stated

'Death after death followed. At last there were scarcely sufficient to bury the dead… There were 94 deaths… those who had been left on Stac an Armin returned mostly to empty houses'.

Fig. 14. Stac an Armin

A small party of three men and eight boys were marooned on Stac an Armin, a 196-metre high sea stack, where they had been taken to collect birds and birds eggs, but also survived through an Atlantic winter till rescued on 13 May 1728 in a little known epic tale of endurance in adversity. A small bothy gave limited shelter, and the group lived off the stack's fresh water supply, birds and their eggs and fish caught with a bent nail, though they were noted to have lost weight. They patched their clothes as well as possible with birds' skins. Although the greater resources of Boreray Island were 100 yards away, the vertical rock face prevented ascent from the water.

A decade earlier Lady Mary Wortley Montagu, wife of the ambassador to Turkey discovered there the procedure of variolation, or intradermal inoculation of smallpox scabs. This caused a moderate infection with a mortality rate of 0.5–2%, compared with a death rate of 10% from the actual disease, She variolated her own children, and though sceptical of physicians there, hoped to introduce the process into England, writing

'I am patriot enough to take pains to bring this useful invention into fashion in England, and I should not fail to write to some of our doctors very particularly about it if I knew any one of them that I thought had virtue enough to destroy such a considerable branch of their revenue for the good of mankind'.

Montagu has an extraordinary coincidental connection with St Kilda. Her sister, Lady Frances Pierrepont, married John Erskine, the 22nd Earl of Mar, a Jacobite general in the 1715 battle of Sherriffmuir. John's brother, Lord Grange married Rachel Chiesley, who soon, perhaps correctly, suspected her husband's infidelity and Jacobite loyalties. Grange imprisoned her, initially on the Monarch Islands, west of North Uist in the Outer Hebrides, and then on St Kilda from 1734 to 1742, where she became the celebrated Lady Grange of St. Kilda. Had Chiesley heard of vaccination from Lady Montagu, her brother-in-law's sister-in-law, and arrived seven years earlier, she may have prevented the epidemic if indeed it was smallpox.

William Heberden differentiated chickenpox from smallpox in 1767; hence 'smallpox' epidemics prior to that date cannot be accepted as indisputable.

Limited biodiversity may well have reduced the population's resistance to this infection, which could equally have been chicken pox. The 1727 epidemic has always been labelled smallpox; however the following points suggest the alternative of chickenpox:

- No clinical features of the 1727 St Kilda epidemic are available;
- The similar exanthemas of smallpox and chickenpox were not distinguished at that time;
- The epidemic spread widely and rapidly among the inhabitants, more like chickenpox than smallpox;
- The epidemic caused a higher death rate among adults than children – again, more like chickenpox than smallpox;
- Chickenpox probably had a high mortality in the 'virgin soil' of the Americas, in a non-immune society, with limited genetic biodiversity of the HLA system, though there are no data on the comparative mortality of smallpox and chickenpox;
- The viability of the smallpox virus is inversely related to its infectivity. Smallpox is only highly contagious in the aerosolised form. Smallpox virions have a low survival rate and smallpox scabs an even lower infectivity in fomites.

9. The boat cough[11]

Another well documented infection peculiar to the inhabitants of St Kilda was called by the native Gaelic speakers '*cnatan-nagall*' or the strangers' cough. [10]

Martin during his visit to the island in 1697, did record details of this boat cough. He wrote:

'They [the islanders] contract a cough as often as any strangers land and stay among them, and it continues for some eight or ten days; they say the very infants on the breast are affected by it.[8]

The Reverend Macaulay arrived in St Kilda in 1758, and related his experience as follows:

'When I landed, all the inhabitants, except two women in child-bed enjoyed perfect health.... On the third day after I landed, some of the inhabitants discovered evident symptoms of a violent cold, such as hoarseness, coughing, discharging of phlegm, etc. and in eight days, they were all infected with this un-common disease, attended in some with severe head-aches and feverish disorders'.[5]

Macaulay noted that once this epidemic had resolved it did not recur without further visitors. Three episodes were once noted to occur within eight weeks following three separate visits from other islands, making the usual suggestion of influenza unlikely, as this organism does not mutate into a new infecting subtypes as quickly as this.
Human rhinoviruses (HRV), with 110 serological types, are the most common worldwide infective viral agents in humans, causing 30–50% of all cases of upper respiratory infection based on viral cultures, or an even higher percentage using improved detection techniques including reverse transcription-PCR. The short incubation period and recurrent infections with severe cough and profuse sputum, plus some cases of pneumonia and rare deaths, strongly support rhinovirus as the cause of the boat cough. Although adults with an averagely robust immune system usually experience one brief episode per year, the prolonged and recurrent bouts experienced by the St Kildans again support some problem in their immune systems. [10]

10. Migration

Unfortunately departure from the island to places of opportunities, a better climate and greater development failed to protect the people of St Kilda.
In 1852, thirty-six of the islanders migrated to Australia. Although they were among the youngest and fittest from the community, they suffered a 50% mortality on the voyage to Melbourne, mainly from measles. In the years immediately after the final evacuation in 1930, several of the young children died of tuberculosis. In these two examples, the factors of poor climate, malnutrition and isolation no longer applied as much, but the factors of low herd immunity and limited biodiversity were the main causes of death and disease in these migrants.

11. Dioxin

Dioxin is a specific organic unsaturated non-antiaromatic six-membered ring compound with a chemical formula of C4H4O2. However, the term dioxin is used generically by most authorities to include chlorinated dioxins with furans and many derivative compounds as a complex of at least 75 ubiquitous and environmentally persistent organochlorine compounds, of variable toxicity.
The incomplete burning of sea water impregnated coastal peat in Scotland well before the industrial revolution has recently been shown to produce dioxins.[9]
Soil containing unburnt peat was taken both from the surface of the St Kilda arable area and excavated from houses down to the time strata of 1800–1850, has been found to contain 114 ng/kg of dioxin. Cultures for the tetanus bacillus at this depth would have been fascinating to ascertain the presence or absence of this micro-organism. Burning peat produced 643 ng/kg of dioxin in the peat smoke and ash. This would give a total dioxin production by the estimated 260,000–420,000 inhabitants of the Scottish highlands and islands in the eighteenth and nineteenth centuries of 1 kg/year, about one fifth of that produced by the whole of the UK in the industrial era.
St Kildans stored peat ash mixed with cow manure on the house floors over winter and in spring fertilised the island's arable land with the mixture, such that more than 70 years after the evacuation there are still high levels of TCDD in the arable area. [11]
The mammalian immune system has been shown to be susceptible to direct damage by low doses of dioxin, and indirectly by damage to the immuno-modulating effect of the hormonal

system. Animal studies have also shown that dioxin toxicity can cause thymic involution, decreased antibody production with thymic dependent and independent antigens reduced function of the HLA system producing some lymphocyte subsets and reduced cytotoxic T-cell function.

The question therefore arises if the dioxin pollution was a contributory factor to the infections experienced by the inhabitants of St Kilda. Comparison with the major known leak of the dioxin, TCDD (2,3,7,8-Tetrachlorodibenzo-p-Dioxin) from Seveso near Milan in 1976, suggests this is unlikely. The soil levels of dioxin in Italy were considerably higher, yet careful follow up over many years in Seveso found no acute increase in the incidence of acute infections, but there was an increase in carcinomas of the gastrointestinal tract and lymphatic systems and an increase in the death rate from respiratory disease. Dioxins are believed to be carcinogenic, but cancers were not a common problem on St Kilda.

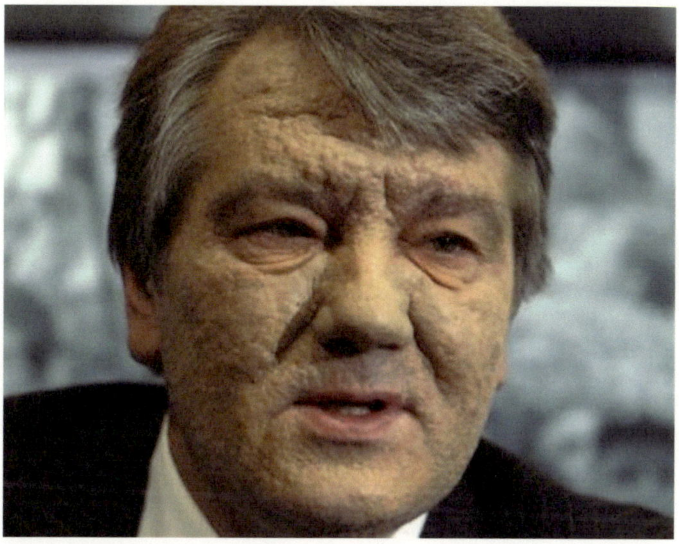

Fig. 15. Chloracne

Chloracne, an acne-like skin eruption, most marked behind the ears, on the cheeks and in the axilla and groin, affected a significant number of Seveso's inhabitants (42 of 214 children in the most contaminated Zone A.) Viktor Yushchenko, past President of the Ukraine, is considered to suffer from this condition, following an acute illness in 2004 in which his face became disfigured, scarred and pockmarked. Levels of dioxin in his blood were reported to be 6,000 times above the safe minimum, but the veracity of these tests are debated by toxicologists, and the possibility of deliberate poisoning is debated in political circles. Chloracne has become the '*sine qua non*' of dioxin poisoning.

Another dioxin leak, probably with TCDD, occurred Germany in 1953. Again chloracne occurred, and acute respiratory tract infections were more common only in the group with severe chloracne. Photographs on the St Kilda population from the late 19th century show no evidence of chloracne in the women and children. Although the men have heavy beards, there is no visible evidence on photographs of chloracne around the neck, nor reports suggesting this dermatological disease. Dioxin toxicity on St Kilda appears a most improbably contributory factor to the infections suffered by the islanders

12. Lord Howe Island – A comparison

The fortunate traveller who has visited both Lord Howe Island in the Pacific Ocean and the island archipelago of St Kilda in the Outer Hebrides will be struck by the many common features of these two remote islands; yet today one is a thriving society and the other was evacuated as a non-sustainable society in 1930. [12]
Lord Howe Island is situated in the Tasman Sea between Australia and New Zealand, 600 kms from the Australian west coast. Lord Howe Island has no detectable trace of human habitation prior to 1788 in spite of the extensive exploratory maritime voyages of the Polynesian people. Its first known sighting was by the HMS Supply captain Lieutenant Henry Lidgbird Ball and his crew on 17 February 1788 while sailing to Norfolk Island. They subsequently landed there on 13 March 1788 during the return journey to Sydney and named the island after Richard Howe, the First Lord of the Admiralty.

Fig. 16. Map of LHI

Fig. 17. Picture of LHI

Both islands are remote even today. Although the distance from Glasgow to St Kilda (main island, Hirte) is only some 340 km as the crow flies, it can take three or four days to get there using sea, land and air transport, including a landing on the 'airstrip' at Barra -the beach at low tide. The journey from Brisbane to Lord Howe Island, a distance of 740 km, in the past took several days by sea, but now a return flight is possible in one day. Both islands are small and have sheer cliffs and high rainfall. Both have UNESCO World Heritage status in which ornithological significance plays a large part, an abundant supply of fish in the surrounding sea and a nearby sea stack renowned for unique bird life. The difference in latitude and ambient temperature were significant factors in the success of one society and the failure of the other.

The earliest settlers on Lord Howe Island had widespread genetic origins, coming from England, Portugal, America, South Africa, Micronesia, New Zealand and Australia, probably with a much greater biodiversity of histo-compatibility antigens creating more resistance to infections than that of the Hebridean people of St Kilda. One of the early settlers, Nathan Chase Thompson, from Somerset, Massachusetts, in the USA, arrived in 1853 with two business partners, George Campbell and Jack Brian, and two women and a girl from the Gilbert Islands (now Kiribati). Thompson initially married Boranga, one of the women, but their only child died aged 11 years and Boranga died soon after. Thompson subsequently married the Gilbertese girl, Bokue, who was by then aged 24. They had five children, two boys and three girls, whose descendants are an important part of today's island population.

Supporting evidence comes from an epidemic of measles, a disease with a mortality of up to 25% in the developing world. In 1868, some inhabitants of the Pitcairn Islands visited Lord Howe Island in the schooner *Pacific* while suffering an outbreak of measles. They landed and recovered on the island, causing an inevitable outbreak of the disease among the islanders. No more details are available, but no deaths in 1868 are to be discovered in the island records or graveyards, implying that poor herd immunity allowed the outbreak of measles but the genetic and environmental background resulted in uneventful recovery.

In contrast to St Kilda, Lord Howe Island appears a paradise of good health and longevity. Regular visits by whalers in the early years of settlement would have helped to reduce isolation and perhaps improve the herd immunity. William Clarson, a visiting teacher, wrote in 1882 that 'sickness is almost unknown'.

A visit to the four island graveyards shows that most inhabitants born more than 100 years ago survived into their 80s, and today Lord Howe Island has 347 permanent residents, with a thriving tourism business.

Medical factors played a major role in the success of Lord Howe Island and the failure of St Kilda. Appendicitis became a manageable problem on Lord Howe Island, with recorded surgery on kitchen tables, but was a final straw leading to the evacuation of St Kilda, when two weeks passed before Mary Gillies with acute appendicitis could be notified and transported to a Glasgow hospital where she died within twenty-four hours. The climate and virgin soil allowed the Pacific islanders a much more beneficial varied diet with food rich in vitamin C. This fact plus the biodiversity of the early settlers and their improved obstetric care protected the young children of Lord Howe Island from infections and guaranteed the survival of the Pacific island society.

13. Conclusion

The inhabitants of St Kilda were an inbred population with limited genetic diversity. They suffered severely from infectious diseases, more than similar inhabitants of nearby islands, who were also exposed to a climate noted for high rainfall, strong winds and cold winters, malnutrition with limited vitamin C intake, dioxin exposure, low herd immunity and lack of health care personnel. These factors are all in stark contrast to the successful society on Lord Howe Island. It is clearly not possible to tease out attributable percentages to each factor, but lack of biodiversity as in North America probably contributed to morbidity and mortality, and ultimately the failure of the society on St Kilda necessitating evacuation.

14. References

[1] Maclean C. St Kilda: island on the edge of the world. Edinburgh: Canongate; 1972.
[2] Steel T. The life and death of St Kilda. London: Fontana; 1975
[3] Harman M. An isle called Hirte. Skye: Maclean Press; 1997.
[4] Robson M. St Kilda Church, visitors and 'natives'. Lewis: Islands Book Trust; 2005.
[5] Macaulay K. The history of St. Kilda. London: Beckett and de Hondt; 1764.
[6] Sykes B Adam's Curse Bantam Press London 2003
[7] Stride P. The St Kilda epidemic of 1727, smallpox or chickenpox? *J R Coll Physicians Edinb* 2009; 39:276–9.
[8] Martin M. Description of the Western Islands of Scotland circa 1695 and a Voyage to St Kilda. Edinburgh: Birlinn; 1999. Available from: http://www.undiscoveredscotland.co.uk/usebooks/martinwesternislands/index.html
[9] Stride P Dioxin, diet and disease on St Kilda *J R Coll Physicians Edinb* 2009; 39:370–4
[10] Stride P. St Kilda, the neonatal tetanus tragedy of the nineteenth century and some twenty-first century answers. *J R Coll Physicians Edinb* 2008; 38:70–7.
[11] Stride P. The St Kilda boat cough under the microscope. *J R Coll Physicians Edinb* 2008; 38:250–60.
[12] Stride P Survival of the fittest: a comparison of medicine and health on Lord Howe Island and St Kilda *J R Coll Physicians Edinb* 2010; 40:368–73

11

Native Tree Species Regeneration and Diversity in the Mountain Cloud Forests of East Africa

Loice M.A. Omoro and Olavi Luukkanen
University of Helsinki
Finland

1. Introduction

Biodiversity is defined by the Convention on Biological Diversity [1] as "the variability among living organisms from all sources including *inter alia,* terrestrial, marine and other aquatic ecosystems, and the ecological complexes of which they are part; this includes diversity within species, between species and of ecosystems". Thus in this definition diversity can also be described by listing of species even though such lists may not show the quality of diversity being described. For plants, assessments of biodiversities are complex because of the magnitude of numbers of species compositions in a setting. In the tropics alone, it has been reported that there are more than 200,000 species of flowering plants, which also include many tree species[2]. Furthermore, variations in species compositions between different geographical areas make biodiversity assessment even more complicated. In tropical African forests for example it has been observed that in an area of about 10^6 km^2, there can be between 30,000 and 120,000 of flowering plant species alone; and in smaller areas such as plots of about 0.1 and 1 hectares this variation has been found to be between 30 and 300 of tree species[3]. Nevertheless, it is acknowledged that patterns and processes of diversity analyses have improved as per FAO forest resource assessments [4,5]. With regard to habitats, it is noted that there is a lack of detailed survey information on habitat types, species or genetic diversity in many forests in the tropics[6]. In addition, the complexity in assessments is exacerbated in the tropics because of the introductions of exotic plant species such as was the case in most of the East African highlands during the last century[7].

The interest in forest biodiversity has increased lately due to the many threats forest ecosystems face in many regions of the world. These threats not only affect the ecosystem co-benefits such as biodiversity; and cultural and aesthetic values but also the service provisions and regulatory functions of the forests. In Kenya, there are nineteen habitats and ecosystems which exclude agricultural and barren lands that are valued for their high biodiversity. One such is the highland moist forests ecosystem which occupies 2% of the land area of the country[8] and occurs in altitudinal ranges of 1400-2000m. Similarly at the global level, the mountain cloud forests have received considerable attention because of their ecological values and threatened stability[9]. The Taita Hills forest ecosystem occurs within the highland moist ecosystems. These forests belong to the only existing mountain cloud forests in East Africa and form a part of the Eastern Arc Mountains which stretch from Tanzania to south-west Kenya. These forests form the northern most parts of these Mountains. The Taita Hills rise from 1400m in the southeastern slopes to above 1700 m altitude in the northwestern slopes[10], and as it is typical with most cloud mountain forests; these forests are the home to a variety of native tree and animal species. The forests host some endemic plant species such as the wild coffee, *Coffea fadenii* Bridson. In all, the plants they host are categorized as follows: 40% of the plant species, 2% of the genera and 13 taxa that are endemic to Taita Hills forests, and additional 22 plants that are endemic only to Kenya and Tanzania[11,12,13]. Due to these diversities, the forests are ranked among the 34 biodiversity hotspots of the world with respect to high ratios of endemic plants and vertebrate species per area[14,15]. Presently, there are only four out of twelve fragments of that are of appreciable coverage in area remaining in these hills and these are: Ngangao (120 ha), Chawia (86 ha), Mbololo (185 ha) and Irizi (47 ha), the largest of the three were investigated (Fig.1).

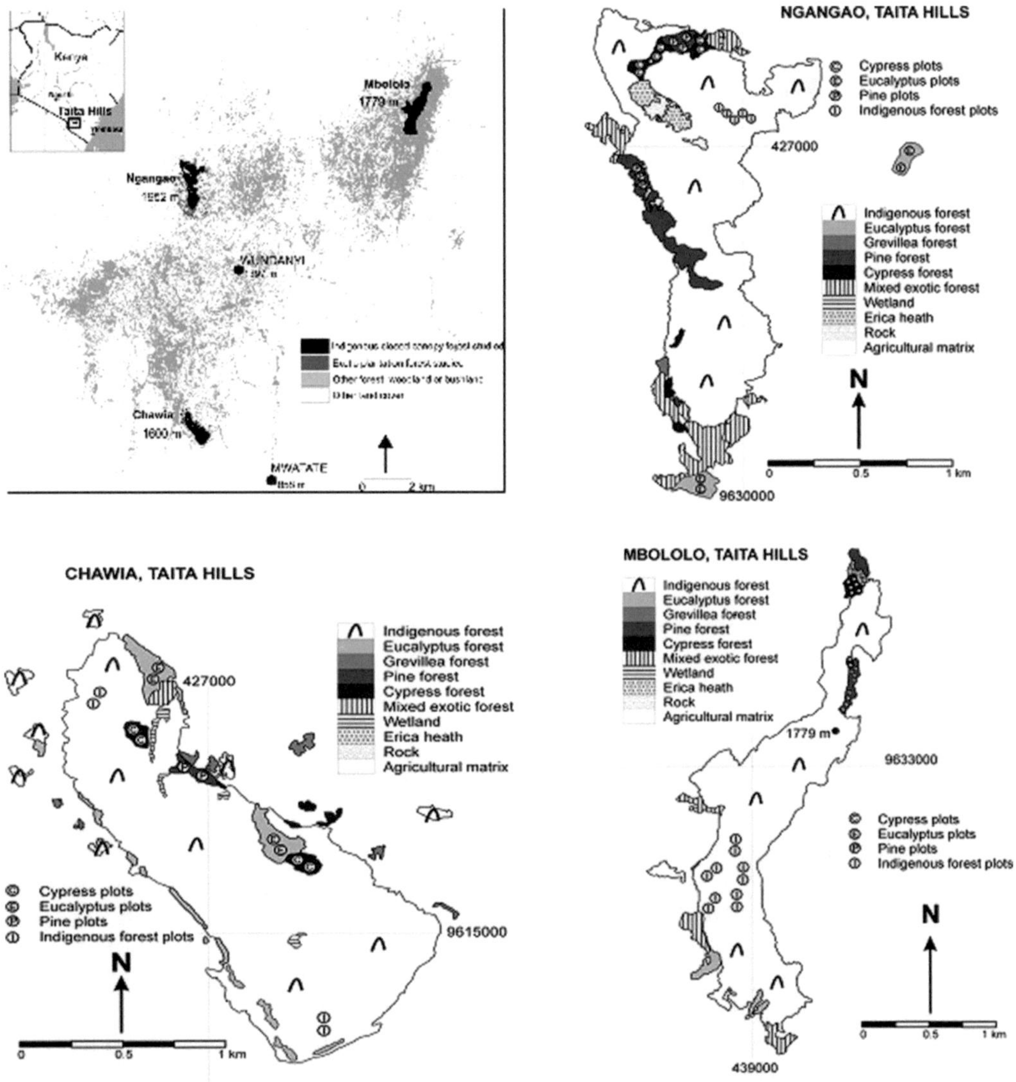

Fig. 1. The location of Taita Hills in Kenya and the three forests studied.
(Source; Omoro et al., 2010)

In this chapter, the observation from research work conducted in three out of the 12 forest fragments considered larger in sizes are discussed. The study sought to assess the diversity of native tree species regenerated within the native and exotic forest plantations. The study assessed this biodiversity by evaluating the species diversity, richness and similarity of the regenerated species between the exotic plantations of pine, eucalyptus and cypress; as well as between native and exotic forests. Other factors such as levels of disturbance were established as not only affecting the ecological functions of these forests but also the loss in some of the species vital for use for medicinal purposes. The perceptions of the communities on how the species composition has affected the ecological functions of the forests was reported as effect on water yield into the streams. Finally, potential management activities that can be instituted to restore the biodiversity are suggested.

1.1 Causes for biodiversity loss

Globally, biodiversity hotspots previously covered an area of 15.7% of the global land surface[14] but currently cover only 2.3% [16] and they are still experiencing different forms of disturbances which constantly threaten their existences. In most cases, these disturbances are due to changes in land use, overexploitation of resources and introductions of exotic species. In addition, ineffective implementation of policies by national institutions exacerbates the losses of biodiversity in many countries, because less attention is given to ensuring that regulations that are in place for the management of these forests are implemented. The Taita Hills forests also faced similar threats until, the 1990s when their conservation was taken seriously. As a result, the disturbances they suffered not only impacted on the biodiversity loss but also on other functions and services of the forests. Farmers in Chawia, for instance, during a Participatory Rural Appraisal exercise by Lekasi[16] recalled that water availability and soil fertility had been reduced from the 1920s to 2005 due to what they perceived as the destruction of forests and the introduction of unsuitable exotic tree species; and that they are presently conserving the forest to improve water yield through reforestation with native species[17].

The sources of disturbances to these forests were anthropogenic and caused mainly by the demographic dynamics. In particular, population growth without corresponding increase in farm holding sizes prompted people to encroach into the forests. As a result, forests were cleared for agricultural production. In addition, other activities such as: firewood collection, charcoal production and grazing[18,19] occurred, which rapidly contributed to the degradation of the forests to the extent that 90% of forest cover loss is estimated to have occurred during the last 200 years[11]. These forests are also red listed by IUCN[20] because 236 of their plants are classified as either endangered or vulnerable. Furthermore, the exotic plantations of pine (*Pinus patula.* Schiede ex Schlecht. & Cham), eucalyptus (*Eucalyptus saligna* R. Baker) and cypress (*Cupressus lusitanica* Miller) that were established between the 1960s and 1980s as forest stands within the native forests and as individual trees for enrichment planting[22] have further led to changes in species composition. These were part of management plans to provide softwood production, protection of the native forests and to mitigate against soil erosion. All these activities inadvertently exacerbated the threats to the biodiversity in these forests[23]. Another reason for the loss in forest cover was the ineffective implementation of the forest policies in place. For instance, despite a national policy in 1977 which banned the logging of the native forests without licence[19], the people continued to encroach and log from these forests; as a result of unsustainable use of forest resources akin to property of the commons[24] and without a well defined forest management strategy, the forests lost much of their ecosystem functions. A study by Himberg[17] shows how the communities living around these forests decry the losses of the services such as reliable water flow and some of the medicinal plants.

1.2 Forest disturbance and its consequences on biodiversity

Several forest degradation types have been documented, at the global level[25,26,27] as well as at the local level[28], and most severe one consists of a total loss of forest cover due to disturbances. The consequences of this loss not only effects the biodiversity but also ecosystem functions[28], including pest control and pollination in agricultural crops[29], seed dispersal [30], and the regulation of water resources[29,31]. While in many instances losses in forest cover have negative impacts, in other circumstances, positive impacts have been noted in which there have been increases in plant species diversity. Such improvement in species diversity has been attributed to resilience which allow the certain plants to regenerate profusely after disturbance or to changed forest conditions such as increased light to forest floor [32] which favour the establishment of pioneer and early successional species whose seeds may have been stored in the soil seed bank. In the Taita Hills, the consequences of the disturbances documented have shown marked changes in tree species compositions[25,33.] The most disturbed forest fragment had a higher degree of changes in tree species composition in which 58 different tree species regenerated with stem densities varying between 10 and 2000 trees per hectare[33]. Differences in species composition occurred between the forest stand types, with native forests showing higher species diversities than the exotic plantations. The loss of biodiversity had also affected the animal species composition[34].

2. Biodiversity assessment

2.1 Sampling of vegetation

Generally for biodiversity assessments vegetation is sampled in designated plots whose designs vary. In the Eastern Arc Mountains, a Y design developed by USDA[35] (Fig.2) and modified for Forest Health Monitoring (FHM) in the Eastern Arc Mountains of Kenya and Tanzania[36,37] have been used. Once the study plots are identified, sampling for vegetation entails enumerating the individual tree found in each plot by species. In order to assess the regeneration dynamics, it is important that during sampling, seedlings and saplings are enumerates as well. In this study, all live juvenile trees of > 5cm in diameter at breast height (dbh) were identified and recorded from each of the two randomly selected subplots. The seedlings were categorised based on height up to 1.3 m in height while sapling were categorized to be higher than 1.3m and with diameters less than 5cm dbh

2.2 Diversity calculations

There are several methods for analyzing plant diversity and its elements[38,39], and these methods provide indices which provide bases for comparison. Lou[40] has gone a step further by deriving Effective Numbers from these indices. These are whole numbers which make comparisons between different diversities easier since the numbers show the magnitude of the differences which makes it easy to perceive. Once the data collected for the diversity analysis is sorted, the diversities can be derived based on several formulae depending on which index is selected. An example is shown for Shannon-Weaver index that has been used in this analysis as Box 1. For practical use however, a Biodiversity Calculator software[39]

and can be used for the calculations. This calculator is freely available online (http://www.columbia.edu/itc/cerc/dunoffburg/MBD/LIMK.html). In this software, values for species derived from the data are fed in the calculator which then provides a range of diversity indices (e.g. Shannon-Weaver, Simpson and Broken stick) which one can select to use. In many cases such as in the Eastern Arc Mountain forest, the Shannon-Weaver index (H') has been commonly used for these calculations[25,33]. This is because it provides an account of both the abundance and evenness[38] and does not disproportionately favour any species as it counts all the species according to their frequency[39,40] unlike indices such as Simpson's which disproportionately highlights common species (in terms of abundance) instead of showing their frequencies (richness) in samples. One documented disadvantage of the Shannon-Weaver index is however, that it requires a large sample size in order to minimize biases[41]. Shannon-Weaver's indices ranges are typically from 1.5 to 3.5 and rarely reach 4.5[42]. Other parameters such as species richness (S) and species evenness (H'E) are also generated from the same calculator.

Box 1. Shannon-Weaver biodiversity Index

Shannon's index, (H') is defined by

$$H' = -\sum_{i=1}^{S} p_i \ln p_i \qquad (1)$$

Where i, is the proportion of the species relative to the total number of species (p_i) multiplied by the natural logarithm of this proportion ($\ln p_i$) and the final product multiplied by -1.

Species richness, the number of species present in an ecosystem (S) was defined by

$$S = \sum n \qquad (2)$$

Where n is number of species in a community.

Species evenness, the proportion of individuals among species in an ecosystem is often assessed by Shannon's equitability index (H'E) which is calculated by

$$H'E = H' / H_{max} \qquad (3)$$

Where H_{max} is defined as ln S. H'E values range from 0 to 1 and 1 indicates complete evenness.

The Shannon-Weaver's indices obtained in the study were converted to *effective numbers* using a method by Lou[40]; this can be done to obtain values that can be used to compare the differences in species diversities. The *effective numbers* are calculated as an exponential of the Shannon's index as:

$$N_{Effect}\ of\ species\ (pi) = exp(-\sum_{i=1}^{S} p_i \ln p_i) \qquad (4)$$

To obtain similarities in species composition between forest fragments and forest stands, the Jaccard's index[43] was used. Jaccard's index (Cj) is defined by

$$Cj = a/a+b+c \quad (5)$$

where a, is the number of species present in both forest types or locations compared, b is the number of species in only one forest type or location; and c is the number of species present in the other forest type or location.

When establishing diversities and studying the regeneration dynamics, densities of native tree species in the study sites and forest stands (e.g., in exotic and native forests) have to be calculated on per area basis as well. In order to show the differences both in diversity and regeneration levels, further statistical analyses have to be done for the indices and densities derived. There are several statistical methods (SAS, STATA, SPSS) which can be used to assess such data sets. In the case of Taita hills, a one way ANOVA was used and SPSS 15 for windows software method adopted for the statistical analyses. The means were separated by applying Tukey's test to test the differences in the diversities and densities between the forest sites and types.

3. Regeneration and species diversities

3.1 General structure of exotic and native forests

A description of the existing trees, seedlings and saplings in all the three exotic and native forests studied in the Mountain cloud forests of Taita Hills is shown in Table 1. Significantly higher tree stem densities were found in the native forests than in the exotic forests. Similarly, the native forests had higher numbers of seedlings and saplings although not significantly different from those in Eucalyptus forests. The numbers of seedlings were not different between the pine and eucalyptus forests.

Forest type	Stand density (tree/ha)	Seedling species (#/ha)	Seedling density (#/ha)	Sapling species (#/ha)	Species richness (S)	Species diversity (H')	Species Evenness (H'E)	Abund-ance (N)
Cypress	765^{b}	1.2^{b}	412^{b}	1.5^{b}	0.85^{b}	0.56^{b}	4.17^{b}	3852
Eucalyptus	897^{b}	2.2^{ab}	962^{ab}	1.7^{b}	0.99^{b}	0.64^{ab}	4.00^{b}	3911
Pine	829^{b}	1.7^{a}	312^{b}	2.8^{ab}	1.14^{b}	0.75^{abc}	5.75^{b}	1667
Native	1016^{a}	6.9^{a}	2575^{a}	4.3^{a}	1.82^{a}	0.78^{a}	11.42^{a}	11015
p-value	≤0.05	≤0.05	≤0.05	≤0.05	≤0.05	0.031	≤0.05	NS

The p-value shows significance levels of a one way ANOVA test for differences between forest types. Values followed by the different letter superscripts are significantly different. Seedlings refer to young trees of < 1.3m height; Saplings ≥1.3.m high and ≤ 5cm dbh.

Table 1. Characteristics of the native tree species regenerated among the four forest types in the Taita Hills

3.2 A typical regeneration and diversity of native species within the exotic and native forests in the three forest fragments in the Taita Hills

In general, the diversity and densities of regenerated native species were higher in the native forests than in the exotic plots forests. The typical native species in cypress plantations were *Macaranga conglomerata, Rapanea melanophloesos, Rytigynia uhligii Tabernaemontana stapfiana* and *Syszygium guineese;* in the eucalyptus plantations were:

Species	*Cupressus lusitanica*			*Eucalyptus saligna*			*Pinus patula*			Native Stands		
	Nga	**Cha**	**Mbo**	**Nga**	**Cha**	**Mbo**	**Nga**	**Cha**	**Mbo**	**Nga**	**Cha**	**Mbo**
**Acacia mearnsii* De wild	1304	15	7	1529		309	29					
Albizia gummifera J.F.Gmel		44	81	0	74		69	88			132	
**Cupressus lusitanica* Miller	637	956	684	574		44						
**Eucalyptus saligna* R. Baker			7	2000	706	1103						
Macarnaga conglomerata Brenan	10	103	15				147	29	15	153		47
Maesa lanceolata Engl.	10		7		29					24		
Newtonia buchananii G.C.C.Gilbert & Boutique		29						294		35		512
Oxyanthus speciosus DC				118			29			82		6
Polyscias fulva Harms	10	29			15					24		18
Phoenix reclinata Jacq		44		15					15		221	
**Phoenix patula* Schiede ex Schlecht.& Cham						29	843	235	485			
Pleiocarpa pycnatha K Schum					29			147	15		44	676
Podocarpus latifolius R.Br. ex Mirb	39	29							15	12		47
Rapanea melanophloeos L. Mez	88	74					10				59	24
Rytigynia uhligii K.schum.& K.Krause	78	235					353	265				82
T.abernamontana stapfiana Britten		309			338			1088		118	176	76
Vangueria volkensii K:Schum	10	15					137	59		24		
Xymalos monospora (Harv.)Baill.		44			29						29	41
Sysygium guineense Willd	108	118			103			88	74	71	15	671

*Denotes exotic species and the woody species include tree, saplings and seedlings regenerated combined

Table 2. Densities all native woody (#/ha) species occurring in at least three plots at each of the three forests fragments of Ngangao (Nga), Chawia (Cha) and Mbololo (Mbo).

Albizia gummifera, Oxyanthus speciosus, T. stapfiana and *S. guineese;* and the pine plantations were A. *gummifera, M. conglomerate, Newtonia buchananii, Pleiocarpa pycnatha, R. uhligii T. stapfiana, Vangueria volkensii* and *S. guineese.* The exotic plantations of Chawia hosted the highest number of native species among all the exotic forests, while cypress and pine forests had more native species than eucalyptus forest.

3.3 Variation in species regenerations and diversity among forest types

The species found naturally regenerated and their diversities are shown in Table 3. With regard to species diversity (H'), richness (S), Evenness (H'E) and abundance (N) of the native species regenerated in the different forests, the native forest had higher average totals than the exotic forests. However, some of the values in native forests were not significantly different from those found in some of the exotic plantations (e.g. H' in native, pine or eucalyptus forests).

Forest Area	Forest type	S	N	H'	H'E	EF
Ngangao	Cypress	4 (2)	2088 (2108)	0.88 (0.22)	0.64 (0.07)	2
	Eucalyptus	3 (1)	9044 (7637)	0.82 (0.89)	0.80(0.099)	2
	Pine	5 (3)	1588 (733)	0.89 (0.26)	0.51 (0.136)	2
	Native	9 (2)	1365 (147)	1.97 (0.16)	0.89 (0.02)	7
Chawia	Cypress	7 (3)	1485 (457)	1.53(0.26)	0.78 (0.216)	5
	Eucalyptus	6 (1)	1176 (825)	1.31 (0.08)	0.78 (0.071)	4
	Pine	11 (1)	2794 (1123)	1.73 (0.15)	0.72 (0.032)	6
	Native	5 (3)	794 (573)	1.07 (0.36)	0.63 (0.056)	3
Mbololo	Cypress	3 (1)	6358 (15609)	0.48 (0.13)	0.39 (0.062)	2
	Eucalyptus	4 (1)	1515 (391)	0.83(0.12)	0.68 (0.041)	2
	Pine	4 (1)	1221 (194)	1.22 (0.16)	0.81 (0.098)	3
	Native	15 (3)	19928 (39748)	2.05 (0.13)	0.78 (0.035)	8

Standard error is shown in the parenthesis for Shannon's index (H') and Shannon's evenness index (H'E); while for ths individual species (N) and species richness (S), the value in parenthesis denotes standard deviation.

Table 3. Number of individual species (N), species richness (S), Shannon's index (H'), Shannon's evenness index (H'E) and effective numbers (EF) observed in four forest types in three forest fragments of the Taita Hills (n=65 plots)

The native forests had a higher species diversity than the exotic plantations. Other studies in East Africa[7] have shown the same trend. The study under discussion however, showed that, the highly disturbed forest site of Chawia had higher tree species diversity in the exotic forests than in the native forests. The reason for this is probably attributed to the fact that native forest land had been cleared for the establishment of the exotic forests instead of bare land. Moreover, some of these exotic plantations were established around the native forests as buffers. Therefore, the high levels of anthropogenic activities at this site[45,46] and a possible presence of soil seed bank[47] associated with the initial clearance of the native forests may explain the high levels of regenerated native species. These results also compare well with those established by Yirdaw and Luukkanen[47] in eucalyptus plantations surrounding a native forest in Ethiopia; in their study, higher species diversity was found close to the native forests as compared to stands that were planted further away.

An analysis of variance showed a significantly higher species richness in the native forests than the exotic plantations (p=0.000). A comparison between the three forest fragments, showed a significant difference in species richness only between the least and intermediately disturbed sites of Mbololo and Ngangao respectively (p=0.002). In general no statistical differences were detected in species abundance (N) between the native and exotic forests types; and between the three fragments. Nevertheless, the native forests in the least disturbed site (Mbololo) had a higher abundance (N) and number of species (S) than the exotic forests at the same fragment. The regenerated species showed higher abundance at this fragment than in either the intermediately disturbed fragment (Ngangao) or the highly disturbed one (Chawia). Among the exotic forests, pine plantations in Chawia had higher species richness than the rest of the exotic forests. A comparison within each fragment showed that in Ngangao the regeneration of trees within the native forest was higher than within the exotic forests (Shannon-Weaver's index 7 and Effective number 7). In contrast, Shannon-Weaver's indices and Effective numbers in the three forest fragments were almost similar. In Chawia, the Shannon-Weaver's indices were higher for the exotic forests than for the native forest. The highest one was that for the pine forest; with a Shannon-Weaver's index of 1.73 and Effective number 6, followed by cypress and eucalyptus forests respectively. Similarly, the pine forest had the highest number of species as well as total density of species regenerated. In Mbololo, the Shannon-Weaver's indices for the regenerated native species were rather similar for all forest types except in the native forest which indicated somewhat higher values. Comparatively, the native species that regenerated in the different forest types in both Ngangao and Mbololo appeared similar again with the exception of those in the native forests where the values were higher. As mentioned above, in Chawia, the regeneration was more pronounced in the exotic forests than in the native forest.

The Shannon-Weaver's diversity indices (H') were higher in the native forests than in the exotic plantations in Ngangao and Mbololo whereas in Chawia this index was lowest in the native forests. A one-way analysis of variance using Tukey's test indicated differences in diversity indices between the native forests and all types of exotic forests. In particular, these were highly significant between the native forests and the cypress or eucalyptus forests (p=000) and between the native forests and pine forests (p=0.001). No significant differences were found between exotic forests. All the exotic forests in Ngangao had almost similar diversities with very slight differences in the regeneration pattern, possibly because the exotic plantations here were established in barren areas[22] which provided fewer opportunities for the regeneration of native species. The differences in regenerations patterns observed between the exotics forests in Mbololo on the other hand and the more at the Chawia and Ngangao fragments were most likely due to a lower level of disturbance Mbololo fragment experienced. This observation is corroborated by the high Shannon-Weaver's indices and effective numbers in the native forests and the presence of few secondary species that are indicators of disturbance (e.g. *P. reclinata M. lanceolata* and *T. stapfiana*). Ngangao, on the other hand, had more species associated with disturbance such as *M. conglomerata, R. uhlighii* and *M. lanceolata* also confirming the earlier observations by Bytebier[48].

The application of effective numbers facilitated distinction of differences in diversity among the exotic forest types. The effective numbers were also consistent with the Shannon-Weaver's indices. The effective numbers were higher in exotic forests in Chawia than in native forests and in all the exotic forests both in Ngangao and Mbololo. The effective numbers were also high in the native forests at Mbololo and Ngangao fragments. The Shannon-Weaver's evenness index (H'E) reached its highest value in the native forests of Ngangao site (0.89) and in pine plantation at Mbololo fragment (0.81). The most uneven forest in terms of diversity was the cypress plantation at Mbololo fragment (0.39). A one way ANOVA for Shannon-Weaver's evenness showed that there was a significant difference between the exotic forests and the native forests in general and also a significant difference in the species evenness between the cypress forest and the native forest (0.031). With respect to the abundance of regenerated individuals at each fragment, the native forest plots in Chawia had the lowest number (794) compared to the other native plots of Ngangao (1365) or Mbololo (19928).

3.4 Similarities of species regenerated at the three sites

Table 4 shows the similarities in species composition between the different forest types and fragments. The highest similarity (77 %) was found between the cypress and pine forests in Chawia; the pine and cypress forests in Ngangao also showed a high species similarity (64%), which almost corresponded with the species diversity values found (11 and 13 species respectively). In Mbololo, the highest species similarity was 59% between the exotic forests of pine and cypress. The cypress forests in Mbololo were twice as diverse with 13 species as the eucalyptus forest and yet they shared a similarity of 50% species, the same as for these exotic forests in Chawia. The majority of the forests studied however, shared less than 30% of the species, while the forests which did not share any species were the eucalyptus forests in Mbololo and the native forests in Ngangao, as well as the eucalyptus and native forests in Mbololo.

		Ngangao				Chawia				Mbololo			
		C	E	P	N	C	E	P	N	C	E	P	N
Nga	C		40	**64**	33	40	32	**52**	18	42	22	24	26
	E			22	7	11	11	25	0	35	36	11	6
	P				26	29	21	38	24	44	29	**50**	29
	N					31	36	43	17	16	0	26	53
Cha	C						**50**	**77**	40	30	29	29	29
	E							**54**	48	37	10	29	29
	P								35	32	21	38	35
	N									25	11	32	21
Mbo	C										**50**	**59**	15
	E											19	0
	P												39
	N												

Table 4. Similarities(%) of species regenerated in four forest types (C=Cypress, E=Eucalyptus, P-Pine and N= Native) from the three forest fragments. Values in bold similarities from 50%

Highest species similarities in the undergrowth of pine and cypress forests in Chawia and Ngangao observed were possibly because these forests were located close to each other. Thus, there were either similar seed dispersal mechanisms or the forests could have had similar soil seed banks. The complete absence of similar species in eucalyptus and native forests in Mbololo and Ngangao implies that the eucalyptus plantations cannot support similar regeneration of species as the native forests, especially if the stem density is high as indeed was the case in these eucalyptus forests. In general, the native forests in Mbololo shared a low species similarity with other forests. A possible explanation is the low level of disturbances in this forest which provided few opportunities for seedling recruitment except through dispersal and gap dynamics. *Ocotea usambarensis* which is extinct in some regions in East Africa[49] and under threat in Tanzania[50], was observed in the native forests of Mbololo with 94 stems, although it is also known to be present in Ngangao. *Coffea fadenii, a* wild coffee species, was only found in Ngangao, indicating the relatively low levels of disturbances experienced by these two sites. In Mbololo, the pine forest had a higher effective number and Shannon-Weaver's index than the eucalyptus and cypress forests plots possibly because the pine forest was located in the middle of the native forest, while cypress and eucalyptus had been planted at the edges.

4. Strategies for diversity restoration

4.1 Stakeholder involvement

Since the advent of the biological diversity convention of 1992, many countries have, as part of their global commitment to sustainable development, paid great attention to ecosystem conservation. Moreover, the realization that some vital biological resources are on the brink of extinction and yet they are vital for social and economic development reinforces the urgency for conservation. Similarly, due to the apparent loss of biodiversity and ecological functions in the forests of the Taita Hills and in many other forests in Kenya, a newly promulgated Forest Act[51] provides several options for managing the forests, including those under threat such as the Taita Hills forests. These options include opportunities for stakeholders, particularly the local communities residing around these forests to participate in their management. As a result, Community Forest Associations have been formed for the different forest fragments in Taita Hills in preparation for their participatory management. Community involvement in management has already entailed participation in forest reforestation activities such as replanting with seedlings of native tree species in order to enhance the replacement of the exotic species. In areas outside the forests and on farms tree planting is being carried out so as to ease pressure off the forest. Scientific approach has been applied to facilitate the restoration by identifying suitable sites for planting. In particular, a GIS-based, least-cost modelling technique has been used to identify such sites[52] and after integrating biological and socio-economic data within the forest corridors[46], a set of exotic plantations with highest priority for restoration activities have been identified by both the government and Non Governmental organizations that are active in the area. The choice of species planted is based on their potential to increase landscape connectivity or on their importance for conservation of critically endangered taxa, although with regard to soil characteristics of the forest fragments, any of the native species is suitable and can be used for restoration exercise[53].

Box 2. Community activities for restoration of forest biodiversity between two fragments

Ngangao and Chawia forest fragments are being linked through a three-step reforestation plan. This includes forest enrichment, agricultural matrix enrichment through on-farm tree planting and conversion of exotic plantations to native forest. As native forest enrichment is being done, it is accompanied by gradual removal of exotic species from the canopy level to increase the light availability for the planted native tree seedlings and to allow faster recruitment of seedlings from the soil seed bank. Local initiatives to enhance these restorations activities include the establishment of tree nurseries to supply seedlings necessary for planting with native tree species (e.g. *Prunus africana.*) and farmer-friendly exotic trees (e.g. *Grevillea robusta*).

4.2 Forest restoration and management for biodiversity enhancement

Forest restoration can be enhanced by the presence of appropriate conditions, some of which include the following: placement of the plantations, edge effect, presence of gaps, seed dispersal mechanisms with mammals as dispersing agents and an existing soil seed bank. In most cases these factors work in tandem. In the study discussed here, fragments which showed some degree of disturbance (Chawia and Ngangao) had stronger edge effects since they were located in the middle of agricultural lands due to their fragmentation, this possibly provided opportunities for increased movements of propagules[54] by fauna from adjacent forest patches especially because high number of rodents and shrews have been reported particularly for the Chawia site[55]. Some of the exotic plantations (e.g. Ngangao) were established on denuded land which eliminated the possibility of the presence of soil seed bank and hence the relative differences in biodiversity observed.

The presence of gaps associated with disturbance is also important in forest ecosystem restoration. The differences in gaps observed at the different study sites showed that at the most disturbed fragement of Chawia, there were growths of secondary native species such as *T. stapfiana, M. lanceolat*a, and *P. reclinata,* which seemed to indicate that disturbance favoured their regeneration. Some studies have shown that forest disturbance does stimulate regeneration of species stipulated to be for intermediate succession stages [54,56].

Disturbances as observed in the forests, could have rendered them to be in early or represent a post-extraction and post-abandonment secondary stage[45], even though some species associated with low disturbance such as *X. monospora, S. guinees* and *R. melanophloeos*[57] were found in the most disturbed site of Chawia. This was possibly an indication of the presence of either seeds in the soil banks or efficient seed dispersals mechanisms. An unexpected observation, however, was the low number of pioneer species which would not be expected with the common presence of gaps in the forest canopy [58] and this was also noted by Rogers et al.[59], *for M. conglomerate* in the native forests of the more disturbed Chawia fragment. This shows that the succession stage had been passed in Chawia and that the native forest had matured and therefore, phased out the pioneer species for the more shade tolerant species[59].

The importance of the presence of soil seed bank is underscored by the observations made at the most disturbed site where, there was a higher regeneration of endemic species namely *Xymalos monospora Rapanea melanophloeos* and *Syzygium guineense* (Willd.) DC which are associated with lower levels of disturbance. This is an indication that a disturbed forest has an inherent potential to regenerate. Thus, without further disturbance the forest can restore itself. This trend has been observed in Nigeria where, a degraded forest recovered to its original status without further disturbances[60]. In the broader context, therefore, disturbance can be considered a key element of landscape diversity, and may be viewed as beneficial to properly functioning systems[25,61]. This would only occur if deliberate efforts are put in place to ensure that no further disturbances occur such as isolating the sites for restoration and eliminating anthropogenic activities which can create the undesired disturbances.

4.3 Species selection

In the study, there were few native species that regenerated in the exotic forest of the relatively less disturbed fragments of Mbololo and Ngangao. Two possible reasons were the stand densities and the inherent physiological characteristics of some of the exotic species. Stand densities in eucalyptus plantations in Ngangao and Mbololo were 2000 and 1103 stems per hacatare, respectively, while in Chawia it was only 706. The densities for pine at the former fragments were 843 and 485 respectively, compared to 235 for Chawia (Table 1). A possible consequence is not only the lack of light for regeneration in a high density stand but also effects of adaptation to particular geochemical characteristics of a given species have been attributed to the exotic species as well. In other studies elsewhere[62] mulches from pine were found to inhibit the germination of seeds. It is thus plausible that the inherent physiological composition of some of these exotic species could not have favoured the regeneration of other species.

5. Conclusions and recommendations

The rich biodiversity of the indigenous forests of Taita Hills has been acknowledged by scientists for decades. The least disturbed native forests have a much higher diversity among the regenerating seedlings and saplings than in the most disturbed forest. In contrast the exotic forests in a highly disturbed fragment seemed to have the highest species biodiversity; an indication that in the absence of further disturbance and by avoiding replanting with exotic species, the forests may regain their diverse native status. Rehabilitation activities to restore the biodiversity of these forests have been initiated with the involvement of the local communities in which there are assisted regeneration efforts.

The high species diversity and high abundance of native tree seedlings and saplings in the exotic plantations in the Taita Hills is very encouraging in terms of conservation efforts. At the highly disturbed site of Chawia, there was a high species diversity in the exotic forests which, with respect to restoration, requires no special management except for elimination of further disturbance. Regeneration was observed of 58 woody plant species with stem densities varying between 10 and 2000 trees per hectare. Marked differences in species diversity were observed between native and exotics forest types. The native forests showed a higher species diversity. Between the forest fragments however, the most disturbed fragment showed greater diversity than either Ngangao or Mbololo, an indication that more regeneration occurred in Chawia fragament than in the less disturbed forests of Mbololo and Ngangao. The implications from the study are that the native tree species diversity in the mountain cloud forests of East Africa is affected by the level of intensity of the disturbances, and in this area, if forest disturbance is properly regulated with regard to forest type; there are possibilities for the original forest ecosystems to be restored.

6. References

[1] Convention on Biological Diversity, 1992. http://sedac.ciesin.org/entri/texts/biodiversity,1992.htm (accessed 22 April 2008)

[2] Prance, G.T., Beentje, H. Dransfield, J., Johns, R. 2000. The tropical flora remains uncollected. Ann. Missouri. Bot. Gardens. 87:67-71

[3] World Conservation Monitoring Centre, 1992. Global Diversity: Status of the Earth's living Resources. Chapman and Hall, London.

[4] FAO, 1993. Forest Resources Assessment 1990 - Tropical countries. FAO Forestry Paper No. 112. Rome.

[5] FAO. 2010. Global Forest Resource Assessment. FAO Foresrty paper 163. Rome

[6] Wilcox, B.A. 1995. Tropical forest resources and biodiversity: assessing the risks of forest loss and degradation. Unasylva 181, No. 46. pp. 43-49.

[7] Fimbel, R.A., C.A. Fimbel., 1995. The role of exotic conifer plantations in rehabilitating degraded tropical forest lands: A case study from the Kibale forest in Uganda. Forest Ecology and Management 81, 215-226.

[8] Dean P.B. and Trump, E.C. 1983. The Biotic Communities and Natural Regions of Kenya. Wildlife Planning Unit, Ministry of Tourism and Wildlife and Canadiaon International Development Agency. 34 p.

[9] Hamilton, L.S. 1995. Mountain Cloud forest conservation and Research: A synopsis. Mountain Research and Development: 15(3): 259-266.

[10] Jaetzold, R., Schmidt, H., 1983. Farm Management Handbook of Kenya, Vol. II. East Kenya. Ministry of Agriculture, Kenya.
[11] Lovett, J.C., 1993. Eastern Arc moist forest flora, In. J.C. Lovette and S.K.Wasser (eds.), Biogeography and Ecology of Eastern African. Cambridge University Press, pp 33-35.
[12] EAWLS, 2001. The Taita Biodiversity Conservation Project. http://www.easterarc.org/org/html/bio.html (accessed 8 February 2008).
[13] GEF, 2002. Project brief: conservation and management of the Eastern Arc mountain forests, Tanzania, GEF Arusha, Tanzania.
[14] Myers, N. Mitterrmeir, R.A., Mittermeir, C.G., Da Fonseca, G.A.B. and Kent, J. 2000. Biodiversity hotspots for conservation priorities. Nature 403: 853-858.
[15] Conservation International, 2005. Biodiversity hotspots. http://www.biodiversityhotspots.org/Pages/default.aspx (accessed 23 November 2008)
[16] Lekasi JK, IV Sijali, P Gicheru, L Gachimbi & MK Nyagw´ara (2005). Agricultural productivity and sustainable land management project report on participatory rapid appraisal for Wusi Sub-location in the Taita Hills catchment. SLM Technical Report No. 10. Kenya Agricultural Research Institute. Nairobi.
[17] Nina Himberg, Omoro, Loice, Pellikka Petri and Luukkanen Olavi. (2009) The benefits and constraints of participation in forest conservation. The case of Taita Hills, Kenya. *Fennia* 187:1, pp 61-76.
[18] Beentje, H.J., 1988. An ecological and floristical study of the forests of the Taita Hills, Kenya. Utafiti 1, 23-66.
[19] Collins, M., Clifton M., 1984. Threatened wildlife in the Taita Hills. Swara 7, 10-14.
[20] IUCN. 2002. The 2002 Red list of threatened species . IUCN, Cambridge and Gland.
[21] Pellikka, P.,Lötjönen,M Siljander, M & Lens, L. 2009. Airborne remote sensing of spatiotemporal change (1955-2004) in native and exotic forest cover in the Taita Hills, Kenya. International Journal of Applied Earth Observations and Geoinformation 11(4):221-232. DOI: 10.1016/j.jag.2009.02.002
[22] Rogo, L., Oguge, N., 2000. The Taita Hills Forest Remnants: a disappearing world heritage. Ambio 29, 522-523.
[23] Hardin, G. 1968. Tragedy of the Commons. Science, 162:1243-1248.
[24] FA0.2009. Forest Resource Assessment. Case Studies on measuring and assessing forest degradation. Working paper No. 16. Rome
[25] Rogers, P.C. 1996. Distrubance ecology and forest management: a review of the literature. Ogden, UT: US.Department of Agriculture, Forest Service, Intermountain Research Station; INT-GTR-336. 16
[26] Krishnaswamy, A., Hanson, A., 1999. Summary Report of the World Commission on Forest and Sustainable Development.
[27] Foley, J.A., Asner, G.P., Costa, M.H., Coe, M.T., DeFries, R., Gibbs, H.K., Howard, E.A., Olson, S., Patz, J., Ramankutty. N., Snyder, P. 2007. Amazonia revealed: forest degradation and loss of ecosystem goods and services in the Amazon Basin. Ecological Society of America, ESA online Journal 5, 25-32.
[28] Kremen, C., Williams, N.E., Aizen, M.A., Gemmill-Herren, B., LeBuhn, G., Minckley, R., Packer, L., Potts, S.G., Roulston, T., Steffan-Dewenter, I., Vázquez, D.P., Winfree, R., Adams, L., Crone, E.E., Greenleaf, S.S., Keitt, T.H., Klein, A.-M, Regetz, J., Ricketts, T.H., 2007. Pollination and other ecosystem services produced by mobile organisms: a conceptual framework for the effects of land-use change. Ecology Letters 10, 299-314.

[29] Laurance, W.F., Nascimento, H.E.M., Laurance, S.G., Andrade, A., Ribeiro, J., Giraldo, J.P., Le Maitre, D.C., van Wilgen, B.W., Gelderblom, C.M., Bailey, C., Chapman, R.A., Nel, J.A., 2002. Invasive alien trees and water resources in South Africa: case studies of the costs and benefits of management. Forest Ecology and Management 160, 143–159.

[30] Howe, H.F. and Smallwood J. 1982, Ecology of seed dispersal. Annual Review of Ecology and Systems 13:201-228

[31] Scott, D.F., Lesch, W., 1997. Streamflow responses to afforestation with *Eucalyptus grandis* and *Pinus patula* and to felling in the Mokobulaan experimental catchments, South Africa. Journal of Hydrology 199:360-377.

[32] Senbeta, F., Teketay, D., Näslund, B-A., 2002. Native woody species regeneration in exotic tree plantations at Munessa-Shashemene forest, Southern Ethiopia. New Forests 24, 131-145.

[33] Loice M.A. Omoro, Petri Pellikka, Paul C. Rogers.2010. Tree species diversity, richness and similarity between exotic and indgenous forest in the cloud mountains of Eastern Arc Mountains, Taita Hills, Kenya. *Journ. of For. Res 21(3):255-264.*

[34] Githiru, M., Lens, L., Creswell, W., 2005. Nest predation in a fragmented Afrotropical forest: evidence from natural and artificial nests. Biological Conservation 123, 189-196.

[35] USDA Forest Service, 2007. Field methods instructions for Phase 2 (Forest Inventory) and Phase 3 (Forest Health) of the national Forest Inventory and Analysis program. Available at: *http://www.fs.fed.us/pnw/fia/publications/fieldmanuals.shtml.* [Cited 8 Dec 2008].

[36] Madoffe, S., Hertel, G.D., Rogers, P., O'Connell, B., & Killenga, R., 2006. Monitoring the health of selected Eastern Arc forests in Tanzania. African Journal of Ecology 44, 171–177.

[37] Magurran, A.E., 1988. Ecological Diversity and Measurement. Princeton University Press, Princeton.

[38] Rogers, P.C., O´Connell, B., Mwangombe, J., Madoffe, S., Hertel, G., 2008. Forest health monitoring in the Ngangao forest, Taita Hills, Kenya: A five year assessment of change. Journal of East African Natural History.

[39] Magurran A.E., 1988. Ecological Diversity and its measurment. Princeton: Princeton Univeristy Press 192pp.
Danoff-burg, J., Xu, C., 2008. Measuring Biological Diversity. http://ww.columbia.edu/itc/cerc/dunoffburg/MBD/LINK.html (accessed 15 April 2008).

[40] Lou, J. 2006. Entropy and Diversity. Oikos 113(2), 363-375. Nordic Ecological Society; Blackwell Publishing.

[41] Liu and Nordheim.2006. Effects of Sampling andSpecies Abundance on the Bias and t test of theShannon-Weaver Diversity Index
http://cbe.wisc.edu/assets/docs/pdf/srp-bio/LiuLrevisedforweb.pdf accessed

[42] Gaines, W.L., Harrod, R.J., Lehmkuhl, J.F., 1999. Monitoring Biodiversity: Quantification and interpretation. USDA Forest Service. Pacific North-west Research Station. General Technical Report PNW-GTR-443.

[43] Krebs, C.J., 1989. Ecological methodology. New York, Harper and Row publishers.

[44]Wandago, B., 2002. Realities and perspectives, Kenya country paper. In FAO/EC LNV/GTZ workshop proceedings on Tropical Secondary forest management in Africa.

[45] Mwangombe, J., 2005. Restoration and increase of forest connectivity in Taita Hills: Surveys and suitability assessment of exotic plantations for restoration. A report

www.cepf.net/xp/static/pdfs/Final_EAWLS_TaitaHills.pdf (Accessed 15th April, 2008).
[46] Wassie, A., Teketay, D., 2005. Soil seed banks in Northern Ethiopia: implications for the conservation of woody plants. Flora 201, 32-43.
[47] Yirdaw, E., Luukkanen, O., 2003. Native woody species diversity in *Eucalyptus globulus* Labill. ssp. *globulus* plantations in the Ethiopian highlands. Biodiversity and Conservation 12, 567-582.
[48] Bytebier, B., 2001. Taita Hills Biodiversity Project report. National Museums of Kenya. Nairobi.
[49] Burgess N, D'Amico Hales J, Underwood E, Dinerstein E, Olson D, Itoua I, Schipper J, Ricketts T, Newman K. 2004. *Terrestrial ecoregions of Africa and Madagascar: a conservation assessment*. Washington, D.C: Island Press, 501pp
[50] FAO. 2009. Country Report on the state of plant genetic resources for food and agriculture.http://www.fao.org/docrep/013/i1500e/United%20Republic%20Tanzania.pdf (Accessed June, 2008)
[51] Government of Kenya, 2005. The Forest Act. Ministry of Environment and Natural Resources. Nairobi, Kenya
[52] Adriaensen F., Githiru, M., Mwangombe, J., Lens, L., 2006. Restoration and increase of connectivity among fragmented forest patches in the Taita Hills, South-east Kenya. CEPF project report, 55 pp.
[53] Loice Mildred Akinyi OMORO, Raija LAIHO, Michael STARR, Petri K. E. PELLIKKA. 2011. Relationships between native tree species and soil properties in theindigenous forest fragments of the Eastern Arc Mountains of the Taita Hills, KenyaFor. Stud. China, 2011, 13(3): 198–210. DOI 10.1007/s11632-011-0303-7(In Press)
[54] Hobbs, R.J., Huenneke, L.F., 1992. Disturbance diversity and invasion: implications for conservation. Conservation Biology 6, 324-337
[55] Githiru M, Lens L, Creswell W. 2005. Nest predation in a fragmented Afro- tropical forest: evidence from natural and artificial nests. *Biological Con- servation,* 123: 189–196.
[56] Chazdon, R.L., 1998. Tropical forests - Log 'Em or Leave 'Em? Science 281,1295-1296.
[57] Chege J, Byteibier B. 2005. Vegetation structure of four small fragments in Taita Hills, Kenya. *Journal of East African Natural History,* 94(1): 231–234.
[58] Brokaw, N.V.L., 1985. Gap-phase regeneration in a tropical forest. Ecology 66, 682-687.
[59] Onyekwelu, J.C., Masandl, R., Sti B., 2007. Tree species diversity and soil status of two forest ecosystems in lowland humid rainforest region of Nigeria. A paper presented at a conference on International Agricultural research for Development. University of Göttingen, Germany.
[60] Franklin, J.F., Spies, T.A., Van Pelt, R., Carey, A.B., Thornburgh, D.A., Berg, D.R., Lindenmayer, D.B., Harmon, M.E., Keeton, W.S., Shaw, D.C., Bible, K., Chen, J., 2002. Disturbances and structural development of natural forest ecosystems with silvicultural implications, using Douglas-fir forests as an example. Forest Ecology and Management 155, 399-423.
[61] Duryea M.L., English, R.J., Hermansen, L.A., 1999. A comparison of landscape mulches: Chemical, allelopathic and Decomposition properties. Journal of Arboriculture 25, 88-98.

12

Destruction of the Forest Habitat in the Tatra National Park, Slovakia

Monika Kopecka
Institute of Geography, Slovak Academy of Sciences
Slovakia

1. Introduction

The dynamically changing land cover configuration and its impact on biodiversity have aroused interest in the study of deforestation and its consequences. Deforestation is generally considered to be one of the most serious threats to biological diversity. Awareness of how different deforestation patterns influence habitat quality of forest patches is essential for efficient landscape-ecological management.

The overall effect of deforestation on the forest patch depends on its size, shape and location. Zipperer (1993) identified the following types of the deforestation pattern:

- Internal deforestation that starts in the forest patch and progresses outwardly;
- External deforestation that starts outside and cuts into the forest patch, including indentation, cropping and removal;
- Fragmentation when the patch is split into smaller parcels.

Forest fragmentation is one of the most frequently cited causes of species extinction making it a crucial contemporary conservation issue. The classic view of a habitat fragmentation is the breaking up a large intact area of a single vegetation type into smaller landscape units or simply the disruption of continuity (Fahrig, 2003; Lord & Norton, 1990). This process represents a transition from being whole to being broken into two or more distant pieces. The outcome is landscape composed of fragments (e. g. forest) with something else (the non-forest matrix) between the fragments. Fragmentation of biotopes affects several ecological functions of landscape, first of all the spatial distribution of selected plant and animal species and associations (Bruna & Kress; 2002, Kurosawa & Askins, 2003; Parker et al., 2005). Fragments of original biotopes with reduced area and increased isolation are not capable of securing suitable conditions for life and reproduction of some organisms. Consequences of fragmentation include the species biodiversity decline, functional changes in ecological processes, for instance disruption of trophic chains (Valladares et al., 2006), and genetic changes in organisms (Cunningham & Moritz 1998; Gibbs, 2001).

The history of research focused on the fragmentation consequences reaches to the 1960s when the theory of insular biogeography was published (MacArthur & Wilson, 1963, as cited in Faaborg et al., 1993). The basis of this theory is the recognition that the smaller and more dispersed islands, the lower number of species is capable to find suitable conditions for their permanent existence. However, the phenomenon is not limited only to islands in the geographic sense. As Madera and Zimová (2004) report similar problems were also

observed in case of "islands" in the sense of fragments of natural biotopes in the "sea" of the agro-industrial landscape.
The most frequently studied fragmentation consequences include assessment of the effects on birds. Betts et al. (2006), who investigated dependence of two bird species on fragmentation of their natural biotopes, confirms the hypothesis that landscape configuration is important for selected species only in case of a too small range area and isolated occurrence of suitable biotopes. Faaborg et al. (1993) pointed to the main fragmentation consequences for the neotropical migrating birds and simultaneously presented his proposals how to minimise the negative effects of fragmentation in landscape management. Parker et al. (2005) studied effects of forest areas and forest edges on distribution of 26 species of singing birds and reports that effects of forest fragmentation is negative for many species while the forest area is more important in terms of bird occurrence than its shape (length of edges). Likewise, Trzcinski et al. (1999) emphasize that the effects of the decline in forest area are more serious than fragmentation alone and that the decline of the forest area cannot be compensated by optimisation of spatial arrangement of the remnant fragments. Kuroshawa and Askins (2003) arrived at a similar conclusion and report that preservation of some species in deciduous forest requires occurrence of sufficiently big forest areas. The same authors also consider the forest acreage a suitable indicator of selected bird species frequency.
Fragmentation not only reduces the area of available habitat but can also isolate populations. As the external matrix is physiognomically and ecologically different from the forest patch, an induced edge is formed. Riitters et al. (2002), leaning on studies of several authors, state that a change in area of forest and an increased fragmentation can affect 80 to 90% of all mammals, birds and amphibians. According to habitat types of matrix, Faaborg et al. (1993) recognize:

- Permanent fragmentation that resulted in islands of forest surrounded by dissimilar habitat types (e.g. urban areas), and
- Temporary fragmentation that occurs through timber harvest practices, which create holes of young forest within a matrix of mature forest.

Large forest areas are rapidly becoming fragmented not only as a result of human activities, but as a result of natural disasters as well. In November 2004, the territory of the Tatra National Park, Slovakia was affected by a calamity whirlwind that destroyed around 12,000 ha of forest at altitudes between 700 m to 1,350 m above sea level and substantially changed the vegetation cover in the whole area. The whirlwind and subsequent logging of damaged timber has radically changed the natural conditions of local fauna and flora.

2. The Tatra National Park

The study area covers the whole of the Slovak part of the Tatra Mountains (High Tatras, Belianske Tatras and West Tatras) and a part of the Podtatranska Basin. The Slovak-Polish frontier runs in the north of the study area (Fig. 1).

2.1 Natural biodiversity

The Tatras form a geomorphological unit at the extreme top of the arching province of the Western Tatras. In terms of exogenous relief-forming processes, the surface of high-mountain landscape is the glacial georelief. The tallest peaks of the Tatras are over 2,600 m

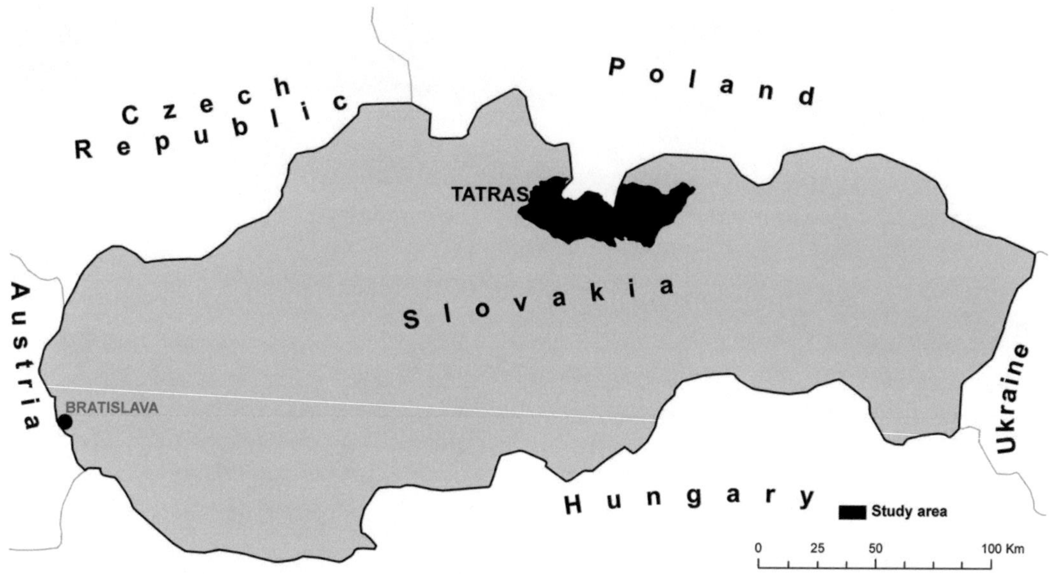

Fig. 1. Location of the study area

a. s. l. The basin is classified as the type of morphostructure of dell grabens and morphotectonic depressions with the relief resting on glacial, glacifluvial and polygenic sediments. The climate of the Tatras is cool to very cool and moist. The mean January temperature in the high-mountain part of the territory is between -7 and -11 degrees of Celsius with the mean annual precipitation totals of 1,200 –2,130 mm. The mean January temperature in the cool subtype oscillates between -6 and -7 °C and the annual precipitation total is between 1,000 and 1,400 mm. The moraine zone in the foreland of the Tatras is classified as the type of cool mountain climate with the mean January temperatures between -5 and -6.5 °C and the mean annual precipitation total between 800 and 1,100 mm. Mean annual air temperatures in the area of the upper timber line in the altitude of 1,600 – 1,800 m above sea level are 2-4 °C; in July it is 10 to 12 °C with the mean annual precipitation totals from 900 to 1,200 mm. The annual course of the air temperature with the minimum in January and the maximum in July prevails in the territory. The lowest temperature instead of occurring in January often moves to February and the highest temperature occurs in August in the highest positions above 2,000 m. n. m. Inversions are typical for the region. The amount of precipitation in the Tatras increases with the increasing sea level altitude. Monthly totals are minimum in winter and maximum in summer. The amplitude of the yearly course depends not only on the sea level altitude but also on exposition of the terrain.

Nature of the Tatras is the unique example of the fully devolved alpine ecosystems on a comparatively small and completely isolated territory lacking any direct links to other alpine mountain ecosystems. It is precisely this feature that makes the Tatras so unique and valuable in terms of natural history not only for Slovakia but also Europe. The great species diversity of fungi, vascular and non-vascular plants in the Tatras is the result of pronounced altitude differences, varied geology but also diverse moist conditions and soils. Endemites of the Tatras, Western Tatras and the Carpathians are the most important representatives of flora. Forest and non-forest plant association linked to five vegetation zones and the substrate exist in this territory.

Fig. 2. Vegetation zones over the Štrbské Pleso Lake in the High Tatras (Photo: M. Kopecká)

The submontane zone covers the lowest part of the region up to the sea level altitude of 800-900 m. The transformed forest is typical while agricultural landscape prevails. The original mixed forests, which once covered the total submontane zone, survive only in inaccessible and mostly wetlogged localities. Spruce-pine and fir-beech woods grow here on acid substrates. The dominant species is the spruce (*Picea abies*). Rare and threatened species like *Ledum palustre, Pedicularis sceptrum-carolinum* and *Iris sibirica* grow in wetland and peat bogs. Among other important species are *Carex lasiocarpa, Carex davalliana, Gymnadenia conopsea, Menyanthes trifoliata, Primula farinosa* and *Pinguicula vulgaris*.

The montane zone is located at the attitude from do 800-900 m a. s. l. to 1,500-1,550 m a.s.l. It includes thick woods with the dominance of *Picea abies*. Broad-leaved forests with dominance of birch and alder trees prevail on the wet soils in the foothills of the Tatras. In altitudes below 1,200 m apart from spruce trees other species like *Pinus sylvestris, Abies alba, Acer pseudoplatanus, Fagus silvatica, Betula pendula,* and *Salix caprea* grow. The most common shrubs include *Lonicera nigra, Lonicera xylosteum, Rosa pendulina* and *Rubus ideaus*. Mountain spruce woods almost exclusively dominate in altitudes above 1,250 m a. s. l. *Sorbus aucuparia, Larix decidua* and *Pinus cembra* thrive on acid soil while *Acer pseudoplatanus* and *Fagus silvatica* prefer calcareous substrates. Peat bogs (*Sphagnum sp.*) with occurrence of many rare species such as *Eriophorum vaginatum, Drosera rotundifolia, Oxycoccus palustris* represent other than forest associations.

Subalpine zone spreads from 1,500-1,800 m. a. s. l. Vegetation consists of continuous growths of *Pinus mugo* and dwarfed trees. In the lower parts of the zone species like *Picea abies, Pinus cembra, Betula carpatica, Ribes alpinum, Ribes petraeum, Sorbus aucuparia* and *Salix silesiaca* occur. Among herbs are *Aconitum firmum* subsp. *firmum, Cicerbita alpina* and *Doronicum austriacum*.

The alpine zone including alpine meadows stretches to 2,300 m a.s. l (Fig. 3). The only wood species resisting the extreme conditions are the low shrubs of *Salix kitaibeliana, Salix alpina, Salix reticulata, Salix herbacea, Juniperus communis* subsp. *alpina, Vaccinium vitis-idaea, Vaccinium myrtillus, Vaccinium gaultherioides, Calluna vulgaris,* and *Empetrum hermaphroditum, Juncus trifidus, Festuca supina, Campanula alpinum, Hieracium alpinum,* and *Pulsatilla scherfelii* dominate on granite substrate. The most exuberant plant associations with the typical representatives like *Dryas octopetala, Festuca versicolor, Saxifraga caesia, Primula auricula* and *Helianthemum alpestre* grow on the base rocks in the Belianske Tatras and in a part of the Western Tatras.

Fig. 3. Biocenoses in the alpine zone (Photo: M. Kopecká)

The subnival zone, as the only in the territory of Slovakia, reaches to the sea level altitude of 2,300 m. Its area in the High Tatras is about 9.6 km^2 (Izakovičová et. al, 2008) and it is located in the core zone of the National Park. It is remarkable for the reduced vegetation period and a very thin soil layer. Lichens, mosses and algae prevail in these conditions while the vascular plants are represented by *Gentiana frigida, Silene acaulis, Saxifraga bryoides, Cerastium uniflorum, Saxifraga retusa, Festuca supina, Poa laxa,* and *Oreochlora disticha*. About forty species of vascular plants also occur in altitudes over 2,600 m. a.s.l. and some of them are glacial relicts.

Endemites that occur only in certain spots are among the extremely rare species of the Tatra flora - 57 species represent the Carpathian endemites in the Tatras. Among them is, for instance, *Aconitum firmum* subsp. *firmum*. The West Carpathian endemites include paleoendemites from the Tertiary Era, for instance, *Saxifraga wahlengergii, Delphinium oxysepalum* and *Dianthus nitidus*. The Tatra endemites cover 36 species of genera *Alchemilla, Thalictrum minus subsp. carpaticum* and *Cochlearia tatrae*. The endemite of the High Tatras is *Ranunculus altitatrensis* and the one of the Belianske Tatras is *Hieracium slovacum*.

In terms of age structure, the young growths at the age below 40 years and three- or multilayer older growths resist best the winds. The planted spruce monoculture at the age of 40-60 years is prone to snow calamities and 60-100 years old stands are susceptible to wind calamities. Ecologically very stable growths at the age of 140-220 year also occur in the subalpine zone. However, the 80-100 years old growths with low resistance due to single layer prevail in the High Tatras and in the Belianske Tatras. In extremes of the valleys of the montane zone of the Western Tatras there are 100-120 year old ecologically stable growths (Izakovičová et al., 2008). The best resistance is observed in a three-layer forest growth, which however, is rare in the Tatras. The whirlwind that struck the mountain range in November 2004 damaged prevailingly single-layer spruce or combined larch/spruce growths.

The varied building of zoocenoses of vertebrates and invertebrates depends on varied types of biotopes in individual zones, while there are several endemites and relic species in the Tatras. The montane zone is the richest in terms of wild life. Thanks to high diversity of biotopes (shrubby vegetation, monocultured woods, mixed woods, thin underwood with grasslands) plenty of animal species live in forests. *Capreolus capreolus* and *Sus scrofa* find food not only in the forest but also in the contiguous farm cultures. Typical field species like *Lepus europaeus* and *Perdix perdix* also live there. *Cervus elaphus* is comparatively common. *Ursus arctos, Lynx lynx, Felis sylvestris, Martes martes and Meles meles* represent carnivorous animals in the forest zone. *Tetrao urogallus, Tetrastes bonasia, Accipiter gentilis, Falco subbuteo,* and *Aquila pomarina* are the bird species that stand out in the fauna of the Tatras. *Lutra Lutra,* an eminent indicator of the water environment quality, is rare and threatened. *Salamandra salamandra, Triturus alpestris* and *Triturus montadoni* are the amphibians worth mentioning. The dwarf pine zone is in fact an intermediate phase between the montane and alpine zones. Chamoix descend to this zone in winter in search for food, while several predators from the forest zone ascend here in summer. Because of harsh living conditions only a limited number of animals lives in the alpine zone, among them *Rupicapra rupicapra tatrica, Marmota marmota latirostris, Pitymys tatricus, Chionomys nivalis, Tichodroma muraria, Anthus spinoletta, Aquila chrysaetos,* and *Oenanthe oenanthe.*

2.2 Disastrous whirlwind of November 2004

19. November 2004 between 15:00 and 20:00 hours, the territory of Slovakia was swept by the whirlwind with almost 200 km/hour gusts. It caused the greatest damage in the territory of the Tatra National Park where in a short time more than 12,000 hectares of forest growths were wrecked (Crofts et al., 2005). It is an area greater than the one annually forested in the total territory of Slovakia. The wind uprooted a continuous belt of forest from Podbanské to Tatranská Kotlina at the altitude from 700 to 1,250-1,350 m a.s.l (Fig. 4). The border between the damaged and undamaged forest was almost straight line following the contour line at the altitude of 1,150 m a.s.l., in the eastern and 1,350 m a.s.l. in the western parts of the territory. In the absolute majority of cases the trees were uprooted, broken trees were rare. Orientation of uprooted trees seen on the aerial photographs and in the terrain confirmed that the damage was caused by the north-eastern to northern winds (Jankovič, 2007).

Representation of individual wood species damaged by the calamity was roughly the same as the wood species composition of growths before the event. The share of spruce trees, of course, dominated with 76%, those of pine, larch, and fir amounted to 8%, 7%, and 15 % respectively while the share of damaged broadleaved wood species was 7.5 %. Estimating by the age, the 60-120 year old specimens with almost 60% share in the total calamity damage were the ones most affected (Fig. 5).

Repeated forest fires that are extremely harmful for biodiversity followed several years after the calamity whirlwind. Fire - either caused by humans or natural - impairs and damages all components of forest ecosystems disrupting the production and other functions of the forest. The biggest fire in the history of the Tatras broke out in a year after the calamity whirlwind (2005) in the calamity area. It damaged 230 hectares of forest biotopes along with about 15,000 cubic metres of unprocessed timber and about 14 hectares of live forest (Fig.6).

Fig. 4. Diminution of forest near the town of Vysoké Tatry a/ in 2000 b/in 2006

Fig. 5. Thousands of hectares of forest were damaged by the 2004 whirlwind (Photo: P. Barabáš)

Fig. 6. Fire of 2005 damaged 230 ha of forest biotopes (Photo: P. Barabáš)

It also damaged the natural undergrowth and artificially restored growth on an area amounting to about 13 hectares. It was additional factor that contributed to the significant fragmentation of forest in the Tatra National Park.

3. Changes in forest fragmentation

3.1 Methodology

An important aspect of fragmentation is the scope and structure of fragments (shape, size, spatial arrangement and the like). These spatial parameters can be assessed using several quantitative methods (D'Eon et al., 2002; Keitt et al., 1997; Kopecká & Nováček, 2008; McGarigal& Marks, 1995; Riitters, 2005; Ritters et al., 2002). In the study of Kummerle et al. (2006), authors used satellite data and compiled land cover maps followed by computation of fragmentation indexes in the boundary regions of Poland, Slovakia and Ukraine. However, actual and reliable information about the land cover and its changes are important input data for any forest fragmentation assessment.

In the early 1990s, the CORINE Land Cover (CLC) database became an essential source of land cover information in the project concerning the majority of the EC countries as well as the PHARE partner countries from Central and Eastern Europe. Standard methodology and nomenclature of 44 classes were applied to mapping and generation of the database in 1:100,000 scale using the 25 ha minimal mapping unit (Feranec & Oťaheľ 2001). The need of updated databases became the impulse for realization of the CLC2000 and CLC2006 projects. All participating countries used a standardized technology and nomenclature to ensure the compatibility of results for the environmental analysis, landscape evaluation and changes. An example of cartographic expression of qualitative changes in forest fragmentation in the selected study area on the regional level related to the years 2000 and 2006 that are based on CLC data assessment is offered here. The applied methodological procedure makes it possible not only to quantify the scope of forest diminishment but also to detect qualitative changes in forest biotopes that survive in the study area.

CLC 2000 and CLC 2006 data layers were used as the input data in the process of forest fragmentation assessment. With the aim to assess the degree of forest fragmentation in the selected model territory, the methodology presented by Vogt et al. (2007) was applied. In the process of morphological image analysis we used the Landscape Fragmentation tool (LFT) developed by Parent and Hurd (2008). LFT is able to perform the fragmentation analysis to clasify a land cover type of interest into four main fragmentation components. Although originally intended for forest fragmentation analysis, the LFT is also aplicable to any land cover type of interest.

CLC data layers are accessible in vector format. For the identification of forest fragmentation, conversion of LFT into the raster format was needed. The preparatory steps consisted of data selection for the model territory and their conversion to the grid reclassification of classes. The module *Polygrid* with 25 m cell size was used for the conversion of the vector format to raster - grid. Cell size was opted taking into account the fact that in interpretation of land cover the LANDSAT 4 TM a LANDSAT 7 ETM satellite images with the resolution capacity of 25 m were used.

As the LFA tool requires a 3 class land cover map as an input, it was necessary to aggregate land cover classes in order to discern forest and other than forest areas, i.e. to reclassify land cover classes so that the grids input into the analysis contains the following values:

- 1 = fragmenting land cover: residential, commercial, urban, pastures, orchards, fallows (on the study area CLC classes 112, 121, 124, 131, 134, 142, 211, 222, 231,242, 243, 321, 322, 324, 333)
- 2 = non-fragmenting land cover: water, rocks, ice, snow, sand (CLC classes 411, 511, 512 and 332)
- 3 = forest (CLC classes 311, 312 and 313)

Forest is classified into four main fragmentation components: *patch, edge, perforated,* and *core* (Fig.7). 'Core forest' is relatively far from the forest/non-forest boundary and 'patch forest' comprises coherent forest regions that are too small to contain core forest. 'Perforated forest' defines the boundaries between the core forest and relatively small perforations, and 'edge forest' includes the interior boundaries with relatively large perforations as well as the exterior boundaries of the core forest regions.

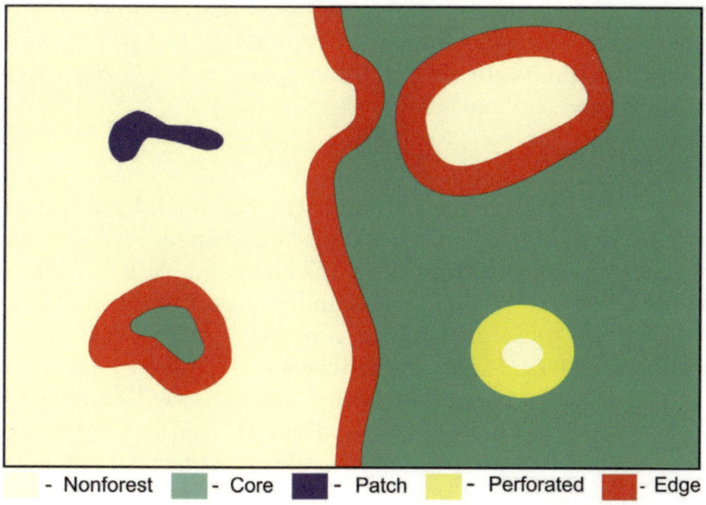

Fig. 7. Illustration of four types of spatial pattern on an artificial map (Vogt et al., 2007)

The forest area classification is based on a specified edge width (Parent & Hurd 2008). The edge width indicates the distance over which other land covers (i.e. urban) can degrade forest. The core pixels are outside the "edge effect" and thus are not degraded from proximity to other land cover types. Edge and perforated pixels occur along the periphery of tracts containing core pixels. Edge pixels make up the exterior peripheries of the tracts whereas perforated pixels make up the interior edges along small gaps in tracts. Patch pixels make up small fragments that are completely degraded by the edge effect.

Changes in forest fragmentation were further assessed according to the following types:

- Type 1: Continuous forest changed into discontinuous forest (Core in forest fragmentation map from 2000 changed into Patch, Perforated or Edge in 2006)
- Type 2: Continuous forest changed into non-forest (Core in forest fragmentation map from 2000 changed into Fragmenting land cover in 2006)
- Type 3: Discontinuous forest changed into non-forest (Patch, Perforated and Edge in 2000 changed into Fragmenting land cover in 2006)

According to experts, the fraze *habitat fragmentation* should be only used in connection with particular plant and animal species regarding the definition of the *habitat*. Franklin et al. (2002) stress that although the notion of *habitat* in connection with a particular species often represents a particular vegetation type, for instance forest interior which can satisfy all ecological demands of the particular species, in many cases it is a combination of several vegetation types (for instance a meadow and a forest while forest provides the living space and the meadow satisfies the reproductive needs of the species). Regarding the above-said, it should be emphasized that CLC databases make it possible to assess fragmentation of selected land cover classes (for instance forest fragmentation) not fragmentation of stands or biotopes of particular species. Another problem in assessment of biotope fragmentation is the fact that the division of the selected area affects species in a different way. Franklin et. al. (2002) use the example of narrow road that can cause fragmentation of the biotope of amphibians but would not alter that of birds of pray. For this reason, the cited authors consider indispensable to define the hierarchic level where the fragmentation is assessed. Fragmentation on the supranational level affects spatial distribution of individual populations, fragmentation on the regional level influences dynamics of population and that on the local level modifies living conditions and reproduction of particular individuals. CLC databases regarding the minimum size of the mapped area (25 ha) makes it possible to assess fragmentation on the regional level.

3.2 Results

Between 1990 and 2000, land cover in the Tatra National Park was relatively stable (Kopecká & Nováček, 2008, 2010). Recorded landscape changes were particularly connected with changes of abandoned agricultural land (pastures, arable land) into woodland scrub and with changes of transitional woodland scrub into forest by natural development. In the period 2000-2006, a remarkable decrease of forestland in the study area was recorded. Decrease of the area of the CLC forest classes (classes 311, 312 and 313) on land cover maps from 2000 and 2006 was connected with an increased number of transitional woodland/shrubs polygons (CLC class 324, see Table 1). This land cover type is represented by the young wood species that are planted after clear-cuts or after calamities of any origin, forest nurseries and stages of the natural development of forest (Feranec & Oťaheľ 2001).

The change of forest into transitional woodland indicates a temporary fragmentation with possible forest regeneration. On the other hand, forest destruction in the National Park facilitated the development of travel and tourism (new hotels, ski parks, etc.). An increased number of construction sites (CLC class 133) indicate that urban sprawl associated with permanent forest fragmentation can be expected in future.

CLC class*	2000		2006		Change 2000 - 2006	
	Number of polygons	Total class area (km2)	Number of polygons	Total class area (km2)	Number of polygons	Total class area (km2)
112 Discontinuous urban fabric	58	37,99	58	38,44	0	0,45
121 Industrial or commercial units	9	6,01	10	6,26	1	0,25
124 Airports	1	1,53	1	1,53	0	0
131 Mineral extraction sites	1	1,26	1	1,26	0	0
133 Construction sites	0	0	5	2,36	5	2,36
142 Sport and leisure facilities	13	10,07	13	10,26	0	0,19
211 Non-irrigated arable land	34	278,02	36	275,01	2	-3,01
222 Fruit trees and berry plantations	1	0,07	1	0,07	0	0
231 Pastures	92	128,49	91	126,98	-1	-1,51
242 Complex cultivation pattern	18	18,04	18	18,04	0	0
243 Land principally occupied by agriculture with significant areas of natural vegetation	69	34,42	69	34,14	0	-0,28
311 Broad-leaved forest	6	3,46	6	3,46	0	**0**
312 Coniferous forest	26	492,66	36	373,45	10	**-119,21**
313 Mixed forest	26	20,01	24	18,01	-2	**-2**
321 Natural grassland	27	81,43	27	81,43	0	0
322 Moors and heathland	38	91,11	38	91,11	0	0
324 Transitional woodland/shrubs	79	51,52	82	174,28	3	122,76
332 Bare rocks	7	60,96	7	60,96	0	0
333 Sparsely vegetated areas	40	40,7	40	40,7	0	0
412 Peatbogs	1	0,56	1	0,56	0	0
511 Water courses	2	1,42	2	1,42	0	0
512 Water bodies	1	0,01	1	0,01	0	0

* CLC classes are described in Feranec & Oťaheľ (2001).

Table 1. CORINE land cover classes on the study area

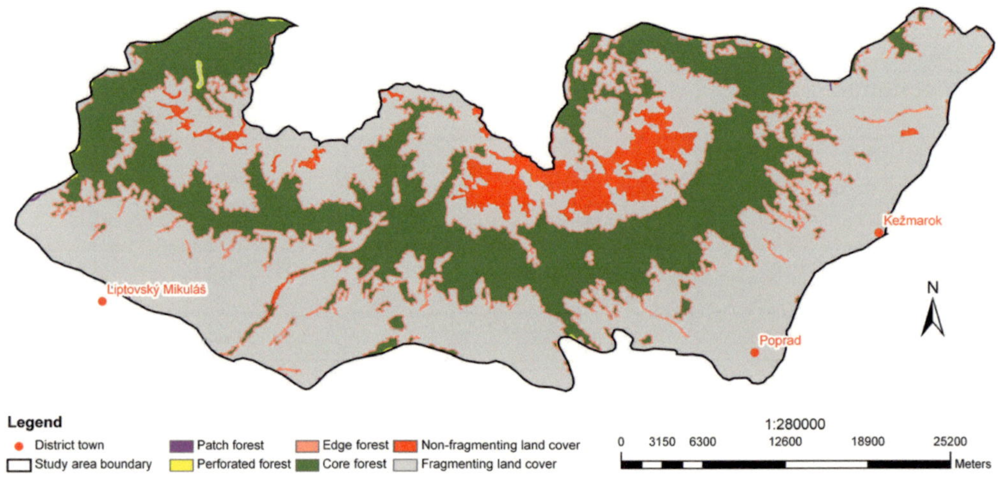

(a)

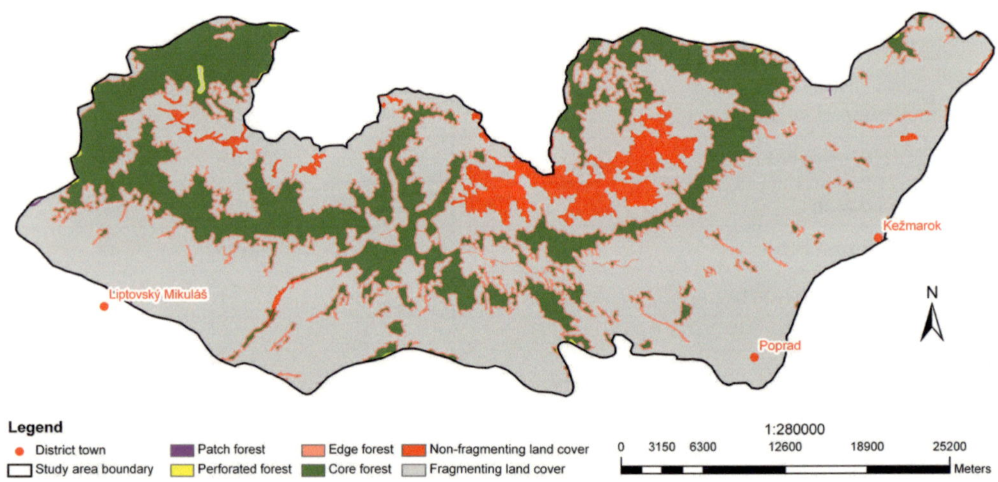

(b)

Fig. 8. Forest fragmentation in Tatra region in a/ 2000, b/2006 (Kopecká & Nováček, 2010)

Fig. 8 and Table 2 demonstrate the decrease of the compact forest areas (Forest core) in 2000 and 2006. On the other side, an increased percentage of disrupted forest areas was observed. Pursuing the applied methodology, these areas were classified as Perforated Forest, Forest Patches and Forest Edge fragmentation components.

	2000		2006		Change 1990-2006	
Fragmentation component	km2	%	km2	%	km2	%
Patch forest	0,964	0,07	0,632	0,05	-0,332	-0,02
Perforated forest	1,646	0,12	1,689	0,12	0,043	0
Edge forest	129,891	9,56	116,652	8,58	-13,239	-0,98
Core forest	357,930	26,32	275,952	20,29	-81,978	-6,03
Non fragmenting land cover	63,695	4,68	62,931	4,63	-0,764	-0,05
Fragmenting land cover	805,635	59,25	901,905	66,33	96,27	7,08
Total	1359,761	100,00	1359,761	100,00	0	0

Table 2. Changes in forest fragmentation in the period 2000 - 2006

The assessment of different types of forest fragmentation (Fig. 9) showed, that the change of continuous forest into the non-forest area was dominant (61%). Discontinuous Forest changed into non-forest area amounted to 22% of the changed territory and the percentage of continuous forest changed into discontinuous forest was 17%.

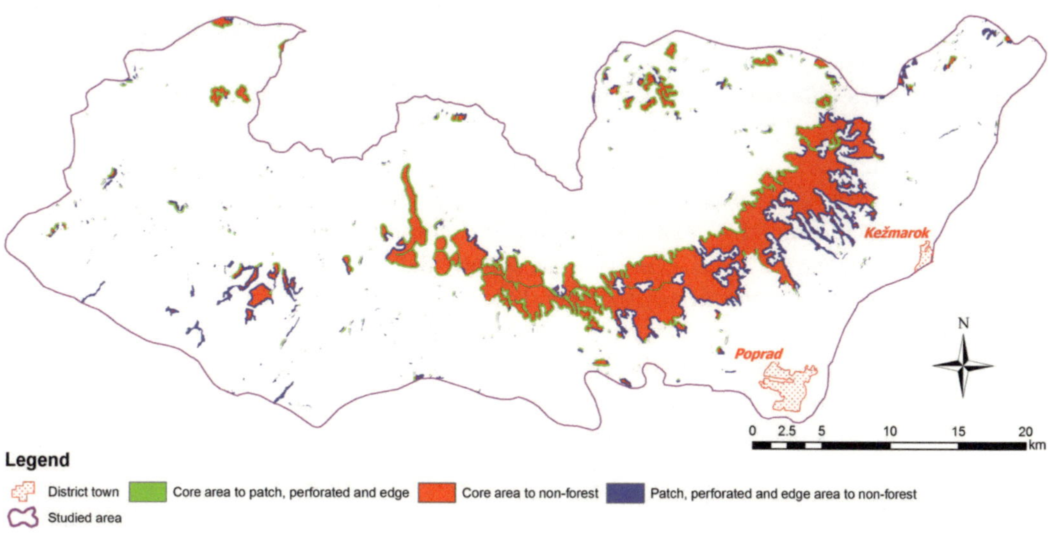

Fig. 9. Map of changes in forest fragmentation in 2000–2006 (Kopecká & Nováček, 2010)

The negative effects on forest biotopes increased when the fallen and broken trees were removed by heavy machinery in order to prevent the large-scale bark-beetle damage. Despite of this, bark beetles destroyed more than 1,700,000 trees before the year 2010 (Fig. 10). This forest habitat changes were not included in the fragmentation analysis based on CLC 2006.

Fig. 10. Bark beetle calamity followed after the windfall (Photo: P. Barabáš)

3.3 Consequences of the forest fragmentation

Ecological succession in an ecosystem represents an organized sequence of association's development including the changes of species composition and processes in time. Succession is considered a dynamic process where the new populations of the same or different species replace the dominant population of one or more species in a particular place. Changes of ecology caused by the sudden damage and subversion of more than 12,000 ha of forest are studied by the international interdisciplinary research often referred to as the "post-calamity" one which concentrates on microclimatic situation, water cycle, bioproduction, succession, species composition, regeneration and restoration processes, biochemical cycles, soil properties, erosion and contamination of the forest ecosystem. The principal aim is to asses the effects of different ways of management applied to damaged growths on the status and development of the model forest association (*Lariceto-Picetum* one), that was most heavily affected and is considered the autochthonous part of the unique anemo-orographic system. Approximately 100 ha areas were delimited with the most similar conditions possible so that they were comparable in terms of properties and features of the forest ecosystem (Fleischer & Matejka, 2009).

Tabular synthesis confirmed 10 types of phytocenoses, mostly secondary, identifiable in the calamity area by mere visual observation (Fig. 11). They are: 1. *Calamagrostis villosa,* 2. *Chamerion angustifolium,* 3. *Calluna vulgaris,* 4. *Vaccinium myrtillus,* 5. *Avenella flexuosa,* 6. *Picea abies,* 7. *Sphagnum magellanicum,* 8. *Carex rostrata,* 9. *Juncus effusus,* and 10. *Veronica officinalis* (Olšavská et al., 2009). Humid substrates are colonized by phytocenoses with the dominating species of *Sphagnum magellanicum* and *Carex rostrata.* The second type of phytocenose is that of associations with increased share of types requiring the higher N level: *Chamaerion angustifolium* and *Veronica officinalis.* Associations with the dominant *Calamagrostis villosa,* are the one most frequent and they form a distinct mosaic along with overgrowths of *Chamaerion angustifolium,* in burnt down places. In future it is expected that *Chamaerion angustifolium,* will be pushed out by *Calamagrostis villosa.* Associations of *Lariceto-Picetum* especially *Vaccinium myrtillus, Avenella flexuosa, Picea abies* and the type of *Calluna vulgaris,* bound to the most acid soil represent the climax stage of forest.

Fig. 11. Natural revitalization of the damaged area (Photo: M. Kopecká)

Some species may be perfectly capable of surviving in a remnant forest many others may not. A forest patch is not the same as a piece of original forest: edge effects may now encroach or even traverse the whole patch. For example, Repel (2008) analysed the breeding bird assemblage structure; nesting, foraging and migrating guilds; bird and habitat relationship and the seasonal dynamics of bird assemblages within four research plots assigned by the management of the Tatra National Park.

- Reference stand, not affected by windstorm calamity
- Plot with extracted wood
- Post wild-fire plot
- Not extracted plot

The average density of breeding bird assemblages in reference stand was much higher than in the plot with extracted wood and wildfire plot. The assemblages on the not extracted plot had the highest average density. The structure of the breeding bird assemblages was most influenced by the portion of the not disturbed forest stands in the plot, number of live standing trees, proportion of dead wood in form of twig heaps, proportion of lying dead wood, and proportion of stones/stone fields in research plots. Kocian et al. (2005) presume that the activity of birds plays an especially significant role in foresting and restoration of forest in the Tatras, as some species (jay, nuthatch nutcracker) propagate natural wood species, such as beech, Swiss pine and hazel.

The use of forest fragmentation indices in the analysis of forest landscapes offers a great potential for integration of spatial pattern information in the landscape-ecological management processes, but requires understanding of the limitations and correct interpretation of results. Further monitoring of forest fragmentation based on remote sensing data together with the terrestrial monitoring of natural vegetation development and dynamics of indicative plant and animal species is necessary to realize the possible revitalization activities and to mitigate negative effects of the calamity windstorm in the Tatra region.

4. Conclusion

Natural forest fragmentation is not a new phenomenon in the Tatras. Windthrows have repeatedly happened in this region in the past (Zielonka et al., 2009) although in a much smaller scale. Urbanization connected with the human-induced deforestation also played an important role in the past because of tourism. The main difference between the old practices and the current deforestation is the difference in scale and rate of increase. In the past, small patches of pastures or damaged forest appeared in the large forested landscape and they quickly grew back upon abandonment. What happened during the bora windstorm in 2004 in the Tatra National Park was precisely the opposite: remnant forest patches were left in the "sea" of the degraded forest landscape.

Anthropogenic disruption of the natural development of the Tatra forest in the past caused that the status of the forest before the calamity did not correspond to the natural development at all. Wood species composition and structure on the greater part of the affected area were not proper for the place. The majority of growths were mature and resembled an economic forest prepared for harvesting. The growths consisted of slender and tall trees with high-situated crowns, which are unstable and highly susceptible to the wind and snow threats. Reasonable and consistent management should insist on growths of different species and age on small areas, which will ensure ecological stability and functionality of forest in an acceptable time horizon and simultaneously provide optimal biotopes for all naturally occurring species. Revitalization of forest affected by the

windstorm is a very complicated process in terms of expertise, organization and economy. The aim of the present monitoring is to observe the process, to identify and assess the results in individual stages. Successful revitalization calls for a new forest-economic concept that is mosaic growths aided by natural processes and natural succession.
Habitat fragmentation not only reduces the area of available habitat but also can also isolate populations and increase edge effects. Whatever the combination of biotic and abiotic changes, the forest patches generally can no longer sustain the production of biodiversity it once had as a part of a larger forest. Understanding of the possible consequences of forest fragmentation remains of great concern to conservationists, biologists and landscape ecologists. The use of forest fragmentation indices in the analysis of forest landscapes offers a great potential for integration of spatial pattern information in the landscape-ecological management processes, but also requires understanding of limitations and correct interpretation of results.

5. Acknowledgment

This paper is one of the outputs of the VEGA Grant Agency Project No 2/0018/10 „Time-spatial analysis of land use: dynamics of changes, fragmentation and stability assessments by application of the CORINE land cover data layers", pursued at the Institute of Geography of the Slovak Academy of Sciences. Author thanks to K2 Studio for the photographs.

6. References

Betts, M. G.; Forbes, G. J., Diamond; A.W.& Taylor, P.D. (2006). Independent Effects of Fragmentation on Forest Songbirds: An Organism-Based Approach. *Ecological Applications*, Vol. 16, No. 3, p. 1076-1089, ISSN: 1051-0761

Bruna, E.M. & Kress, J.W. (2002). Habitat Fragmentation and the Demographic Structure of an Amazonian Understory Herb (Heliconia acuminata). *Conservation Biology*, Vol. 16, No. 5, p.1256-1266, ISSN: 1523-1739

Crofts, R.; Zupancic-Vicar, M.; Marghescu, T. & Tederko, Z., 2005. *IUCN Mission to Tatra National Park, Slovakia, April 2005*. IUCN – the World Conservation Union: 43 pp. 04. 04. 2011 Available from: http://www.wolf.sk/files/dokumenty/IUCN_EN_zaverecna_sprava_2005.pdf

Cunningham, M. & Moritz, C. (1998). Genetic effects of forest fragmentation on rainforest restricted lizard (Scincidae: Gnypetoscincus queenslandiae). *Biological Conservation*, Vol. 83, No. 1, pp. 19 – 30, ISSN 0006-3207

D'Eon, R. G.; Glenn, S. M.; Parfitt, I. & Fortin M. J. (2002). Landscape connectivity as function of scale and organism vagility in a real forested landscape. *Conservation Ecology*. Vol. 6, No.2, ISSN 1195-5449 31. 03. 2011 Available from: http://www.ecologyandsociety.org/vol6/iss2/art10/print.pdf

Faaborg, J.; Brittingham, M.; Donovan, T. & Blake, J. (1993). Habitat Fragmentation in the Temperate Zone: A perspective for managers. In: *Status and management of neotropical migratory birds. General Technical Report RM- 229*, Finch, D.M., Stangel, P.

W. (eds.), p. 331 - 338, Rocky Mountains Forest and Range Expert Station, U. S. Department of Agriculture, Forest Service, Fort Collins, Colorado, USA

Fahrig, L. (2003). Effects of habitat fragmentation on biodiversity. *Annual Rewiew of Ecology, Evolution and Systematics*, Vol. 34, (November 2003), pp. 487-515, ISSN 1546-2069

Feranec, J. & Oťaheľ, J. (2001). *Land cover of Slovakia*. Veda , ISBN 80-224-0663-5, Bratislava, Slovakia

Fleisher, P. & Matejka, F. (2009.) *Windfall research in TANAP-2008*. Geophysical Institute of the Slovak Academy of Sciences, Research Station of the TANAP, State Forest of TANAP, ISBN 978-80-85754-20-9, Bratislava, Slovakia

Franklin, A. B.; Noon, B. R. & George L. T. (2002).What is habitat fragmentation? *Studies in Avian Biology*, Vol. 25, pp. 20 - 29, ISSN 0197-9922

Gibbs, J. P. (2001). Demography versus habitat fragmentation as determinants of genetic variation in wild populations. *Biological Conservation*, Vol. 100,No 1, pp.15 - 20, ISSN 0006-3207

Izakovičová, Z.; Boltižiar M.; Celer, S.; David, S.; Ditě, D.; Gajdoš, P.; Hreško, J.; Ira, V.; Grotkovská, L.; Kenderessy, P.; Kozová, M.; Oszlányi, J.; Petrovič, F.; Válkovcová, Z. & Vološčuk, I.(2008). *Krajinnoekologicky optimálne priestorové a funčné využitie územia biosférickej rezervácie Tatry*. Veda, ISBN 978-80-224-0998-8, Bratislava, Slovakia

Jankovič, J.(2007). *Projekt revitalizácie lesných ekosystémov na území Vysokých Tatier postihnutom veternou kalamitou dňa 19. 11. 2004*. National Forest Centre, Zvolen , Slovakia, 31. 3. 2011 Available from:
http://www.nlcsk.sk/files/57.pdf

Keitt, T. H.; Urban, D. L.& Milne, B. T. (1997). Detecting critical scales in fragmented landscapes. *Conservation Ecology*, Vol. 1, No.1, ISSN 1195-5449 Available from:
http://www.ecologyandsociety/vol1/iss1/art4

Kocian, Ľ.; Topercer, J.; Baláž, E. & Fiala, J. (2005). Breeding birds and their habitat requirements in the windthrown area of the Tatra National Park (Slovakia). *Folia faunistica Slovaca*, Vol. 10, No. 9, pp. 37 - 43, ISSN 1336-4529

Kopecká, M. & Nováček, J. (2008). Hodnotenie fragmentácie krajinnej pokrývky na báze dátových vrstiev CORINE Land Cover. *Geografický časopis*, Vol 60, No. 1, pp. 31-43, ISSN 0016-7193

Kopecká, M. & Nováček, J. (2010). Natural forest fragmentation: An Example from the Tatra Region, Slovakia. In: *Land Use/Cover Changes in Selected Regions in the World*. Bičík, I., Himiyama, Y., Feranec, J. (eds.), pp. 51- 56, International Geographical Union Commission on Land Use and Land Cover Change, ISBN 978-4-907651-05-9 Asahikawa, Japan

Kummerle, T.; Radeloff, V. C.; Perzanowski, K. & Hostert, P. (2006). Cross-border comparison of land cover and landscape pattern in Eastern Europe using a hybrid classification technique. *Remote Sensing of Environment*, Vol. 103, No. 4, pp.449-464, ISSN 0034-4257

Kurosawa, R. & Askins, R. A. (2003). Effects of Habitat Fragmentation on Birds in Deciduous Forests in Japan. *Conservation Biology*, Vol. 17, No. 3, p.695 - 707, ISSN 1523-1739

Lord, J. M. & Norton, D. A. (1990). Scale and the spatial concept of fragmentation. *Conservation Biology*, Vol. 4, No. 2, pp. 197 -202, ISSN 1523-1739

Madera, P. & Zimová, E. (2004). *Metodické postupy a projektování lokálního ÚSES*. Ústav lesnické botaniky, denrologie a geobiocenologie, Brno, Czech republic , 31. 03. 2011, Available from: http://www2.zf.jcu.cz/public/departments/kpu/vyuka/tvok/metodika_uses.pdf

McGarigal, K. & Marks., B. J. (1995). *FRAGSTATS: Spatial Pattern Analysis Program for Quantifying Landscape Structure*. United States Department of Agriculture, Forest Service, Pacific Northwest Research Station, Portland, USA , 31. 03. 2011, Available from: http://www.fs.fed.us/pnw/pubs/pnw_gtr351.pdf

Olšavská, G.; Križová E. & Šoltés, R. (2009). Pokalamitý vývoj vegetácie na trvalo monitorovacích plochách vo Vysokých Tatrách. In: *Windfall research in TANAP-2008*. Fleisher, P. & Matejka, F. (eds.), pp. 172 - 182, Geophysical Institute of the Slovak Academy of Sciences, Research Station of the TANAP, State Forest of TANAP, ISBN 978-80-85754-20-9, Bratislava, Slovakia

Parker, T. H.; Stanberry, B. M.; Becker, C. D. & Gipson, P. S. (2005). Edge and Area Effects on the occurrence of Migrant Forest Songbirds. *Conservation Biology*, Vol. 19, No. 4, pp. 1157-1167, ISSN 1523-1739

Parent, J., & Hurd, J. (2008). *An improved method for classifying forest fragmentation*. CLEAR, 02. 04. 2011 Available from: http://clear.uconn.edu/publications/research/presentations/parent_nearc2008_ff.ppt

Repel, M. (2008). *Diverzita, denzita a potravné vzťahy zoskupení vtákov vo Vysokých Tatrách postihnutých vetrovou kalamitou*. PhD. thesis, Technická univerzita, Zvolen, Slovakia

Riitters, K. H.; Wickham, J. D.; O'Neil, R.V.; Jones, K. B.; Smith, E. R.; Coulston, J. W.; Wade, T. G. & Smith, J. H. (2002). Fragmentation of Continental United States Forests. *Ecosystems*, Vol. 5, No. 8, pp. 815 - 822, ISSN 1435-0629

Riitters, K. H. (2005). Downscaling indicators of forest habitat structure from national assessments. *Ecological Indicators*, Vol. 5, No. 4, pp. 273 - 279, ISSN 1470-160X

Trzcinski, M. K.; Fahrig, L.& Merriam, G. (1999). Independent effects of forest core and fragmentation on the distribution of forest breeding birds. *Ecological Applications*, Vol. 9, No. 2., pp. 586-593, ISSN 1051-0761

Valladares, G.; Salvo, A. & Cagnolo, L. (2006). Habitat Fragmentation Effects on Trophic Processes of Insect-Plant Food Webs. *Conservation Biology*, Vol. 20, No.1, pp. 212-217, ISSN 1523-1739

Vogt, P., Riitters, K. H.; Estreguil C.; Kozak, J.; Wade, T. G. & Wickham, J. D. (2007). Mapping spatial patterns with morphological image processing. *Landscape Ecology*, Vol. 22, No. 2, pp. 171 - 177, ISSN 1572-9761

Zielonka, T.; Holeksa, J.; Malcher, P. & Fleischer, P., (2009). A two-hundred year history of spruce - larch stand in the Slovakian High Tatras damaged by windstorm in 2004. In: *Windfall research in TANAP-2008*. Fleisher, P. & Matejka, F. (eds.), pp. 269 - 274, Geophysical Institute of the Slovak Academy of Sciences, Research Station of the TANAP, State Forest of TANAP, ISBN 978-80-85754-20-9, Bratislava, Slovakia

Zipperer, W., C. (1993). Deforestation patterns and their effects on forest patches. *Landscape Ecology,* Vol. 8, No. 3, pp. 177–184, ISSN 1572 - 9761

13

Modern Methods of Estimating Biodiversity from Presence-Absence Surveys

Robert M. Dorazio[1], Nicholas J. Gotelli[2] and Aaron M. Ellison[3]
[1]*U.S. Geological Survey, Southeast Ecological Science Center, Gainesville, Florida*
[2]*University of Vermont, Department of Biology, Burlington, Vermont*
[3]*Harvard University, Harvard Forest, Petersham, Massachusetts*
USA

1. Introduction

Communities of species are often sampled using so-called "presence-absence" surveys, wherein the apparent presence or absence of each species is recorded. Whereas counts of individuals can be used to estimate species abundances, apparent presence-absence data are often easier to obtain in surveys of multiple species. Presence-absence surveys also may be more accurate than abundance surveys, particularly in communities that contain highly mobile species.

A problem with presence-absence data is that observations are usually contaminated by zeros that stem from errors in detection of a species. That is, true zeros, which are associated with the absence of a species, cannot be distinguished from false zeros, which occur when species are present in the vicinity of sampling but not detected. Therefore, it is more accurate to describe apparent presence-absence data as detections and non-detections, but this terminology is seldom used in ecology.

Estimates of biodiversity and other community-level attributes can be dramatically affected by errors in detection of each species, particularly since the magnitude of these detection errors generally varies among species (Boulinier et al. 1998). For example, bias in estimates of biodiversity arising from errors in detection is especially pronounced in communities that contain a preponderance of rare or difficult-to-detect species. To eliminate this source of bias, probabilities of species occurrence and detection must be estimated simultaneously using a statistical model of the presence-absence data. Such models require presence-absence surveys to be replicated at some – but not necessarily all – of the locations selected for sampling. Replicate surveys can be obtained using a variety of sampling protocols, including repeated visits to each sample location by a single observer, independent surveys by different observers, or even spatial replicates obtained by placing clusters of quadrats or transects within a sample location. Information in the replicated surveys is crucial because it allows species occurrences to be estimated without bias by using a model-based specification of the observation process, which accounts for the errors in detection that are manifest as false zeros.

Several statistical models have been developed for the analysis of replicated, presence-absence data. Each of these models includes parameters for a community's incidence matrix (Colwell et al. 2004, Gotelli 2000), which contains the binary occupancy state (presence or absence) of each species at each sample location. The incidence matrix is only partially observed owing to species- and location-specific errors in detection; however, the incidence matrix can be estimated by fitting these models to the replicated, presence-absence data. Therefore, any function of the incidence matrix – including species richness, alpha diversity, and beta diversity (Magurran 2004)– also can be estimated using these models.

Models for estimating species richness – and other measures of biodiversity – from replicated, presence-absence data were first developed by Dorazio & Royle (2005) and Dorazio et al. (2006). By including spatial covariates of species occurrence and detection probabilities in these models, Kéry & Royle (2009) and Royle & Dorazio (2008) estimated the spatial distribution (or map) of species richness of birds in Switzerland. Similarly, Zipkin et al. (2010) showed that this approach can be used to quantify and assess the effects of conservation or management actions on species richness and other community-level characteristics. More recently, statistical models have been developed to estimate *changes* in communities from a temporal sequence of replicated, presence-absence data. In these models the dynamics of species occurrences are specified using temporal variation in covariates of occurrence (Kéry, Dorazio, Soldaat, van Strien, Zuiderwijk & Royle 2009) or using first-order Markov processes (Dorazio et al. 2010, Russell et al. 2009, Walls et al. 2011), wherein temporal differences in occurrence probabilities are specified as functions of species- and location-specific colonization and extinction probabilities. The latter class of models, which includes the former, is extremely versatile and may be used to confront alternative theories of metacommunity dynamics (Holyoak & Mata 2008, Leibold et al. 2004) with data or to estimate changes in biodiversity. For example, Dorazio et al. (2010) estimated regional levels of biodiversity of butterflies in Switzerland using a model that accounted for seasonal changes in species composition associated with differences in phenology of flight patterns among species. Russell et al. (2009) estimated the effects of prescribed forest fire on the composition and size of an avian community in Washington.

In the present paper we analyze a set of replicated, presence-absence data that previously was analyzed using statistical models that did not account for errors in detection of each species (Gotelli & Ellison 2002). Our objective is to illustrate the inferential benefits of using modern methods to analyze these data. In the analysis we model occurrence probabilities in assemblages of ant species as a function of large-scale, geographic covariates (latitude, elevation) and small-scale, site covariates (habitat area, vegetation composition, light availability). We fit several models, each identified by a specific combination of covariates, to assess the relative contribution of these potential sources of variation in species occurrence and to estimate the effect of these contributions on geographic differences in ant species richness and other measures of biodiversity. We also provide the data and source code used in our analysis to allow comparisons between our results and those obtained using alternative methods of analysis.

2. Study area and sampling methods

2.1 Ant sampling

The data in our analysis were obtained by sampling assemblages of ant species found in New England bogs and forests. The initial motivation for sampling was to determine the extent of the distribution of the apparent bog-specialist, *Myrmica lobifrons*, in Massachusetts and Vermont. Bogs are not commonly searched for ants, but in 1997 we had identified *M. lobifrons* as a primary component of the diet of the carnivorous pitcher plant, *Sarracenia purpurea*, at Hawley Bog in western Massachuestts. This was the first record for *M. lobifrons* in Massachusetts. At the time the taxonomic status of this species was being re-evaluated (Francoeur 1997), and it was largely unknown in the lower (contiguous) 48 states of the United States. In addition to our interest in *M. lobifrons*, we also wanted to explore whether bogs harbored a distinctive ant fauna or whether the ant faunas of bogs were simply a subset of the ant species found in the surrounding forests. Thus, at each of the sites selected for sampling, we surveyed ants in the target bog and in the upland forest adjacent to the bog (Gotelli & Ellison 2002).

At each of 22 sample sites, we established two 8 $\times$ 8 m sampling grids, each containing 25 evenly spaced pitfall traps. One sampling grid was located in the center of the bog; the other was located within intact forest 50-500 m away from the edge of the bog. Each pitfall trap consisted of a 180-ml plastic cup (95 mm in diameter) that was filled with 20 ml of dilute soapy water. Traps were buried so that the upper lip of each trap was flush with the bog or forest-soil surface, and left in place for 48 hours during dry weather. At the end of the 48 hours, trap contents were collected, immediately fixed with 95% ethanol, and returned to the laboratory where all ants were removed and identified to species. Traps were sampled twice in the summer of 1999, and the time between each sampling period was 6 weeks (42 days); therefore, we consider the two sampling periods as early- and late-summer replicates. Locations of traps were flagged so that pitfall traps were placed at identical locations during the two sampling periods.

2.2 Measurement of site covariates

The geographic location (latitude (LAT) and longitude (LON)) and elevation (ELEV, meters above sea level) of each bog and forest sample site was determined using a Trimble Global Positioning System (GPS). At each forest sample site we also estimated available light levels beneath the canopy using hemispherical canopy photographs, which were taken on overcast days between 10:00 AM and and 2:00 PM at 1 m above ground level with an 8 mm fish-eye lens on a Nikon F-3 camera. Leaf area index (LAI, dimensionless) was determined from the subsequently digitized photographs using HemiView software (Delta-T, Cambridge, UK). Because there was no canopy over the bog, the LAI of each bog was assigned a value of zero. To compute a global site factor (GSF, total solar radiation) for each forest sample site (Rich et al. 1993), we summed weighted values of direct site factor (DSF, total direct beam solar radiation) and indirect site factor (total diffuse solar radiation). GSF values are expressed as a percentage of total possible solar radiation (i.e., above the canopy) during the growing season (April through October), corrected for latitude and solar track. The GSF of each bog was assigned a value of one.

Digital aerial photographs were obtained for each sampled bog from state mapping authorities, or, when digital photographs were unavailable (five sites), photographic prints (from USGS-EROS) were scanned and digitized. Aerial photographs were used to construct a set of data layers (Arc-View GIS 3.2) from which bog area (AREA) was calculated. The area of the surrounding forests was not measured, as the forest was generally continuous for at least several km^2 around each bog.

3. Statistical analysis

We analyzed the captures of ant species observed at our sample sites using a modification of the multi-species model of occurrence and detection that includes site-specific covariates (Kéry & Royle 2009, Royle & Dorazio 2008). This modification allows a finite set of candidate models to be specified and fit to the data simultaneously such that prior beliefs in each model's utility can be updated (using Bayes' rule) to compute the posterior probability of each model. The resulting set of posterior model probabilities can be used to select a single ("best") model for inference or to estimate scientifically relevant quantities while averaging over the posterior uncertainty of the models (Draper 1995).

To compare our results with previous analyses (Gotelli & Ellison 2002), we analyzed the data observed in bogs and forests separately. These two habitats are sufficiently distinct that differences in species occurrence – and possibly capture rates – are expected a priori. Furthermore, the potential covariates of occurrence differ between the two habitats, adding another reason to analyze the bog and forest data separately.

3.1 Hierarchical model of species occurrence and capture

We summarize here the assumptions made in our analysis of the ant captures. Let $y_{ik} \in \{0, 1, \ldots, J_k\}$ denote the number of pitfall traps located at site k that contained the ith of n distinct species of ants captured in the entire sample of $R = 22$ sites. At each site 25 pitfall traps were deployed during each of 2 sampling periods (early- and late-season replicates); therefore, the total number of replicate observations per site was constant ($J_k = 50$). While constant replication among sites simplifies implementation of the model, it is not required. However, it *is* essential that $J_k > 1$ for some (ideally all) sample sites because information from within-site replicates allows both occurrence and detection probabilities to be estimated for each species. In the absence of this replication these two parameters are confounded.

The observed data form an $n \times R$ matrix $\boldsymbol{Y}_{obs}$ of pitfall trap frequencies, so that rows are associated with distinct species and columns are associated with distinct sample sites. Note that n, the number of distinct ant species observed among all R sample sites, is a random outcome. In the analysis we want to estimate the total number of species N that are present and vulnerable to capture. Although N is unknown, we know that $n \leq N$, i.e., we know that the number of species observed in the samples provides a lower bound for an estimate of N.

To estimate N, we use a technique called parameter-expanded data augmentation (Dorazio et al. 2006, Royle & Dorazio 2011), wherein rows of all-zero trap frequencies are added to the observed data $\boldsymbol{Y}_{obs}$ and the model for the observed data is appropriately expanded to analyze the augmented data matrix $\boldsymbol{Y} = (\boldsymbol{Y}_{obs}, \boldsymbol{0})$. The technical details underlying this technique are described by Royle & Dorazio (2008, 2011), so we won't repeat them here. Briefly, however, the idea is to embed the unobserved, all-zero trap frequencies of the $N - n$ species in the community within a larger data set of fixed, but known size (say, M species, where $M > N$) for the purpose of simplifying the analysis. The conventional model for the community of N species is necessarily modified so that each of the $M - n$ rows of augmented data can be estimated as either belonging to the community of N species (and containing sampling zeros) or not (and containing structural zeros). In particular, we add a vector of parameters $\boldsymbol{w} = (w_1, \ldots, w_M)$ to the model to indicate whether each species is a member of the community ($w = 1$) or not ($w = 0$). The elements of $\boldsymbol{w}$ are assumed to be independentally and identically

	Site k								
	Observed				Partially observed				
species i	1	2	$\cdots$	R	1	2	$\cdots$	R	w_i
1	y_{11}	y_{12}	$\cdots$	y_{1R}	z_{11}	z_{12}	$\cdots$	z_{1R}	w_1
2	y_{21}	y_{22}	$\cdots$	y_{2R}	z_{21}	z_{22}	$\cdots$	z_{2R}	w_2
$\vdots$	$\vdots$	$\vdots$		$\vdots$	$\vdots$	$\vdots$		$\vdots$	$\vdots$
n	y_{n1}	y_{n2}	$\cdots$	y_{nR}	z_{n1}	z_{n2}	$\cdots$	z_{nR}	w_n
$n+1$	0	0	$\cdots$	0	$z_{n+1,1}$	$z_{n+1,2}$	$\cdots$	$z_{n+1,R}$	w_{n+1}
$\vdots$	$\vdots$	$\vdots$		$\vdots$	$\vdots$	$\vdots$	$\vdots$		$\vdots$
N	0	0	$\cdots$	0	z_{N1}	z_{N2}	$\cdots$	z_{NR}	w_N
$N+1$	0	0	$\cdots$	0	$z_{N+1,1}$	$z_{N+1,2}$	$\cdots$	$z_{N+1,R}$	w_{N+1}
$\vdots$	$\vdots$	$\vdots$		$\vdots$	$\vdots$	$\vdots$		$\vdots$	$\vdots$
M	0	0	$\cdots$	0	z_{M1}	z_{M2}	$\cdots$	z_{MR}	w_M

Table 1. Conceptualization of the supercommunity of M species used in parameter-expanded data augmentation. $\boldsymbol{Y}$ comprises a matrix of n rows of observed trap frequencies and $M - n$ rows of unobserved (all-zero) trap frequencies. $\mathbf{Z}$ denotes a matrix of species- and site-specific occurrence parameters. $\boldsymbol{w}$ denotes a vector of parameters that indicate membership in the community of N species vulnerable to sampling.

distributed (iid) as follows:

$$w_i \overset{iid}{\sim} \text{Bernoulli}(\Omega)$$

where the parameter Ω denotes the probability that a species in the augmented data set is a member of the community of N species that are present and vulnerable to capture. Note that the community's species richness N is not a formal parameter of the model. Instead, N is a derived parameter to be computed as a function of $\boldsymbol{w}$ as follows: $N = \sum_{i=1}^{M} w_i$. Therefore, estimation of Ω and $\boldsymbol{w}$ is essentially equivalent to estimation of N (Royle & Dorazio 2011).

The incidence matrix of the community (Colwell et al. 2004, Gotelli 2000) is a parameter of the model that is embedded in an $M \times R$ matrix of parameters $\mathbf{Z}$, whose elements indicate the presence ($z = 1$) or absence ($z = 0$) of species i at sample site k. Although $\mathbf{Z}$ is treated as a random variable of the model, each element associated with species that are not members of the community is equal to zero because z_{ik} is defined conditional on the value of w_i as follows:

$$z_{ik}|w_i \sim \text{Bernoulli}(w_i \psi_{ik}) \tag{1}$$

where ψ_{ik} denotes the probability that species i is present at sample site k. Thus, if species i is not a member of the community, then $w_i = 0$ and $\Pr(z_{ik} = 0|w_i = 0) = 1$; otherwise, $w_i = 1$ and $\Pr(z_{ik} = 1|w_i = 1) = \psi_{ik}$. For purposes of computing estimates of community-level characteristics, $\mathbf{Z}$ may be treated as the incidence matrix itself because the $M - N$ rows associated with species not in the community contain only zeros and make no contribution to the estimates.

The matrix of augmented data $\boldsymbol{Y}$ and the parameters $\boldsymbol{Z}$ and $\boldsymbol{w}$ may be conceptualized as characteristics of a supercommunity of M species (Table 1). This supercommunity includes N species that are members of the community vulnerable to sampling and $M - N$ other species that are added to simplify the analysis. The parameters $\boldsymbol{Z}$ and $\boldsymbol{w}$ are paramount in terms of estimating measures of biodiversity. We have shown already that estimates of $\boldsymbol{w}$ are used to compute estimates of species richness N (a measure of gamma diversity). Similarly, $\boldsymbol{Z}$ may be used to estimate measures of alpha diversity, beta diversity, and other community-level characteristics. For example, summing the columns of $\boldsymbol{Z}$ yields the number of species present at each sample site (alpha diversity). Similarly, different columns of $\boldsymbol{Z}$ may be compared to express differences in species composition among sites (beta diversity). For example, the Jaccard index, a commonly used measure of beta diversity (Anderson et al. 2011), is easily computed from $\boldsymbol{Z}$. The Jaccard index requires the number of species from two distinct sites, say k and l, that occur at both sites. Off-diagonal elements of the $R \times R$ matrix $\boldsymbol{Z}'\boldsymbol{Z}$ contain the numbers of species shared between different sites. Therefore, the proportion of all species present at two sites, say k and l, that are common to both sites is

$$J_{kl} = \frac{z_k' z_l}{z_k'\mathbf{1} + z_l'\mathbf{1} - z_k' z_l}$$

where $\mathbf{1}$ denotes a $M \times 1$ vector of ones, and z_k and z_l denote the kth and lth columns of $\boldsymbol{Z}$. Note that J_{kl} is a measure of the similarity in species present at sites k and l; its complement, $1 - J_{kl}$, corresponds to the dissimilarity – or beta diversity – between sites.
In Section 4 we provide estimates of gamma diversity, alpha diversity, and beta diversity in our analyses of the ant data sets. In these analyses we assume that the community of ants contains a maximum of $M = 75$ species in the forest habitat and a maximum of $M = 25$ species in the bog habitat. The lower maximum is based on five years of collecting ants in New England bogs that yielded only 21 distinct species (Ellison and Gotelli, *personal observations*). The total number of ant species in all of New England is somewhere between 130 and 140 (Ellison et al. 2012); however, many of these species are field or grassland species, and six species, which are not indigenous to New England, are restricted mainly to warm indoors. By excluding these species and those found only in bogs, we obtain the upper limit for the number of ant species in the forest habitat.

3.1.1 Modeling species occurrence probabilities

Equation 1 implies that each element of the incidence matrix is assumed to be independent given ψ_{ik}, the probability of occurrence of species i at sample site k. Let $\boldsymbol{x}_k = (x_{1k}, x_{2k}, \ldots, x_{pk})$ denote the observed value of p covariates at site k. We assume that each of these covariates potentially affects the species-specific probability of occurrence at site k. Naturally, the effects of these covariates may differ among species, so their contributions are modeled on the logit-scale as follows:

$$\text{logit}(\psi_{ik}) = b_{0i} + \delta_1 b_{1i} x_{1k} + \cdots + \delta_p b_{pi} x_{pk} \quad (2)$$

where b_{0i} denotes a logit-scale, intercept parameter for species i and b_{li} denotes the effect of covariate x_l on the probability of occurrence of species i ($l = 1, \ldots, p$). If each covariate is centered and scaled to have zero mean and unit variance, b_{0i} denotes the logit-scale probability

of occurrence of species i at the average value of the covariates. This scaling of covariates also improves the stability of calculations involved in estimating $\boldsymbol{b}_i = (b_{0i}, b_{1i}, \ldots, b_{p_i})$.
The additional parameter $\boldsymbol{\delta} = (\delta_1, \ldots, \delta_p)$ in Eq. 2 is used to specify whether each covariate is ($\delta = 1$) or is not ($\delta = 0$) included in the model. Specifically, we assume

$$\delta_l \stackrel{iid}{\sim} \text{Bernoulli}(0.5)$$

which implies an equal prior probability (0.5^p) for each of the 2^p distinct values of $\boldsymbol{\delta}$. This approach, originally developed by Kuo & Mallick (1998), allows several regression models to be considered simultaneously and yields the posterior distribution of $\boldsymbol{\delta}$. After all models have been considered (as described in Section 3.2), the posterior probability $\Pr(\boldsymbol{\delta}|\boldsymbol{Y}, \boldsymbol{X})$ of each model (vis a vis, each distinct value of $\boldsymbol{\delta}$) can be computed. In our analyses the model with the highest posterior probability is used to compute estimates of species occurrence and biodiversity.

3.1.2 Modeling species captures

We assume a relatively simple model of the pitfall trap frequencies y_{ik}, owing to the simplicity of our sampling design. Specifically, we assume that if ants of species i are present at site k (i.e., $z_{ik} = 1$), their probability of capture p_{ik} is the same in each of the J_k replicated traps. This assumption implies the following binomial model of the pitfall trap frequencies:

$$y_{ik}|z_{ik} \sim \text{Binomial}(J_k, z_{ik}p_{ik})$$

where p_{ik} denotes the conditional probability of capture of species i at site k (given $z_{ik} = 1$). Note that if species i is absent at site k, then $\Pr(y_{ik} = 0|z_{ik} = 0) = 1$. In other words, if a species is absent at sample site k, then none of the J_k pitfall traps will contain ants of that species under our modeling assumptions.
None of the covariates observed in our samples is thought to be informative of ant capture probabilities; therefore, rather than using a logistic-regression formulation of p_{ik} (as in Eq. 2), we assume that the logit-scale probability of capture of each species is constant:

$$\text{logit}(p_{ik}) = a_{0i}$$

at each of the R sample sites.

3.1.3 Modeling heterogeneity among species

In order to estimate the occurrences of species not observed in any of our traps, a modeling assumption is needed to specify a relationship among all species-specific probabilities of occurrence and detection. Therefore, we assume that the ant species in each community are ecologically similar in the sense that these species are likely to respond similarly, but not identically, to changes in their environment or habitat, to changes in resources, or to changes in predation. The assumption of ecological similarity seems reasonable for the species we sampled owing to their overlapping diets, habitats, and life history characteristics. As a point of emphasis, we would *not* assume ecological similarity if our assemblage had included species of tigers and mice! The idea of ecological similarity has been used previously to analyze assemblages of songbird, butterfly, and amphibian species (Dorazio et al. 2006,

Kéry, Royle, Plattner & Dorazio 2009, Walls et al. 2011); however, this idea is not universally applicable. For example, if the occurrence of one species depends on the presence or absence of another species (as might occur between a predator and prey species or between strongly competing species), then ecological similarity would not be a reasonable assumption. In this case a model must be formulated to specify the pattern of co-occurrence that arises from interspecific interactions (MacKenzie et al. 2004, Waddle et al. 2010). The formulation of statistical models for inferring interspecific interactions in communities of species is an important and developing area of research (Dorazio et al. 2010).

In assemblages of ecologically similar species, it seems reasonable to use distributional assumptions to model unobserved sources of heterogeneity in probabilities of species occurrence and detection. For example, occurrence probabilities may be low for some species (the rare ones) and high for others, but all species are related in the sense that they belong to a larger community of ecologically similar species. By modeling the heterogeneity among species in this way, the data observed for any individual species influence the parameter estimates of every other species in the community. In other words, inferences about an individual species do not depend solely on the observations of that species because the inferences borrow strength from the observations of other species. A practical manifestation of this multispecies approach is that the estimate of a parameter (e.g., occurrence probability) of a single species reflects a compromise between the estimate that would be obtained by analyzing the data from each species separately and the average value of that parameter among all species in the community. In the statistical literature this phenomenon is called "shrinkage" (Gelman et al. 2004) because each species-specific estimate is shrunk in the direction of the estimated average parameter value. Of course, the amount of shrinkage depends on the relative amount of information about the parameter in the observations of each species versus the information about the mean value of that parameter. An important benefit of shrinkage is that it allows parameters to be estimated for a species that is detected with such low frequency that its parameters could otherwise not be estimated. Such species are often the rarest members of the community, and it is crucial that these species be included in the analysis to ensure that estimates of biodiversity are accurate.

In the present analysis we use a normal distribution

$$\begin{bmatrix} b_{0i} \\ a_{0i} \end{bmatrix} \overset{iid}{\sim} \text{Normal}\left(\begin{bmatrix} \beta_0 \\ \alpha_0 \end{bmatrix}, \begin{bmatrix} \sigma_{b_0}^2 & \rho\,\sigma_{b_0}\sigma_{a_0} \\ \rho\,\sigma_{b_0}\sigma_{a_0} & \sigma_{a_0}^2 \end{bmatrix} \right), \quad (3)$$

to specify the variation in occurrence and detection probabilities among ant species. The parameters σ_{b_0} and σ_{a_0} denote the magnitude of this variation, and ρ parameterizes the extent to which species occurrence and detection probabilities are correlated.

We also use the normal distribution to specify variation among the species-specific effects of covariates on occurrence. Specifically, we assume $b_{li} \overset{iid}{\sim} \text{Normal}(\beta_l, \sigma_{b_l}^2)$ (for $l = 1, \ldots, p$), so that the effects of different covariates are assumed to be mutually independent and uncorrelated.

3.2 Parameter estimation

The hierarchical model described in Section 3.1 would be impossible to fit using classical methods owing to the high-dimensional and analytically intractable integrations involved

		Posterior probability	
Habitat	Covariates	Uniform prior	Jeffreys' prior
Forest	LAT, LAI, GSF, ELEV	0.818	0.767
Forest	LAT, LAI, ELEV	0.177	0.229
Forest	LAT, ELEV	0.005	0.003
Forest	LAT, GSF, ELEV	< 0.001	0.001
Bog	ELEV	0.424	0.416
Bog	None	0.342	0.412
Bog	LAT	0.082	0.070
Bog	AREA, ELEV	0.060	0.034
Bog	LAT, ELEV	0.045	0.029
Bog	AREA	0.038	0.036
Bog	LAT, AREA	0.006	0.003
Bog	LAT, AREA, ELEV	0.004	0.001

Table 2. Posterior probabilities of models containing different covariates of species occurrence probabilities. Covariates include latitude (LAT), leaf area index (LAI), light availability (GSF), elevation (ELEV), and bog area (AREA). Models with less than 0.001 posterior probability are not shown.

in evaluating the marginal likelihood function. We therefore adopted a Bayesian approach to inference and used Markov chain Monte Carlo methods (Robert & Casella 2004) to fit the model. In the appendix (Section 7) we describe our choice of prior distributions for the model's parameters. We also provide the data and the computer code that was used to calculate the joint posterior distribution of the model's parameters. All parameter estimates and credible intervals are based on this distribution.

4. Results

4.1 Effects of covariates on species occurrence

The posterior model probabilities calculated in our analysis of forest and bog data sets are only mildly sensitive to our choice of priors for the logit-scale parameters of the model (Table 2). Recall that these parameters are of primary interest in assessing the relative contributions of geographic- and site-level covariates. Regardless of the prior distribution used (Uniform or Jeffreys' (see appendix)), the model with highest probability includes all four covariates (LAT, LAI, GSF, ELEV) in the analysis of data observed at forest sample sites and a single covariate (ELEV) in the analysis of data observed at bog sample sites. However, the model without any covariates has nearly equal probability to the favored model of the bog data, and the combined probability of these two models far exceeds the probabilities of all other models. These results suggest that occurrence probabilities of ant species found in the bog habitat are not strongly influenced by the LAT or AREA covariates, either alone or in combination with other covariates.

Each of the four covariates used to model species occurrences in the forest habitat has an average, negative effect on occurrence probabilities. Estimates of β_l and 95% credible intervals are as follows: LAT, -0.717 $(-1.217, -0.257)$; LAI, -0.850 $(-1.302, -0.440)$; GSF, -0.494, $(-0.916, -0.098)$; ELEV, -0.662 $(-1.014, -0.339)$. However, as illustrated in Figure 1, there is considerable variation among species in the magnitude of these effects . Similarly, the estimated occurrence probabilities of ants in the bog habitat decrease with ELEV ($\hat{\beta}_1 = -0.500$ $(-1.019, -0.098)$), and there is considerable variation among species ($\hat{\sigma}_{b_1} = 0.320$ $(0.014, 1.000)$) in the magnitude of ELEV effects.

4.2 Estimates of biodiversity

Our pitfall trap surveys revealed $n = 34$ distinct species of ants at the forest sample sites and $n = 19$ species at the bog sample sites. The estimated species richness of ants found in the forest habitat ($\hat{N} = 43$ (95% interval = $(37, 70)$) is nearly twice the estimated richness of ants in the bog habitat ($\hat{N} = 25$ (95% interval = $(21, 25)$); however, the estimate of forest ant richness is relatively imprecise and the estimate of bog ant richness is strongly influenced by the upper bound ($M = 25$ species).

The numbers of species found in forest and bog communities are perhaps better compared using estimates of species richness at the sample sites. These measures of alpha diversity are plotted against each site's elevation in Figure 2, which also includes the number of ant species actually captured. The estimated richness at sites in the forest habitat usually exceeds that at sites in the bog habitat when the effects of elevation on species occurrences are taken into account. Note also that a site's estimated species richness can be much higher than the numbers of species captured because capture probabilities are much lower than one for most species (Tables 3 and 4).

Site-specific estimates of beta diversity between bog and forest communities of ants are relatively high, ranging from 0.71 to 1.0 (Figure 3). These estimates also generally exceed the beta diversities between ants from different sites within each habitat (Figure 4), adding further support for the hypothesis that composition of ant species differs greatly between forest and bog habitats.

5. Discussion

5.1 Analysis of ant species

It is interesting to compare the results of our analyses with the results reported by Gotelli & Ellison (2002), who analyzed the same data but did not account for errors in detection of species. Gotelli & Ellison (2002) used linear regression models to estimate associations between the number of observed species (which was referred to as "species density") and environmental covariates. For bog ants Gotelli & Ellison (2002) reported a significant association between species density and latitude ($P = 0.041$) and a marginally significant association between species density and vegetation structure (as measured by the first principal-component score; $P = 0.081$). Collectively, these two variables accounted for about 30% of the variation in species density. In the present analysis of the bog data, the best fitting model included the effect of a single covariate (ELEV) on ant species occurrence probabilities, though a model without any covariates was a close second (Table 2). In the analysis of forest ants Gotelli & Ellison (2002) reported significant positive associations between species

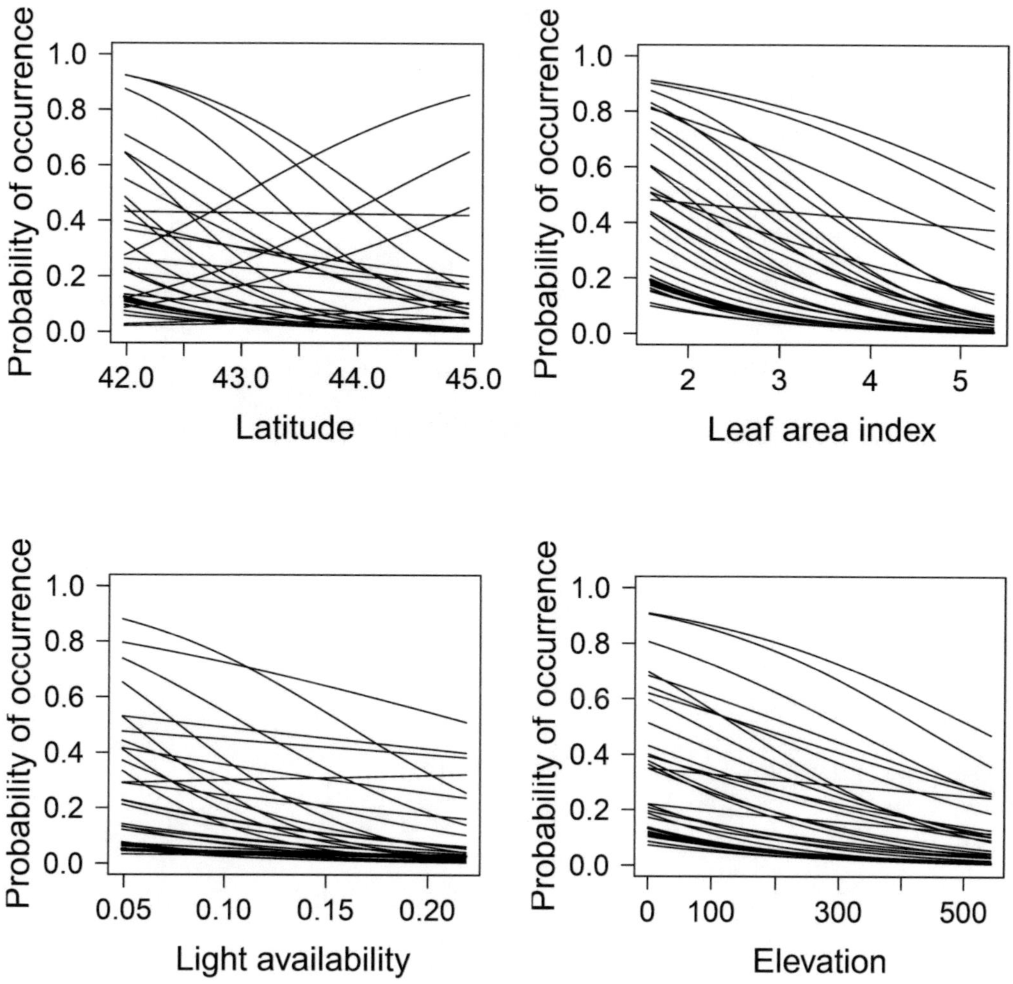

Fig. 1. Estimated effects of covariates on occurrence probabilities of ant species in forest habitat.

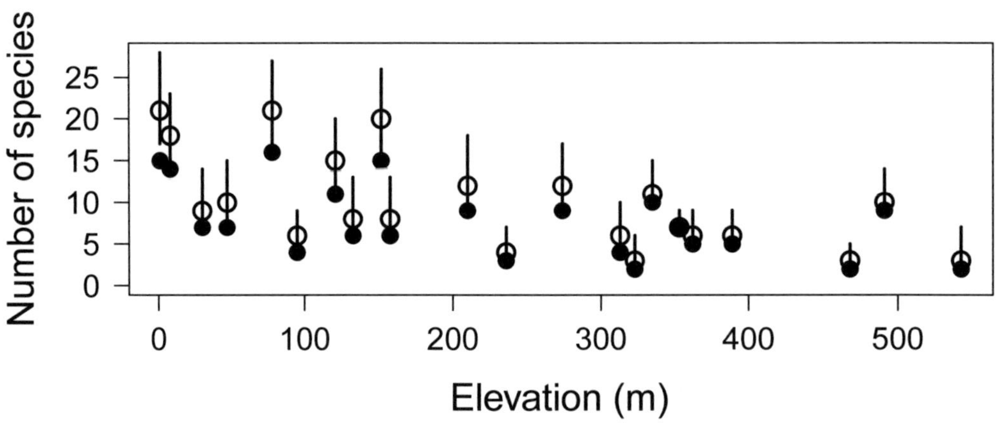

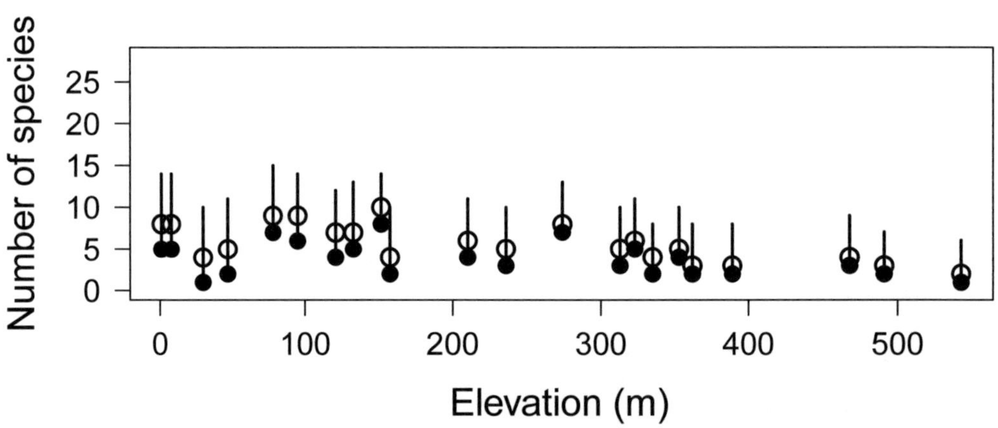

Fig. 2. Estimates of site-specific species richness (open circles with 95% credible intervals) for ants in forest habitat (upper panel) and bog habitat (lower panel) versus elevation. Number of species captured at each site (closed circles) is shown for comparison.

Species	Capture probability			Occurrence probability		
	Median	2.5%	97.5%	Median	2.5%	97.5%
Amblyopone pallipes	0.028	0.008	0.073	0.043	0.005	0.237
Aphaenogaster rudis (species complex)	0.237	0.209	0.269	0.779	0.539	0.927
Campnnotus herculeanus	0.090	0.062	0.123	0.255	0.104	0.482
Campnnotus nearcticus	0.035	0.013	0.074	0.083	0.014	0.316
Campnnotus novaeboracensis	0.017	0.008	0.037	0.454	0.121	0.897
Campnnotus pennsylvanicus	0.131	0.107	0.158	0.587	0.322	0.819
Dolichoderus pustulatus	0.011	0.002	0.053	0.042	0.003	0.389
Formica argentea	0.011	0.001	0.053	0.044	0.003	0.411
Formica glacialis	0.012	0.002	0.055	0.045	0.003	0.413
Formica neogagates	0.096	0.049	0.163	0.038	0.005	0.166
Formica obscuriventris	0.010	0.001	0.051	0.046	0.003	0.448
Formica subaenescens	0.051	0.029	0.081	0.229	0.085	0.476
Formica subintegra	0.166	0.083	0.284	0.029	0.003	0.140
Formica subsericea	0.248	0.184	0.320	0.059	0.009	0.218
Lasius alienus	0.053	0.035	0.075	0.499	0.260	0.761
Lasius flavus	0.011	0.002	0.051	0.043	0.003	0.397
Lasius neoniger	0.036	0.013	0.076	0.097	0.020	0.333
Lasius speculiventris	0.012	0.003	0.040	0.080	0.009	0.502
Lasius umbratus	0.017	0.007	0.037	0.429	0.109	0.931
Myrmecina americana	0.011	0.002	0.052	0.042	0.003	0.398
Myrmica detritinodis	0.078	0.049	0.117	0.169	0.055	0.378
Myrmica lobifrons	0.056	0.036	0.082	0.299	0.118	0.568
Myrmica punctiventris	0.248	0.218	0.279	0.739	0.474	0.911
Myrmica species 1 ("AF-scu")	0.102	0.078	0.131	0.368	0.152	0.642
Myrmica species 2 ("AF-smi")	0.064	0.039	0.097	0.148	0.036	0.385
Prenolepis imparis	0.012	0.002	0.054	0.031	0.002	0.334
Stenamma brevicorne	0.017	0.005	0.046	0.103	0.014	0.526
Stenamma diecki	0.030	0.014	0.056	0.302	0.097	0.725
Stenamma impar	0.049	0.026	0.081	0.168	0.052	0.396
Stenamma schmitti	0.013	0.005	0.030	0.252	0.046	0.753
Tapinoma sessile	0.023	0.010	0.047	0.171	0.035	0.552
Temnothorax ambiguus	0.056	0.015	0.138	0.031	0.003	0.150
Temnothorax curvispinosus	0.057	0.022	0.113	0.037	0.005	0.169
Temnothorax longispinosus	0.086	0.062	0.114	0.333	0.141	0.587

Table 3. Estimated probabilities of capture and occurrence (with 95% credible intervals) for ant species captured in forest habitat. Probabilities are estimated at the average value of the covariates observed in the sample.

Species	Capture probability			Occurrence probability		
	Median	2.5%	97.5%	Median	2.5%	97.5%
Camponotus herculeanus	0.014	0.002	0.050	0.190	0.040	0.731
Camponotus novaeboracensis	0.066	0.043	0.094	0.348	0.172	0.571
Camponotus pennsylvanicus	0.007	0.001	0.040	0.134	0.017	0.723
Dolichoderus plagiatus	0.015	0.002	0.073	0.105	0.016	0.515
Dolichoderus pustulatus	0.090	0.071	0.112	0.701	0.491	0.863
Formica neorufibarbis	0.007	0.001	0.040	0.126	0.015	0.691
Formica subaenescens	0.353	0.308	0.402	0.371	0.194	0.580
Formica subsericea	0.014	0.004	0.037	0.295	0.083	0.774
Lasius alienus	0.020	0.006	0.054	0.191	0.051	0.550
Lasius speculiventris	0.050	0.010	0.138	0.077	0.014	0.263
Lasius umbratus	0.008	0.001	0.034	0.210	0.037	0.766
Leptothorax canadensis	0.007	0.001	0.039	0.142	0.018	0.764
Myrmica lobifrons	0.559	0.529	0.589	0.916	0.748	0.984
Myrmica punctiventris	0.006	0.001	0.039	0.150	0.018	0.783
Myrmica species 1 ("AF-scu")	0.015	0.002	0.073	0.102	0.015	0.486
Myrmica species 2 ("AF-smi")	0.008	0.001	0.034	0.231	0.041	0.826
Stenamma brevicorne	0.007	0.001	0.041	0.149	0.019	0.772
Tapinoma sessile	0.167	0.133	0.207	0.356	0.184	0.561
Temnothorax ambiguus	0.007	0.001	0.042	0.127	0.017	0.697

Table 4. Estimated probabilities of capture and occurrence (with 95% credible intervals) for ant species captured in bog habitat. Probabilities are estimated at the average value of the covariates observed in the sample.

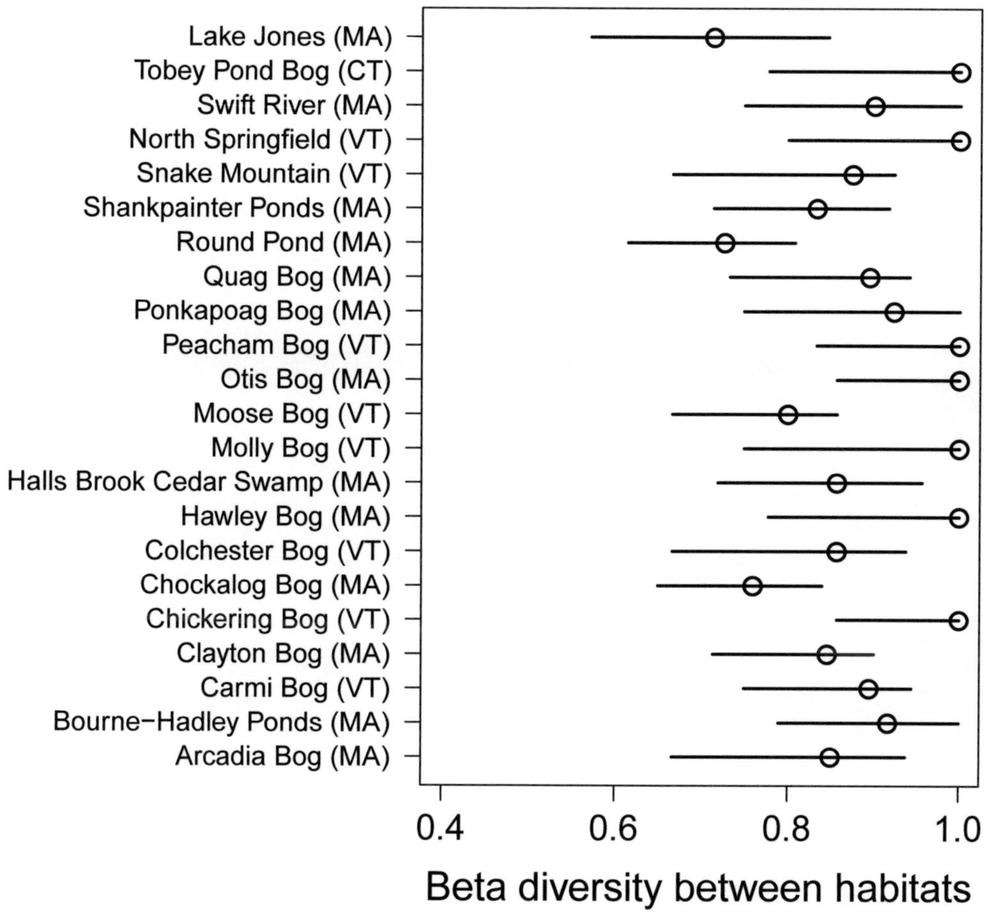

Fig. 3. Estimates of beta diversity (open circles with 95% credible intervals) between ant communities present in bog and forest habitats at each sample location.

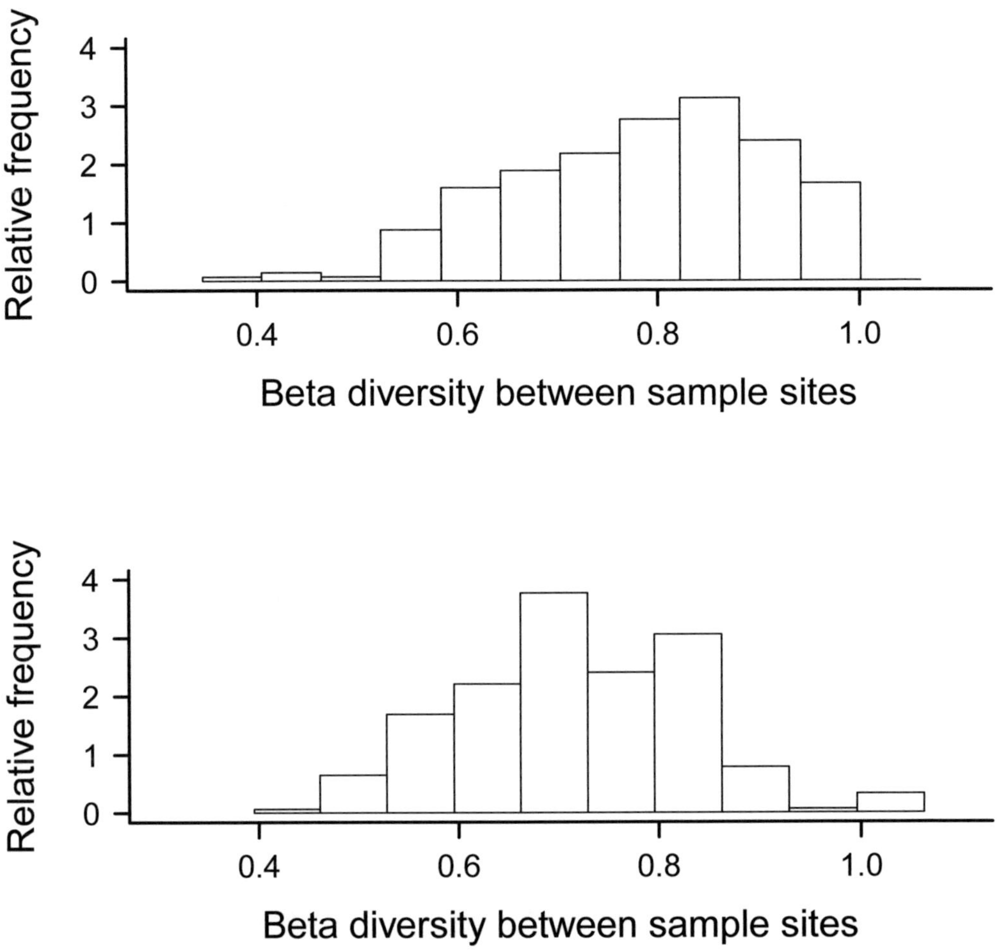

Fig. 4. Distribution of estimates of beta diversity computed for all pairwise combinations of samples collected in forest habitat (upper panel) or bog habitat (lower panel).

density and the first two principal components of vegetation structure, and they reported significant negative associations between species density and four other covariates (LAT, LAI, GSF, and ELEV). Collectively, these six regressors accounted for 83% of the variation in species density. In the present analysis of forest data, the best-fitting model included the effects of four covariates (LAT, LAI, GSF, and ELEV), and the estimated effects of these covariates were all significantly negative, which agrees qualitatively with the regression results of Gotelli & Ellison (2002), though principal components of vegetation structure were not included in the present analysis.

In comparing the results obtained using the linear regression model (Gotelli & Ellison 2002) and the hierarchical model of species occurrences and captures, we note that while both models revealed the same set of negative predictors of ant occurrence in forest habitat (Figure 1), the regression model's associations between species density of bog ants and two predictors (latitude and vegetation structure) are not supported by the hierarchical model. Part of the difference in these results may be attributed to the fact that slightly different data sets were used in the two analyses. Species detected using tuna baits, hand collections, and leaf-litter sorting (in forest habitats) were included in the regression analysis, whereas only species captured in pitfall traps were used in the present analysis. However, these differences in data are relatively minor because the alternative sampling methods used by Gotelli & Ellison (2002) added only a few rare species to their analysis. Instead, we believe the different results stem primarily from differences in the underlying assumptions of these two models. The regression model assumes (1) that the effects of environmental covariates are identical for each species and are linearly related to species density and (2) that residual errors in species density are normally distributed and do not distinguish between measurement errors and heterogeneity among species in their response to covariates. In contrast, the hierarchical model assumes that the effects of environmental covariates differ among species (Figure 1) and that occurrence probabilities and capture probabilities can be estimated separately for each species (Tables 3 and 4) owing to the replicated sampling at each site.

The estimated probabilities of occurrence and capture of each species are of great interest in themselves and highlight differences in species compositions between ants found in bog and forest habitats. For example, the forest species with the highest occurrence probability was *Aphaenogaster rudis* (species complex) ($\hat{\psi} = 0.779$). This species is taxonomically unresolved and currently includes a complex of poorly differentiated species across its geographic range (Umphrey 1996). *Myrmica punctiventris* had the second highest occurrence probability ($\hat{\psi} = 0.739$). Both of these species are characteristic of forest ant assemblages in New England. *A. rudis* (species complex) was never captured in bogs and the occurrence probability of *M. punctiventris* in bogs was only 0.150, almost a fivefold difference between the two habitats.

In bogs the highest occurrence probabilities were estimated for the bog specialist, *Myrmica lobifrons* ($\hat{\psi} = 0.916$), and for *Dolichoderus pustulatus* ($\hat{\psi} = 0.701$), a generalist species that sometimes builds carton nests in dead leaves of the carnivorous pitcher plant *Sarracenia purpurea* (A. Ellison and N. Gotelli, personal communication). Occurrence probabilities of these species in forests were only 0.299 (*M. lobifrons*) and 0.042 (*D. pustulatus*), a 3- to 16-fold difference. These pronounced differences in the occurrence probabilities of the most common species in each habitat suggest that the two habitats support distinctive ant assemblages, a conclusion also supported by the relatively high estimates of beta diversity between habitats (Figure 3).

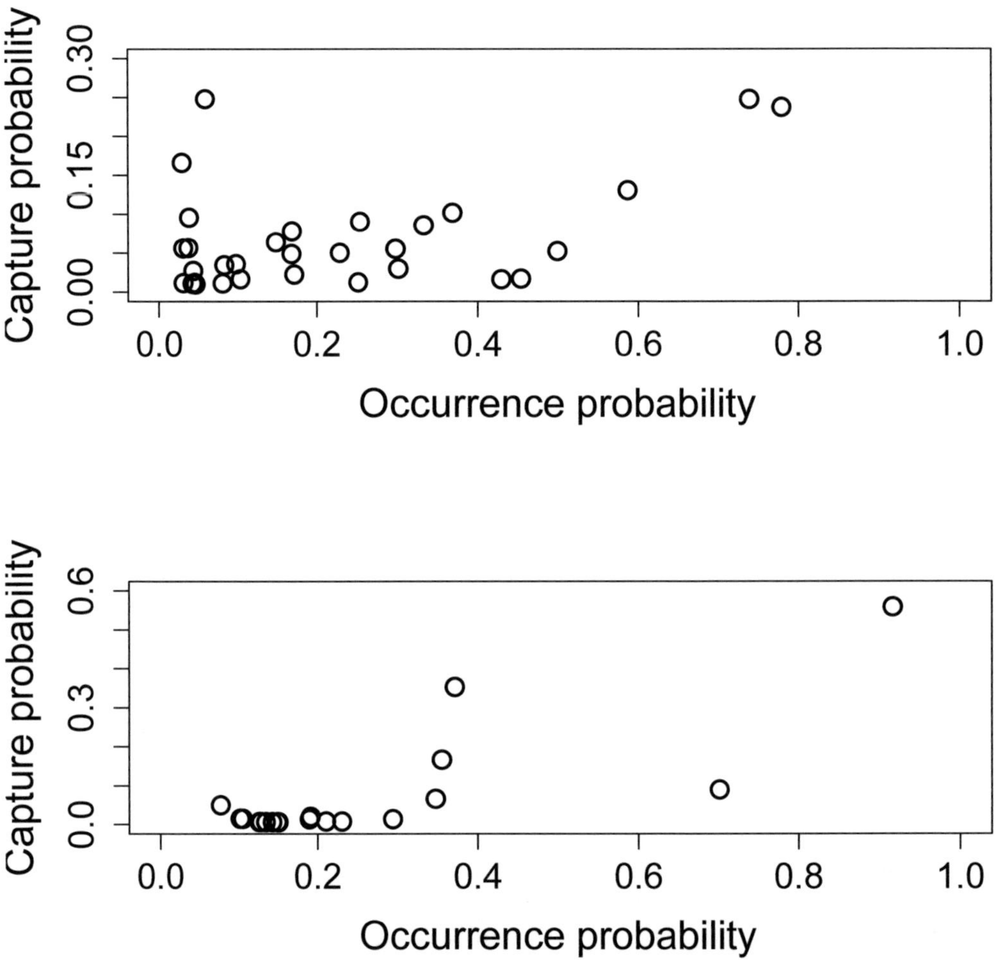

Fig. 5. Estimates of species-specific capture probability versus occurrence probability for ants in forest habitat (upper panel) and bog habitat (lower panel). Note difference in scale between ordinates of upper and lower panels.

Although occurrence and capture probabilities were positively correlated among species (Figure 5), a few rare forest species (*Formica subintegra* and *Formica subsericea*) had relatively high capture probabilities. In the forest habitat the two species with the highest capture probabilities were *F. subsericea* ($\hat{p} = 0.248$) and *Myrmica punctiventris* ($\hat{p} = 0.248$). In bogs these species had capture probabilities of only 0.014 (*F. subsericea*) and 0.006 (*M. punctiventris*), a 17- to 41-fold difference. The two species with the highest capture probabilities in the bog habitat were *Myrmica lobifrons* ($\hat{p} = 0.559$), the bog specialist, and *Formica subaenescens* ($\hat{p} = 0.353$). In the forest habitat these species had capture probabilities of only 0.056 (*M. lobifrons*) and 0.051 (*F. subaenescens*), a 7- to 9-fold difference.

The estimated probabilities of occurrence of most species in the forest habitat decreased with latitude (Figure 1), which is consistent with previous regression analyses of species density (Gotelli & Ellison 2002, figure 1). However, the occurrence probabilities of three species (*Camponotus herculeanus*, *Lasius alienus*, and *Myrmica detritinodis*) significantly increased with latitude. Two of these species, *C. herculeanus* and *M. detritinodis*, are boreal, cold-climate specialists (Ellison et al. 2012), whereas *L. alienus* has a more widespread distribution. Under climate change scenarios of increasing temperatures at high latitudes, species whose occurrence probabilities currently increase with latitude might disappear from New England as their ranges shift northward; other species in the assemblage might show no change in distribution, or might increase in occurrence.

To summarize the comparisons between our results and those reported by Gotelli & Ellison (2002), we note that within-site replication of presence-absence surveys allowed us to estimate species-specific probabilities of capture and occurrence and species-specific effects of environmental covariates. These results represent a considerable advance over traditional regression analyses of observed species density. Using a hierarchical approach to model building, we were able to infer sources of variation in measures of biodiversity – such as the effect of elevation on site-specific species richness (Figure 2) and the effect of habitat on beta diversity (Figure 3) – and to determine how these community-level patterns were related to differences in occurrence of individual species. Although many macroecological data sets collected at large spatial scales do not include within-site replicates, regional studies often use replicated sampling grids of traps or baits (Gotelli et al. 2011) that are ideal for the kind of analysis we have described. We therefore recommend that within-site replication be used in presence-absence surveys of communities, particularly when surveys are undertaken to assess levels of biodiversity.

5.2 Benefits and challenges of hierarchical modeling

Our analysis of the ant data illustrates the benefits of using hierarchical models to estimate measures of biodiversity and other community-level characteristics. By adopting a hierarchical approach to model building, an analyst actually specifies two models: one for the ecologically relevant parameters (or state variables) that are usually of primary interest but are not directly observable, and a second model for the observed data, which are related to the ecological parameters but are influenced also by sampling methods and sampling errors. This dichotomy between models of ecological parameters and models of data is extremely useful and has been exploited to solve a variety of inference problems in ecology (Royle & Dorazio 2008).

In our hierarchical model of replicated, presence-absence surveys, the parameter of primary ecological interest is the community's incidence matrix. This matrix is only partially observable because a species may be present at a sample location but not observed in the surveys. We use a binomial sampling model to specify the probability of detection (or capture) of each species and thereby to account for detection errors in the observed data. In this way estimates of the community's incidence matrix are automatically adjusted for the imperfect detectability of each species.

In our approach, measures of biodiversity are estimated indirectly as functions of the estimated incidence matrix of the community. Thus, species richness and measures of alpha or beta diversity depend on a set of model-based estimates of species- and site-specific occurrences. This approach differs considerably with classes of statistical models wherein species richness is treated as a single random variable – usually a discrete random variable – that represents the aggregate contribution of all species in the community. This "top-down" view of a community may yield incorrect inferences if heterogeneity in detectability exists among species or if the effects of environmental covariates on occurrence differ among species, as illustrated in our analysis of the ant data.

The inferential benefits of using hierarchical models to estimate measures of biodiversity are not free. As described earlier, the price to be paid for the ability to estimate probabilities of species occurrence and species detection is replication of presence-absence surveys within sample locations. In our opinion the improved understanding acquired in modeling the community at the level of individual species and the versatility attained by having accurate estimates of a community's incidence matrix far outweigh the cost of additional sampling. That said, there are other, perhaps less obvious, costs associated with these hierarchical models. Specifically, estimates of species richness and other community-level parameters may be sensitive to the underlying assumptions of these models, and these assumptions can be difficult to test using standard goodness-of-fit procedures. For example, the choice of distributions for modeling heterogeneity among species or sites may exert some influence on estimates of species richness. We assumed a bivariate normal distribution for the distribution of logit-scale, mean probabilities of occurrence and detection, but other distributions – even multimodal distributions – also might be useful. In single-species models of replicated, presence-absence surveys, estimates of occurrence are sensitive to the distribution used to specify heterogeneity in detection probabilities among sample sites (Dorazio 2007, Royle 2006); therefore, similar sensitivity can be expected in multispecies models, though this aspect of model adequacy has not been rigorously explored.

Another assumption of our model that is difficult to test is absence of false-positive errors in detection. In other words, if a species is detected (or captured), we assume that its identify is known with certainty. However, in surveys of avian or amphibian communities where species are detected by their vocalizations, misidentifications of species can and do occur (McClintock et al. 2010a,b, Simons et al. 2007). These misidentifications are even more common in circumstances where surveys are conducted by volunteers whose identification skills are highly variable (Genet & Sargent 2003). If ignored, false-positive errors in detection induce a positive bias in estimates of species occurrence because species are incorrectly "detected" at sites where they are absent. While it is possible to construct statistical models of presence-absence data that include parameters for both false-positive and false-negative detection errors (Royle & Link 2006), these models are prone to identifiability problems. To reduce these problems, Royle & Link (2006) recommended that the model's parameters be constrained to ensure that estimates of misclassification probabilities are lower than estimates of detection probabilities. This constraint, though sensible, does not provide a solution when the probabilities of misclassification and detection are nearly equal (McClintock et al. 2010b, Royle & Link 2006). The development of statistical models of species occurrence that include both false-positive and false-negative errors in detection, as well as unobserved sources of heterogeneity in both occurrence and detection probabilities, is an active area of research owing to the difficulties associated with aural detection methods.

The conceptual framework described in this paper is broadly applicable in ecological research and in assessments of biodiversity. Hierarchical, statistical models of multispecies, presence-absence data can be used to estimate current levels of biodiversity, as illustrated in our analysis of the ant data, or to assess changes (e.g., trends) in communities over time (Dorazio et al. 2010, Kéry, Dorazio, Soldaat, van Strien, Zuiderwijk & Royle 2009, Russell et al. 2009, Walls et al. 2011). The models of community change are especially relevant in ecological research because they provide an analytical framework wherein data may be used to confront alternative theories of metacommunity dynamics (Holyoak & Mata 2008, Leibold et al. 2004). Although a few classes of statistical models have been developed to infer patterns of co-occurrence among species (MacKenzie et al. 2004, Waddle et al. 2010), models for estimating the dynamics of interacting species (e.g., competitors or predators) from replicated, presence-absence data have not yet been formulated. Such models obviously represent an important area of future research.

6. Acknowledgments

Collection of the original ant dataset was supported by NSF grants 98-05722 and 98-08504 to AME and NJG, respectively, and by contract MAHERSW99-17 from the Massachusetts Natural Heritage and Endangered Species Program to AME. Additional support for AME's and NJG's research on the distribution of ants in response to climatic change is provided by the U.S. Department of Energy through award DE-FG02-08ER64510. The statistical modeling and analysis was conducted as a part of the Binary Matrices Working Group at the National Institute for Mathematical and Biological Synthesis, sponsored by the National Science Foundation, the U.S. Department of Homeland Security, and the U.S. Department of Agriculture through NSF Award #EF-0832858, with additional support from The University of Tennessee, Knoxville.

Any use of trade, product, or firm names is for descriptive purposes only and does not imply endorsement by the U.S. Government.

7. Appendix: Technical details

7.1 Model fitting and software

Here we describe methods for fitting our hierarchical model using the Markov chain Monte Carlo (MCMC) algorithms implemented in the software package, JAGS (Just Another Gibbs Sampler), which is freely available at the following web site: `http://mcmc-jags.sourceforge.net`. This software allows the user to specify a model in terms of its underlying assumptions, which include the distributions assumed for the observed data and the model's parameters. The latter distributions include priors, which are needed, of course, to conduct a Bayesian analysis of the data (see below). Part of the reason for the popularity of JAGS is that it allows the model to be specified and fitted without requiring the user to derive the MCMC sampling algorithms used in computing the joint posterior. That said, naive use of JAGS may yield undesirable results, and some experience is needed to ensure the accuracy of the results.

We prefer to execute JAGS remotely from R (R Development Core Team 2004) using functions defined in the R package RJAGS (`http://mcmc-jags.sourceforge.net`). In this way R is used to organize the data, to provide inputs to JAGS, and to receive outputs (results) from JAGS. However, the model's distributional assumptions must be specified in the native language of JAGS. The data files and source code needed to fit our model are provided below.

In our analysis of each data set, the posterior was calculated by initializing each of 5 Markov chains independently and running each chain for a total of 250,000 draws. The first 50,000 draws of each chain were discarded as "burn-in", and every 50th draw in the remainder of each chain was retained to form the posterior sample. Based on Gelman-Rubin diagnostics of the model's parameters (Brooks & Gelman 1998), this approach appeared to produce Markov chains that had converged to their stationary distribution. Therefore, we used the posterior sample of 20,000 draws to compute estimates of the model's parameters and 95% credible intervals.

7.2 Prior distributions

Our prior distributions were chosen to specify prior indifference in the magnitude of each parameter. For example, we assumed a Uniform(0,1) prior for Ω, the probability that a species in the augmented data set is a member of the N species vulnerable to capture. It is easily shown that this prior induces a discrete uniform prior on N, which assigns equal probability to each integer in the set $\{0, 1, \ldots, M\}$. We also used the uniform distribution for the correlation parameter ρ; specifically, we assumed a Uniform(-1,1) prior for ρ, thereby favoring no particular value of ρ in the analysis.

Each of the heterogeneity parameters $(\sigma_{a_0}, \sigma_{b_0}, \sigma_{b_l})$ was assigned a half-Cauchy prior (Gelman 2006) with unit scale parameter, which has probability density function

$$f(\sigma) = 2/[\pi(1+\sigma^2)].$$

Gelman (2006) showed that this prior avoids problems that can occur when alternative "noninformative" priors are used (including the nearly improper, Inverse-Gamma(ϵ, ϵ) family).

Currently, there is no consenus choice of noninformative prior for the logit-scale parameters of logistic-regression models (Gelman et al. 2008, Marin & Robert 2007). To specify a prior for the logit-scale parameters of our model $(\alpha_0, \beta_0, \beta_l)$, we used an approach described by Gelman et al. (2008). Recall that the covariates of our model are centered and scaled to have mean zero and unit variance; therefore, we seek a prior that assigns low probabilities to large effects on the logit scale. The reason for this choice is that a difference of 5 on the logit scale corresponds to a difference of nearly 0.5 on the probability scale. Because shifts in the value of a standardized covariate seldom, in practice, correspond to outcome probabilities that change from 0.01 to 0.99, the prior of a logit-scale parameter should assign low probabilities to values outside the interval (-5,5). The family of zero-centered t-distributions with parameters σ (scale) and ν (degrees of freedom) can be used to specify priors with this goal in mind.

For example, Gelman et al. (2008) recommended a t-distribution with $\sigma = 2.5$ and $\nu = 1$ as a "robust" alternative to a t-family approximation of Jeffreys' prior ($\sigma = 2.5$ and $\nu = 7$). However, when the logit-scale parameter (say, θ) is transformed to the probability scale ($p = 1/(1 + \exp(-\theta))$), both of these priors assign high probabilities in the vicinity of $p = 0$ and $p = 1$, which is not always desirable. As an alternative, we used a t-distribution with $\sigma = 1.566$ and $\nu = 7.763$ as a prior for each logit-scale parameter of our model. This distribution approximates a Uniform$(0, 1)$ prior for p and assigns low probabilities to values outside the interval (-5,5).

Given our choice of priors and the amount of information in the ant data, parameter estimates based on a single model are unlikely to be sensitive to the priors used in our analysis. However, it is well known that the distributional form of a noninformative prior can exert considerable influence on posterior model probabilities (Kadane & Lazar 2004, Kass & Raftery 1995). Because these probabilities are used to select a single model for inference, we examined the sensitivity of the model probabilities to our choice of priors. In particular, we considered a t-family approximation of Jeffreys' prior ($\sigma = 2.482$ and $\nu = 5.100$) as an alternative for the logit-scale parameters of our model. As described earlier, Jeffreys' prior is commonly used in Bayesian analyses of logistic-regression models.

7.3 Data files and source code

The following files were used to fit our hierarchical model to the ant data sets.

`AntDetections1999.csv` – species- and site-specific capture frequencies of ants in bog and forest habitats (format is comma-delimited with first row as header)

`GetDetectionMatrix.R` – R code for reading capture frequencies of ants from data file and returning a species- and site-specific matrix of capture frequencies of ants collected in a specified habitat ('Forest' or 'Bog')

`GetSiteCovariates.R` – R code for reading covariates from data file

`MultiSpeciesOccModelAve.R` – R and JAGS code for defining and fitting the hierarchical model

`SiteCovariates.csv` – site-specific values of covariates (format is comma-delimited with first row as header)

8. References

Anderson, M. J., Crist, T. O., Chase, J. M., Vellend, M., Inouye, B. D., Freestone, A. L., Sanders, N. J., Cornell, H. V., Comitka, L. S., Davies, K. F., Harrison, S. P., Kraft, N. J. B., Stegen, J. C. & Swenson, N. G. (2011). Navigating the multiple meanings of β diversity: a roadmap for the practicing ecologist, *Ecology Letters* 14: 19–28.

Boulinier, T., Nichols, J. D., Sauer, J. R., Hines, J. E. & Pollock, K. H. (1998). Estimating species richness: the importance of heterogeneity in species detectability, *Ecology* 79: 1018–1028.

Brooks, S. P. & Gelman, A. (1998). General methods for monitoring convergence of iterative simulations, *Journal of Computational and Graphical Statistics* 7: 434–455.

Colwell, R. K., Mao, C. X. & Chang, J. (2004). Interpolating, extrapolating, and comparing incidence-based species accumulation curves, *Ecology* 85: 2717–2727.

Dorazio, R. M. (2007). On the choice of statistical models for estimating occurrence and extinction from animal surveys, *Ecology* 88: 2773–2782.

Dorazio, R. M., Kéry, M., Royle, J. A. & Plattner, M. (2010). Models for inference in dynamic metacommunity systems, *Ecology* 91: 2466–2475.

Dorazio, R. M. & Royle, J. A. (2005). Estimating size and composition of biological communities by modeling the occurrence of species, *Journal of the American Statistical Association* 100: 389–398.

Dorazio, R. M., Royle, J. A., Söderström, B. & Glimskär, A. (2006). Estimating species richness and accumulation by modeling species occurrence and detectability, *Ecology* 87: 842–854.

Draper, D. (1995). Assessment and propagation of model uncertainty (with discussion), *Journal of the Royal Statistical Society, Series B* 57: 45–97.

Ellison, A. M., Gotelli, N. J., Alpert, G. D. & Farnsworth, E. J. (2012). *A field guide to the ants of New England*, Yale University Press, New Haven, Connecticut.

Francoeur, A. (1997). Ants (Hymenoptera: Formicidae) of the Yukon, *in* H. V. Danks & J. A. Downes (eds), *Insects of the Yukon*, Survey of Canada (Terrestrial Arthropods), Ottawa, Ontario, pp. 901–910.

Gelman, A. (2006). Prior distributions for variance parameters in hierarchical models (Comment on article by Browne and Draper), *Bayesian Analysis* 1: 515–534.

Gelman, A., Carlin, J. B., Stern, H. S. & Rubin, D. B. (2004). *Bayesian data analysis, second edition*, Chapman and Hall, Boca Raton.

Gelman, A., Jakulin, A., Pittau, M. G. & Su, Y.-S. (2008). A weakly informative default prior distribution for logistic and other regression models, *Annals of Applied Statistics* 2: 1360–1383.

Genet, K. S. & Sargent, L. G. (2003). Evaluation of methods and data quality from a volunteer-based amphibian call survey, *Wildlife Society Bulletin* 31: 703–714.

Gotelli, N. J. (2000). Null model analysis of species co-occurrence patterns, *Ecology* 81: 2606–2621.

Gotelli, N. J. & Ellison, A. M. (2002). Biogeography at a regional scale: determinants of ant species density in New England bogs and forests, *Ecology* 83: 1604–1609.

Gotelli, N. J., Ellison, A. M., Dunn, R. R. & Sanders, N. J. (2011). Counting ants (Hymenoptera: Formicidae): biodiversity sampling and statistical analysis for myrmecologists, *Myrmecological News* 15: 13–19.

Holyoak, M. & Mata, T. M. (2008). Metacommunities, *in* S. E. Jorgensen & B. D. Fath (eds), *Encyclopedia of Ecology*, Academic Press, Oxford, pp. 2313–2318.

Kadane, J. B. & Lazar, N. A. (2004). Methods and criteria for model selection, *Journal of the American Statistical Association* 99: 279–290.

Kass, R. E. & Raftery, A. E. (1995). Bayes factors, *Journal of the American Statistical Association* 90: 773–795.

Kéry, M., Dorazio, R. M., Soldaat, L., van Strien, A., Zuiderwijk, A. & Royle, J. A. (2009). Trend estimation in populations with imperfect detection, *Journal of Applied Ecology* 46: 1163–1172.

Kéry, M. & Royle, J. A. (2009). Inference about species richness and community structure using species-specific occupancy models in the national Swiss breeding bird survey MHB, *in* D. L. Thomson, E. G. Cooch & M. J. Conroy (eds), *Modeling demographic processes in marked populations, series: environmental and ecological statistics, volume 3*, Springer, Berlin, pp. 639–656.

Kéry, M., Royle, J. A., Plattner, M. & Dorazio, R. M. (2009). Species richness and occupancy estimation in communities subject to temporary emigration, *Ecology* 90: 1279–1290.

Kuo, L. & Mallick, B. (1998). Variable selection for regression models, *Sankhya* 60B: 65–81.

Leibold, M. A., Holyoak, M., Mouquet, N., Amarasekare, P., Chase, J. M., Hoopes, M. F., Holt, R. D., Shurin, J. B., Law, R., Tilman, D., Loreau, M. & Gonzalez, A. (2004). The metacommunity concept: a framework for multi-scale community ecology, *Ecology Letters* 7: 601–613.

MacKenzie, D. I., Bailey, L. L. & Nichols, J. D. (2004). Investigating species co-occurrence patterns when species are detected imperfectly, *Journal of Animal Ecology* 73: 546–555.

Magurran, A. E. (2004). *Measuring biological diversity*, Blackwell, Oxford.

Marin, J.-M. & Robert, C. P. (2007). *Bayesian Core*, Springer, New York.

McClintock, B. T., Bailey, L. L., Pollock, K. H. & Simon, T. R. (2010a). Experimental investigation of observation error in anuran call surveys, *Journal of Wildlife Management* 74: 1882–1893.

McClintock, B. T., Bailey, L. L., Pollock, K. H. & Simon, T. R. (2010b). Unmodeled observation error induces bias when inferring patterns and dynamics of species occurrence via aural detections, *Ecology* 91: 2446–2454.

R Development Core Team (2004). *R: A language and environment for statistical computing*, R Foundation for Statistical Computing, Vienna, Austria. ISBN 3-900051-07-0.
URL: *http://www.R-project.org*

Rich, P. M., Clark, D. B., Clark, D. A. & Oberbauer, S. F. (1993). Long-term study of solar radiation regimes in a tropical wet forest using quantum sensors and hemispherical photography, *Agricultural and Forest Meteorology* 65: 107–127.

Robert, C. P. & Casella, G. (2004). *Monte Carlo Statistical Methods (second edition)*, Springer-Verlag, New York.

Royle, J. A. (2006). Site occupancy models with heterogeneous detection probabilities, *Biometrics* 62: 97–102.

Royle, J. A. & Dorazio, R. M. (2008). *Hierarchical modeling and inference in ecology*, Academic Press, Amsterdam.

Royle, J. A. & Dorazio, R. M. (2011). Parameter-expanded data augmentation for Bayesian analysis of capture-recapture models, *Journal of Ornithology* 123: in press.

Royle, J. A. & Link, W. A. (2006). Generalized site occupancy models allowing for false positive and false negative errors, *Ecology* 87: 835–841.

Russell, R. E., Royle, J. A., Saab, V. A., Lehmkuhl, J. F., Block, W. M. & Sauer, J. R. (2009). Modeling the effects of environmental disturbance on wildlife communities: avian responses to prescribed fire, *Ecological Applications* 19: 1253–1263.

Simons, T. R., Alldredge, M. W., Pollock, K. H. & Wettroth, J. M. (2007). Experimental analysis of the auditory detection process on avian point counts, *Auk* 124: 986–999.

Umphrey, G. (1996). Morphometric discrimination among sibling species in the *fulva - rudis - texana* complex of the ant genus *Aphaenogaster* (Hymenoptera: Formicidae), *Canadian Journal of Zoology* 74: 528–559.

Waddle, J. H., Dorazio, R. M., Walls, S. C., Rice, K. G., Beauchamp, J., Schuman, M. J. & Mazzotti, F. J. (2010). A new parameterization for estimating co-occurrence of interacting species, *Ecological Applications* 20: 1467–1475.

Walls, S. C., Waddle, J. H. & Dorazio, R. M. (2011). Estimating occupancy dynamics in an anuran assemblage from Louisiana, USA, *Journal of Wildlife Management* 75: in press.

Zipkin, E., Royle, J. A., Dawson, D. K. & Bates, S. (2010). Multi-species occurrence models to evaluate the effects of conservation and management actions, *Biological Conservation* 143: 479–484.

14

Isolation and Identification of Indigenous Microorganisms of Cocoa Farms in Côte d'Ivoire and Assessment of Their Antagonistic Effects Vis-À-Vis *Phytophthora palmivora*, the Causal Agent of the Black Pod Disease

Joseph Mpika, Ismaël B. Kebe and François K. N'Guessan
Laboratoire de Phytopathologie, CNRA, Divo, Côte d'Ivoire

1. Introduction

The black pod disease due to *Phytophthora* spp is a destructive disease of cocoa. Worldwide, yield losses have been estimated to 30% (Lass, 1985). Côte d'Ivoire, the first cocoa producing country in the world, with 44% of the world market (ICCO, 2000) is also concerned by this disease. Several species of *Phytophthora* are involved in the disease. In Africa, two species, *P. palmivora* and *P. megakarya*, are the most damaging. The first species, which is the most common, causes damage in all the cocoa producing countries in the world, with yield losses between 20 to 30% ; the second, endemic to central and west Africa, is the most aggressive. This pathogen may cause the loss of the whole pod production in some countries (Flood, 2006). In Côte d'Ivoire, since the discovery of *P. megakarya* in the western region in the 90s, the black pod disease problem became more serious (Koné, 1999; Kouamé, 2006). Yield losses increased from an average of 10% to 35-40% (Kébé *et al.*, 1996). Thus, the control of the disease became also a priority.
Although chemical control was developed by the research scientists, the dissemination of this method to the farmers was little successful. The low level of adoption of this technology by the farmers could be explained by the high cost of the fungicides as well as the difficulties related to the provision of water and the application of the fungicides. In addition, the requirements of the international market in terms of bean quality, environmental constrains, health issues for the consumers, and the different moratoriums in this area from the market partners (Anonyme, 2006), are numbers of constraints that do not facilitate the development of the chemical control method.
The strategy adopted in Côte d'Ivoire to control the black pod disease is based on integrated management, which is cost effective and environmentally friendly. This approach combines the use of agronomic practices, resistant cocoa varieties and natural antagonistic microorganisms of *Phytophthora*. Research works continue in order to improve agronomic practices and varietal resistance. The use of natural antagonistic microorganisms of

Phytophthora in the control of black pod disease is a new area of investigation explored by research scientists in several cocoa producing countries. Thus, some species in the genera *Trichoderma* and *bacillus* have been described by several scientific teams as potential biological agent for the control of *Phytophthora* spp. on coca (Bong *et al.*, 1996; Krauss *et al.*, 2003; Mpika, 2002). On other crops, fungi and bacteria belonging to several genera including *Pseudomonas, Burkholderia, Streptomyces, Serratia, Penicillium, Geniculosporiun, Gliocladium, Aspergillus, Coniothyrium, Ampelomyces, Phytophthora, Botrytis, Colletotrichum, Pythium, Rhizoctonia, Fusarium, Gaeunannomyces* and *verticillium* have been described as antagonists of many fungi, pathogens of plants. Members of these genera are pathogens of plants such tomato, rice, cucumber, maize, cotton and beans (Hebber *et al.*, 1998 ; Benhamoun *et al.*, 2000 ; Singh *et al.*, 1999 ; de Cal *et al.*, 1999 ; Paulitz and Linderman, 1991 ; Gerlagh *et al.*, 1999 ; Vidhyasekaran and Muthamilan, 1999 ; Bong and Stephen, 1999 ; Tondje *et al.*, 2006a,b).

During this study, the biodiversity was explored in the cocoa ecosystems. Microorganisms, potential antagonists of *Phytophthora* spp., were collected from pods and soils of cocoa farms. A collection of microorganisms was established. The antagonistic effect of these microorganisms on *Phytophthora* was assessed in the laboratory and in the field on the cacao trees.

2. Materials and methods

2.1 Samples and culture media

The microorganisms used in this study were isolated from the cocoa ecosystems, either from soils or pods. The soil samples were collected from cocoa farms in Abengourou in the the Eastern region of the country, and in Divo in the West-central region. In each of the locations, the soil samples were taken in 8 cocoa farms grouped in two categories according to the age of the trees. The first category was made up by 4 young plots with trees being 3 to 5 years old. These newly established plots still have open canopies with little soil litters. The second category was made up by 4 olds farms (25 to 30 years old). These farms, with very mature and fully bearing trees have closed canopies and abundant litters on the soil. In each plot, a bulk sample of 800 g of soil was made-up with 4 samples taken at the base of 4 cacao trees bearing many healthy pods and selected randomly in the plot.

Before taking each soil sample, the litter was totally removed. The samples were then taken in the superficial zones colonized by the fine root system because of fertilizer application. It is in this horizon of 30 to 40 cm deep that the samples were taken (Davet and Rouxel, 1997). Each bulk sample was carefully mixed and divided in two equal parts which were put in plastic containers. In one of them, baits for the antagonists made-up by fragments of pod plugs infected by *Phytophthora palmivora,* were buried in the soil sample (Tim *et al.*, 2003). The other was left without any bait. In order to obtain a good colonization of the pod plugs, the soil samples were kept in the laboratory at 20°C for 30 days.

The pods used for the isolation of the microorganisms were collected in the main cocoa producing regions of Côte d'Ivoire. Thus, 390 healthy pods were collected from the 13 cocoa producing regions. In each region, pods were collected in 10 farms, at a rate of 3 pods per farm. The pods were kept in plastic bags labeled with information on the samples and brought to the laboratory for isolation of endophytes.

The diversity of fungi and bacteria in the soil of cocoa farms and in the pods were evaluated on selective culture media. For the isolation of bacteria, two selective media, including the PCAT medium (P. cepacia Azeaic acid tryptamine), specific to *Pseudomonas* and *Burkhoderia*

(Burbage, 1982), and the NYD medium (nutrient yeast dextrose) (Guizzardi and Pratella, 1996), adapted to a larger spectrum of bacteria were used. For the isolation of fungi, the TME (*Trichoderma* medium E), specific to fungi belonging to the genera *Trichoderma* and *Gliocladium* (Papavizas and Lumsden, 1982) were used. The PDA (potato dextrose agar), adapted to a larger spectrum of fungi was used for the isolation of the other fungi.

2.2 Isolation of the microorganisms

2.2.1 Pod endophytes

The surface of the pods was beforehand washed with tap water, and then underwent a series of disinfection in ethanol at 95 % for 30 seconds, in sodium hypochlorite at 10 % for 2 minutes, and again in ethanol at 75 % for 2 minutes, in order to eliminate the microorganisms present on the husk. The pods were rinsed three times in sterile distilled water to eliminate any trace of disinfectant (Arnold, 1999; Evans, et *al.*, 2003; Rubini et *al.*, 2005).

The sampling zone is chosen and the superficial tissues were removed using a sterile scalpel. Ten cubic shape fragments of 5 to 7 mm were taken per pod in the husk. The samples taken were put in culture on the selective media contained in Petri dishes. The incubation was made in the dark in a steam room, at 26 °C for 2 days for the bacteria and 7 days for the fungi.

2.2.2 Soil microorganisms

The microorganisms were obtained by direct isolation from the soil according to the method described by Davet and Rouxel (1997) and from fragments of pod husk buried in the soil samples. The soil samples were beforehand dried, ground and calibrated by sieving. The fragments of husk were ground in a porcelain mortar to separate each living propagule, because of the gelatinous consistency of the decomposing pod husk.

In both cases, 10 g of ground soil were transferred in 90 ml of sterile distilled water contained in an Erlenmeyer. The mixture is then put in agitation for 30 minutes to obtain a good separation of the particles. To obtain a variable concentration of propagules and facilitate the enumeration of the colonies, a series of dilution was performed from the initial solution whose concentration was 10^{-1} (Rapilly, 1968). To obtain a solution of 10^{-2}, 1 ml of the initial solution was mixed in 9 ml of sterile distilled water. Thus, a series of dilution from 10^{-2} to 10^{-9} was performed in hemolytic tubes. For each dilution, 100µl was pipetted and spread onto the surface of the culture media in Petri dishes. For each dilution and for each medium, 4 Petri dishes were inoculated. The incubation was also made in the dark in a steam room at 26 °C for 2 days for the bacteria and 7 days for the fungi.

2.3 Conservation and identification of the microorganisms

The microorganisms isolated were first purified by two or three successive monospore transplantings on specific culture media. Once purified, each isolate was designated by a code number. For strains of bacteria, the code numbers were preceded by the letter B, followed by an order number. The nomenclature of the fungi isolates begins with one or several first letters of the name of the genus, followed by an order number. The conservation of the microorganisms was then made in a freezer at a temperature of - 80°C for the bacteria, and - 10°C for the fungi. In both cases, agar disks taken near the edge of the purified culture were transferred to 1.5 ml sterile Eppendorf microtubes containing glycerol at 50 %. The identification of the isolated microorgamisms was based on the macroscopic, microscopic, biochemical and molecular characters. The molecular characterization was made by the method developed by Druzhinina *et al.* (2005) to identify the various species of fungi isolated. This method uses baits specific targeting genes encoding for translation elongation factor 1-alpha (*tef1*) obtained on sequences IST 1 and 2 of the DNAr.

2.4 Evaluation of the antagonistic effect of the microorganisms

The antagonistic effect of the isolated microorganisms against *P. palmivora* was evaluated using three tests. The first test conducted *in vitro,* is a test of direct confrontation between *P. palmivora* and the microorganism in a mixed culture. The second test was realized *in vivo* on leaf disks of cacao tree. The test consists in measuring the leaf susceptibility to *P. palmivora* in the presence of the microorganism according to the scoring scale of Blaha which varies from 0 to 5 (Nyassé *et al.*, 1997). The third test was carried out in the field on cacao trees.

The confrontation between *Trichoderma* and *P. palmivora* was realized in Petri dishes containing agar culture media made of potato broth (PDA medium). A fragment of mycelium, 6 mm in diameter, was taken around the edge of the cultures of each fungus. These fragments were transplanted face to face in the same Petri dish, 2 cm from the center of the dishes (Benhanou and Chet, 1996). The controls were monocultures of each of the 2 fungi being confronted. In this case, the fragment of mycelium was placed in the center of the Petri dishes. Each treatment was conducted in 6 replications. The incubation was done in the dark in a cryptogrammic steam room. Twenty four hours after the start of the cultures, the mycelial growth of each fungus was measured daily until the Petri dish was full with the fungal development. Seven days after the mycelial strands of both fungi have met, the survival of the spores of *P. Palmivora* was evaluated by taking fragment of mycelium in the Petri dishes containing the mixed culture according to an axis which passes by the center and the transplanting sites of both fungi. Two consecutive samples were taken 0,5 cm apart. Thus, nine samples were taken in every Petri dish of mixed culture. The fragments of mycelium taken were placed in lesions made on healthy pods collected from trees of the same clone, making sure the pods were not attacked by *Phytophthora* sp. These inoculated pods were placed in crystallizers in which the humidity was maintained by a plug of sterile cotton wool soaked with sterile distilled water. The incubation was done at the ambient temperature of the laboratory. The survival of *Phytophthora* was evaluated by recording the number of brown spots on the surface of the inoculated pods after 15 days of daily observation. The presence of *Phytophthora* in the spots was confirmed by microscopic observations.

The effect of the microorganisms on *P. palmivora* was evaluated through the leaf disk test. This test was performed on disk of cocoa leaves 15 mm in diameter. The disks were beforehand dipped into the bacterial suspension or the suspension of *Trichoderma* for 1 min, and then arranged in containers on a plate of foam soaked with water. Each disk received 10 µl of a suspension of zoospores of *P. palmivora* calibrated to 3.10^5 zoospores / ml. The controls were not dipped in bacterial suspension or suspension of *Trichoderma* before receiving the suspension of zoospores of *P. palmivora.* The leaf disks were obtained from 3 cocoa clones, the reaction of which to *Phytophthora* sp. is known (Tahi et *al.*, 2000). Thus the susceptible clone (IFC5), the moderately resistant clone (P7) and the resistant clone (SCA6) were tested. For each clone, 40 leaf disks were inoculated and placed in 4 containers, each representing a replicate. The incubation was done in the dark at 26°C for 7 days. The results were scored according to the scale of Blaha (Nyassé *et al.*, 1995).

In the field, the random target method was used to assess the effect of the microorganisms on *P. palmivora.* This method consisted in following the development of the disease on groups of 100 Cacao trees with different treatments. The 100 trees corresponding to each group were chosen randomly in the field, numbered and marked with the same color. However, care was taken so that each test tree was surrounded by 8 border trees. The *Trichoderma* based biological fungicides were applied using a knapsack sprayer at the concentration of 10^7 conidia ml^{-1}. The entire cocoa tree was treated. Six applications at 21 days interval were made. Over the duration of the trial, weekly count were made for healthy mature pods, rotting pods, wilting pods and pods damaged by squirrels. During the trial, the survival of *Trichoderma* on treated pods and flower cushions was evaluated. Thus, pod samples were taken on the treated trees every 15 days.

2.5 Data analyses

The SAS program (Statistical Analysis System, SAS Institute, Cary, NC) was used for all the statistical analyses. For the isolation of the microorganisms and the leaf disk test, the analyses of variances (ANOVA) were performed on the mean number of microorganisms colonies counted on the culture media and the mean rating score of leaf susceptibility to *Phytophthora* in the presence of bacteria and fungi. The normality of residuals and the homogeneity of the variances were verified. The mean comparisons were realized with the Student Newman and Keuls test at 5%.

3. Results

3.1 Microorganisms isolated

The exploration of the biodiversity of microorganisms obtained from soils of cocoa field and endophytes of pods revealed 2 categories of microorganisms: fungi and bacteria. On the pods, the fungi represented 66.31% of the positive isolation, against 33.04 for the bacteria. In the soil, the two categories of microorganisms were present in inverted proportion: 55.8% for the bacteria and 29.8% for the fungi. In the group of the fungi, the yeast that represented 11.6% of the population on the pods, represented only 3.7% in the soil. Among the bacteria, the Actinomycetes which represented only 0.64% on the pods, reached 10.5% in the soil (Fig.1).

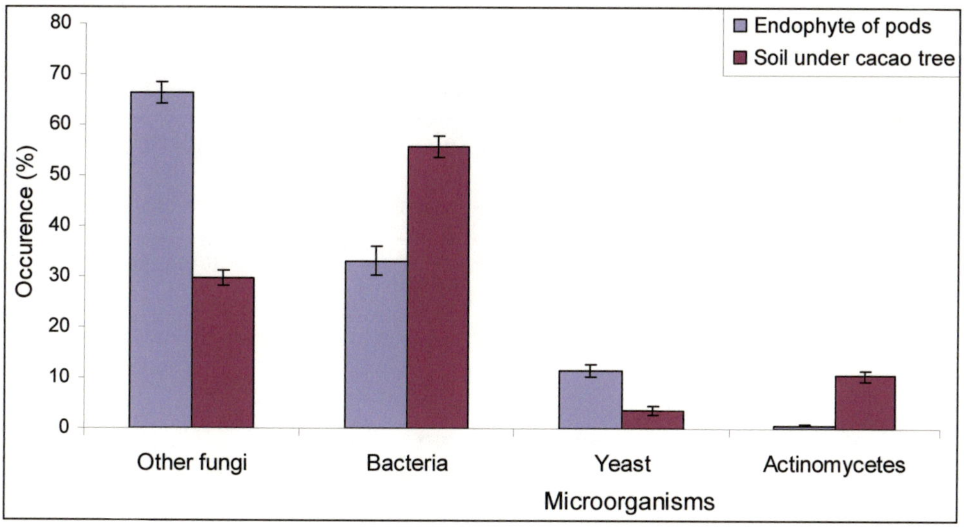

Fig. 1. Group of micro-organisms and their relative importance in the soil under cacao-plantation and the pod cortex

The results of the statistical analyses showed that the mean number of microorganisms colonies counted per gram of soil varied significantly ($P<0.05$) with the isolation method (with baits or no bait). This result revealed that the use of fragment of pods infected by *Phytophthora* as bait, significantly improved the isolation of the microorganisms on PDA, TME and PCAT media (Table 1).

	Culture media			
Isolation methods	NYD	PDA	TME	PCAT
Direct isolation	$3.27\ 10^{11}$ **a** ± $2.14\ 10^{11}$	$4.72\ 10^{6}$ **a** ±$1.42\ 10^{6}$	$4.64\ 10^{5}$ **a** ± $1.60\ 10^{5}$	$5.10\ 10^{5}$ **a** ± $2.50\ 10^{5}$
Baiting isolation	$3.80\ 10^{11}$ **a** ± $2.35\ 10^{10}$	$2.43\ 10^{9}$ **b** ± $1.41\ 10^{9}$	$6.10\ 10^{6}$ **b** ± $1.60\ 10^{6}$	$7.64\ 10^{7}$ **b** ± $3.67\ 10^{7}$

Means within the same column followed by the same letter are not significantly different according to Newman & Keuls's test at 5 % probability

Table 1. Mean numer of colonies for unit (cfu/g) according to the isolation methods and the culture media

For the samples taken in Divo and Abengourou, the statistical analyses did not show any significant differences ($P>0.05$) between regions with regard to the mean number of microorganisms colonies for the different culture media except for the NYD. There was no clear relationship between the density of microorganisms isolated and the age of field except for the bacteria on the NYD medium in Abengourou (Table 2).

		Culture media			
Locations	Age of the farms	NYD	PDA	TME	PCAT
Abengourou	Young farms	$1.9\ 10^{7}$ **a** ± $1.42\ 10^{7}$	$1.75\ 10^{6}$ **a** ± 0	$4.67\ 10^{6}$ **a** ± $2.2\ 10^{6}$	$1.2\ 10^{8}$ **a** ± $6.96\ 10^{8}$
	Old farms	$1.38\ 10^{12}$ **b** ± $4.31\ 10^{10}$	$1.15\ 10^{7}$ **b** ± $1.11\ 10^{6}$	$5.97\ 10^{6}$ **a** ± $3.22\ 10^{6}$	$2.77\ 10^{7}$ **a** ± $1.52\ 10^{7}$
Divo	young farms	$5.81\ 10^{7}$ **a** ± $4.78\ 10^{7}$	$2.05\ 10^{8}$ **a** ± $1.42\ 10^{8}$	$1.42\ 10^{6}$ **a** ± $7.96\ 10^{5}$	$5.75\ 10^{6}$ **a** ± $4\ 10^{4}$
	Old farms	$3.2\ 10^{10}$ **a** ± $2.15\ 10^{10}$	$3.45\ 10^{9}$ **a** ± $1.99\ 10^{9}$	$1\ 10^{6}$ **a** ± $5.76\ 10^{5}$	$5.78\ 10^{4}$ **a** ± $3.77\ 10^{4}$

Means within the same column followed by the same letter are not significantly different according to Newman & Keuls's test at 5 % probability

Table 2. Mean number of colonies for unit (cfu/g) according to the locations and the age of the farms

3.2 Identification of the microorganisms isolated

Regarding the pods, 313 fungi isolates were purified. Among the purified isolates, 58 isolates belonging to 9 genera were identified. These were *Penicillium* sp (6), *Fusarium* sp (7), *Botrytis* sp. (9), *Pestalotia* sp. (24). The remaining isolates identified belong to the genera *Nigrospora* (2), *Physoderma* (1), *Polynema* (1), and *Botryodiplodia* (8). The other isolates (255) belong to diverse species or genera, but the relationship has not been established yet.

One hundred and two (102) colonies of bacterial endophytes of pods were identified. These bacteria belong to two groups based on Gram-coloration response to chemicals: 56 Gram-positive bacteria (45 bacilli and 11 cocci) and 46 gram-negative bacteria (9 bacilli and 37 cocci). The bacterial colonies B105 and B116 were identified as belonging to the genus *Bacillus*. Finally, 55 yeast strains and 2 isolates of actinomycetes were identified.
With regard to the soil, amongst 455 isolates collected and purified, 254 bacteria, 136 fungi, 48 actinomycetes and 17 yeasts were identified. In the group of the fungi, 44 isolates belonging to the genus *Trichoderma* were identified. These were *T. virens* (32 isolates), *T. harzianum* (4 isolates), *T. spirale* (6 isolates) and *T. asperellum* (2 isolates). Three isolates belonging to the genus Clonostachys were also identified. The other isolates have not been identified yet.

3.3 Effect of the microorganisms on *Phytophthora palmivora*
3.3.1 Effect of *Trichoderma* on the mycelial growth of *P. palmivora*
The direct confrontation tests realized *in vitro*, between the isolates of *Trichoderma* sp. and of those of *Phytophthora palmivora* revealed an inhibitory effect of *Trichoderma* on *P. palmivora* in mixed culture. After 3 days of confrontation, the inhibition of the growth speed became very high and the growth of *P. palmivora* practically stops (Fig.2).

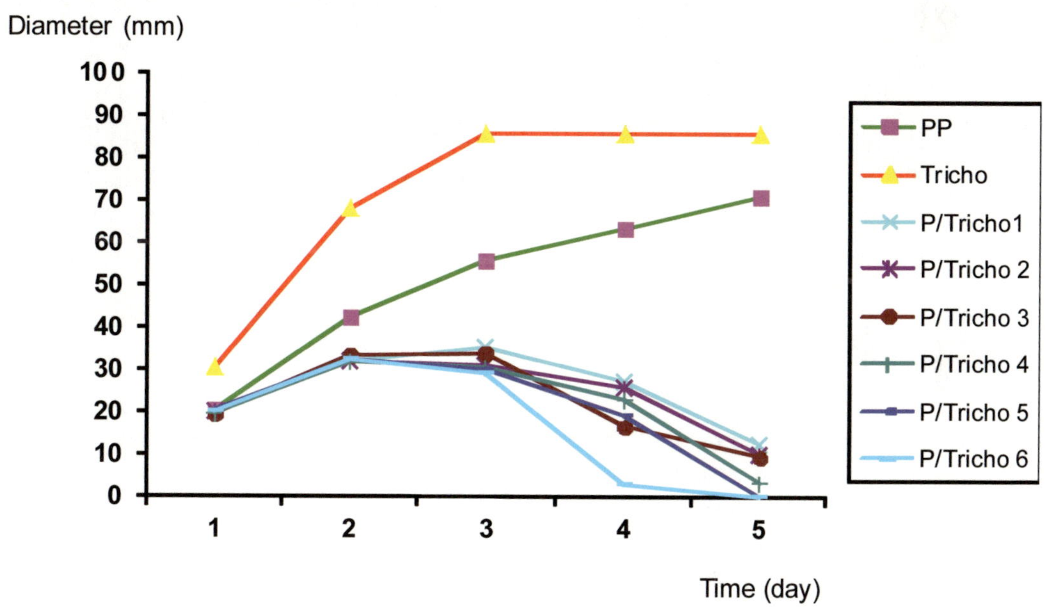

Fig. 2. Influence of *Trichoderma* on the mycelial growth of *Phytophthora palmivora*. Legend. PP : *Phytophthora palmivora*, Tricho: *Trichoderma*, P/Tricho : *Phytophthora/Trichoderma*

The capacity of *Trichoderma* to stop the mycelial growth of *P. palmivora* reveals a deep fungistatic effect. From the fourth day, we note a progressive disappearance of the mycelium of *P. palmivora*. This degradation of the mycelium of *P. palmivora* which is more accentuated at the fifth day, with all the isolates, reveals a mycoparasitic effect of

Trichoderma. After 7 days of confrontation, in mixed culture, the survival of the spores of *P. palmivora* was assessed. The influence of *Trichoderma* sp. varied according to isolates. The percentage of survival of *Phytophthora* varied from 90 to 50 % respectively with *Trichoderma* isolates 3 and 6. This rate falls to 30 to 10 % with *Trichoderma* 2 and 4. Finally, the presence of *Trichoderma* 1 and 5 in mixed culture has a very clear fungicidal effect with a percentage of survival of *P. palmivora* equal to 0.

3.3.2 Influence of the bacteria and *Trichoderma* on the leaf susceptibility to *P. palmivora*

The effect of 37 strains of bacteria on *P. palmivora* was evaluated using the leaf disk test. The results showed that, for the resistant clone (SCA 6) of cacao tree, the rating scores of the leaf susceptibility varied from 3.3 to 0.59, respectively with strains B104 and B105 (Fig. 3).

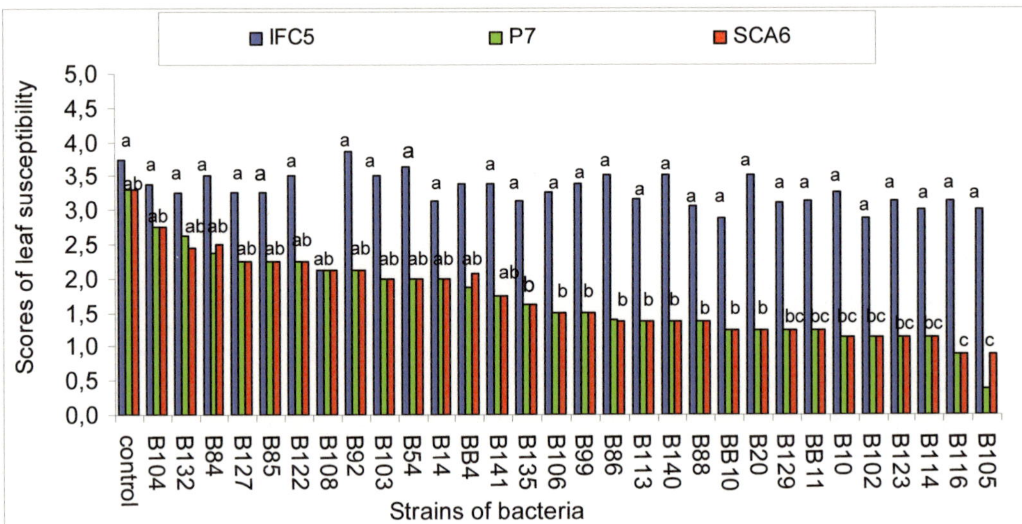

Fig. 3. Effect of the bacterial strains on the leaf discs susceptibility to *P. palmivora* for clones IFC5, P7 and SCA6.

With the moderately resistant clone (P7), the rating scores varied in the same proportions. In both cases, the analysis of variance revealed a significant ($P<0.05$) effect of bacteria and the Student Newman & Keuls test revealed 5 homogeneous groups of bacteria (a, ab, b, bc, and c). The best results were obtained with bacteria BI05 and B116 (group c). With the susceptible clone IFC 5, we note that the reduction of the rating scores of leaf susceptibility is relatively low. The scores varied from 3.8 to 2.8. The statistical analyses did not reveal any significant ($P>0.05$) differences between the treatments and the untreated control.
Similarly, the effect of 57 *Trichoderma* isolates on *P. palmivora* was evaluated using leaf disk of three cocoa clones. For the susceptible clone (IFC5), the scores of leaf susceptibility varied from 2.4 to 0.02 respectively with the isolates T39 of *T. spirale* and T55 of *T. virens*. With the moderately resistant clone (P7), we note a reduction of the scores, which varied from 2.04 to 0.03 respectively with the isolates T39 of *T. spirale* and T28 of *T. virens*. With the resistant clone (SCA6), the scores were less than 1.5 for all the isolates of *Trichoderma* (Fig. 4).

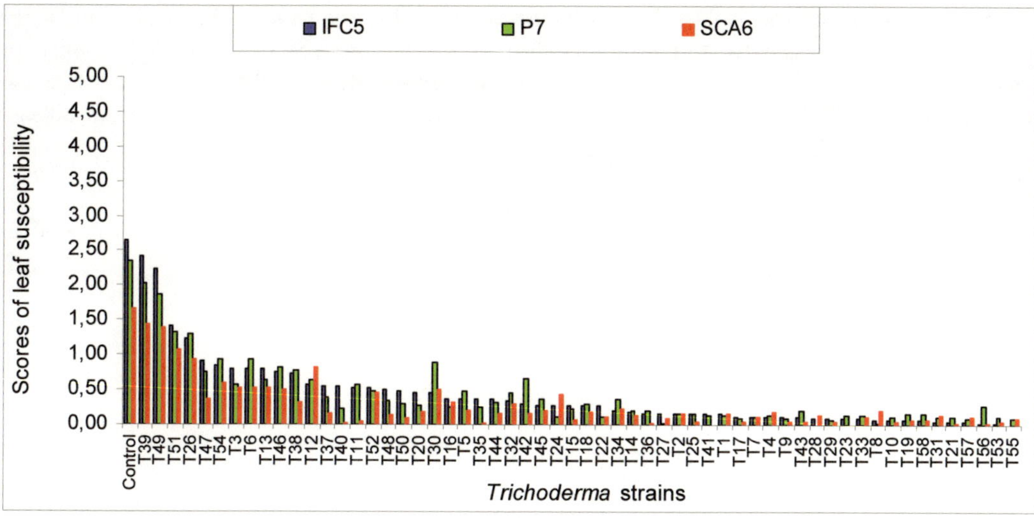

Fig. 4. Effect of *Trichoderma* strains on the leaf discs susceptibility to *P. palmivora* for clones IFC5, P7 and SCA6.

3.3.3 Field efficacy of *Trichoderma* in the control of the black pod disease due to *P. palmivora*

The effects of 4 species of *Trichoderma*, applied to the cacao trees were compared. The results obtained are presented in Table 3. The final percentages of rotten pods for year 1, were 11.75, 8.44, 7.34, and 3.6 respectively with the isolates T4 of *T. spirale*, the isolate T40 of *T. harzianum*, the isolate T7 of *T. virens* and the isolate T54 of *T. asperellum*. These percentages were lower than 18.33 obtained with the untreated control, indicating a reduction of the yield losses from 36 to 80 %. For year 2, the final percentages of diseased pods were lower than that recorded in year 1. For both years of study, the calculated efficacy index was

	Treatments	Rotten pods (%)	Reduction of loss in relation to untreated control (%)	Slope of the epidermic curve	Efficacy index (%)
Year1	Untreated control	18.33	-	15.6	
	T. spirale T4	11.75	36	8.96	42
	T. harzianum T40	8.44	54	8.82	43
	T. virens T7	7.34	60	5.29	66
	T. asperellum T54	3.6	80	3.29	79
Year2	Untreated control	7.46	-	12.04	
	T. harzianum T40	4.16	44	6.73	44
	T. virens T7	4.23	43	6.08	49
	T. spirale T4	2.46	67	4.81	60
	T. asperellum T54	2.21	70	4.68	61

Table 3. Reduction of the final losses due to the black pod disease and efficacy (%) of 4 species of *Trichoderma* during 2 years of field trials after the application

higher than 60 % for the isolate T54 of *T. asperellum*. The analysis of the evolution of the epidemic curves in each treated plot reveals that the isolate T54 of *T. asperellum* and the isolate T7 of *T. virens* substantially reduced the losses due to the black pod disease. Furthermore, these isolates delayed the onset of the epidemic of the black pod disease compared to the untreated control (Fig. 5).

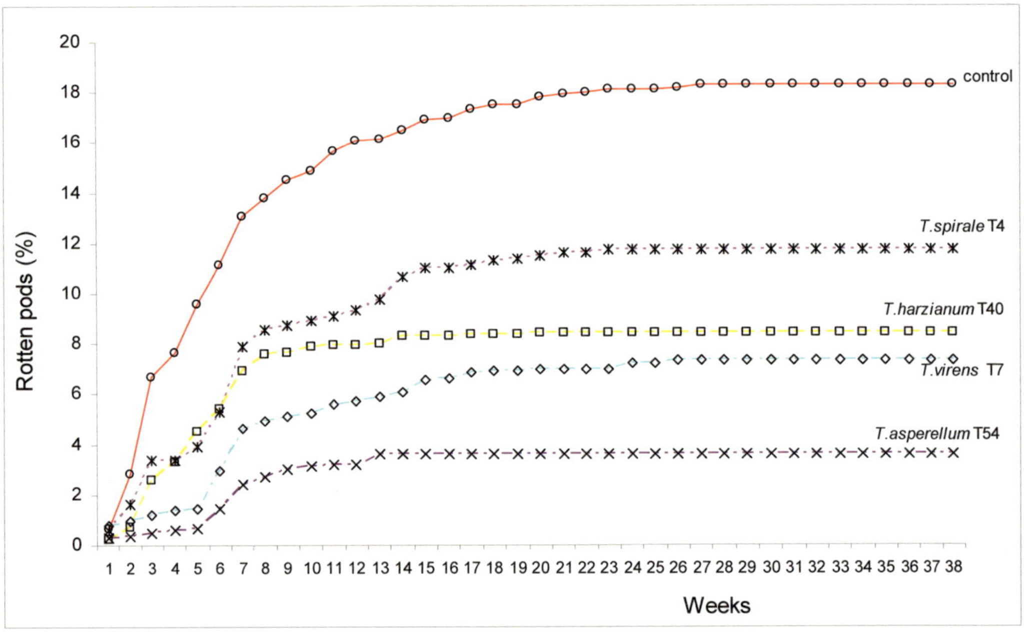

Fig. 5. Effect of the applications of biofungicide containing various species of *Trichoderma* on the evolution of black pod disease in the field.

4. Discussion

This study revealed the existence of a high biodiversity within the microorganisms' populations in the cocoa ecosystem. Regarding the two areas investigated, the soil microorganisms and pod endophytes, the biodiversity varied quantitatively and qualitatively. Endophytes are microorganisms that colonize plant tissues without causing visible symptoms in normal conditions (Carroll, 1998). Since the beginning of the 2000s, these microorganisms are usually studied in the tropical plants for their use in biological control or for the production of substances having pharmacological properties (Azevedo, 2002; Peixoto-Neto, 2002). The isolation carried out on cocoa pods showed a predominance of fungi (66.31%), against 33.04% of bacteria. Similar results were obtained by Evans *et al.* (2003) and Rubini *et al.* (2005). Regarding the soil, the results also showed a high proliferation of both categories of microorganisms with a greater proportion of bacteria. The soils of cocoa field and the pods are therefore the preferred sites of indigenous microorganism, potential antagonists of *P palmivora*, occupying the same ecological niche. The results showed that the use of baits, made by fragments of cocoa pod, infected by *P. palmivora*, significantly improved the isolations of microorganisms in the soil. This improvement could be explained by an affinity between *P. palmivora* and the collected microorganisms. An analysis of the list of the collected microorganisms revealed the existence of fungi and bacteria identified by several authors as having antagonistic effect against pathogens responsible for plant diseases. This is the case of the genus *Trichoderma* in which several species have been tested for the control of cocoa diseases (Sanogo *et al.*, 2002; Krauss and Soberanis, 2001; Tondje *et al.*, 2007; Samuels, 1996). In the group of the bacteria, the antagonistic effect of the genus *Bacillus* was demonstrated by Shari Fuddin, 2000. The direct confrontation tests carried out *in vitro* between *P. palmivora* and the *Trichoderma* isolates revealed an antagonistic effect going from the inhibition of mycelia growth (fungistatic effect), to the degradation and the disappearance of the mycelium of *P. palmivora* (mycoparasitic and fungicidal effect). Similar results were obtained with *Trichoderma harzianum* on *Fusarium oxysporum* (Hibar *et al.*, 2005). In Central America, similar results were also obtained with *Trichoderma stromaticum* in the control of witch's brooms disease of cacao and with *Trichoderma virens* against the black pod disease (Krauss and Soberanis, 2002). This study thus allowed to isolate and to purify several potential antagonists of *Phytophthora* sp, susceptible to be used in the control the black pod disease.

The evaluation of the effect of the bacterial strains on *P. palmivora* using the leaf disk test revealed a reduction of the leaf susceptibility to *P. palmivora* for resistant clones (SCA 6) and moderately resistant clones (P 7). On the other hand, this effect is less perceptible with the sensitive clone (IFC5). Similar results were obtained by Maurhofer et *al.* (1994) on tobacco, Duijff et *al.* (1997) on tomato and Chen et *al.* (1998) on cucumber. These results could reveal an increase in the level of the intrinsic resistance of the plant by the bacterial strains. The highest effects were obtained with bacterial strains B 105 and B 116, which makes them potential candidates for the biological control against *P. palmivora*. These two bacteria belong to the genus *Bacillus* endowed with an ability to sporulate, suggesting a good ability of dissemination within the framework of a biological control program. Also, on leaf disks, the effect of *Trichoderma* revealed a reduction of the size and the frequency of the necrotic lesions due to *P. palmivora*. Similar results were obtained by Bowers *et al.* (2001b) on leaf disk with *Phytophthora megakarya*. This effect of reducing leaf susceptibility results from the germination of the spores of *Trichoderma,* on the underside of the leaves, which probably inhibits or hampers the germination of the zoospores of *Phytophthora*. This germination would stimulate the mechanisms of defense, and consequently would strengthen the resistance to the penetration and the dissemination of the parasite. Similar results were obtained by Bigirimana et *al.* (1997), Howell et *al.* (2000), Sid Ahmed et *al.*, 2000 and Harman et *al.* (2004) on bean, cotton, hot pepper and corn inoculated by *T. virens* and *T. harzianum,* subjected to the attacks of *Rhizoctonia* sp., *Colletrotrichum* sp. and *Phytophthora* sp.

The effects of the isolates T4 of *T. spirale*, T7 of *T. virens* and T40 of *T. harzianum* in the field and the isolate T54 of *T. asperellum* on *P. palmivora*, showed a reduction of the incidence of the black pod disease due to *P. palmivora*. However, this effect is more striking on cacao trees treated with *T. asperellum*. Similar results were obtained by Tondje et *al.* (2007) in Cameroon after evaluating the potential effect of the isolates of *T. asperellum* on the incidence of the black pod disease due to *P. megakarya*. This reduction would be due to the mycoparasitic effect of *T. asperellum* against *Phytophthora* sp. Indeed, *T. asperellum* penetrates and destroys the propagules of *Phytophthora* on the flower cushions, thus reducing the quantity of inoculums of this parasite. Besides the mycoparasitic effect on the propagules (mycelium and sporocystes) of *P. palmivora*, the production of cellulase by *T. asperellum* in the presence of *Phytophthora* sp. would also be determining. The synergy of the modes of action in the efficicacy of *Trichoderma* to achieve the control of the disease was demonstrated by Chet *et al.* (1997); Howell (2003) and Benitez *et al.* (2004).

5. Conclusion

The exploration of the biodiversity of microorganisms in the soils collected from cocoa farms and in cocoa pods showed that it is possible to know natural populations of microorganisms and beneficial interactions for the cacao tree as well as to identify potential antagonists for *Phytophthora* sp. For the two components investigated (soil and pod), the biodiversity appeared to be highly variable both qualitatively and quantitatively. The results obtained in this study showed that it is possible to exploit the biodiversity for the control of the black pod disease due to *Phytophthora palmivora*. Several isolates of *Trichoderma* and some bacterial strains showed a deep antagonistic effect against *P. palmivora* in the laboratory. The study allowed to demonstrate the efficacy of *Trichoderma asperellum* in the control of the black pod disease. This result arouses a great hope for the cocoa farmers whose revenues constantly decline because of this disease. At national level, appropriate measures have been taken to allow a large scale use of *T. asperellum* for the control of the black pod disease in Côte d'Ivoire.

6. References

Anonyme 2006. New EU regulation on maximum residue levels of pesticides in food : minimising the impact on the cocoa sector. *In*: proceedings of the 15th International Cocoa Research Conference. 9-10 October 2006. San Jose (Costa Rica). pp. 1565-1571.

Arnold B.1999. Fungal endophytes of tropical trees: Methods and potential for biological control of fungal pathogens of cocoa. Research methodology in biocontrol of plant diseases with special reference to fungal diseases of cocoa. Worshop manual, CATIE, Turrialba, Costa Rica 28 june – 4 july, 1999. Edited by Ulrike Krauss & Prakash Hebbar. pp 44 -54.

Azevodo J.L. 2002. Microrganismos endofiticos e seu papel em plantas tropicais. *In: Serafini LA, et al. Ed.Biotecnologia : avanços na agricultura e na agroindùstria.* Caxias do sul : EDUCS, 2002 : 233-268.

Benhamou N., Chet I. 1996. Parasitism of sclerotia of *Sclerotium rolfsii* by *Trichoderma harzianum* : ultrastructural and cytochemical aspects of the interaction. Phytopathology 86:405-416

Benhamou N., Gagné S., Quéré D.I., Dehbi I. 2000. Bacterial-mediated induced resistance in cucumber : beneficial effect of the endophytic bacterium *Serratia phymuthica* on the protection against infection by *Pythium ultimum*. Phytopathology 90: 45-56

Benitez, T., Rincon, A.M., Limo,, M. C., Codon, A. C. 2004. Biocontrol mechanisms of Trichoderma strains. Review article. International Microbiology 7 : 249 – 260.

Bigirimana J., De Meyer G., Poppe J., Elad Y., Hofte M. 1997. Induction of systemic resistance on bean (*Phaseolus vulgaris*) by *Trichoderma harzianum*. Med. Fac. Landbouww. Univ. Gent 62 : 1001-1007.

Bong, C. L., Wong, A., Chu, F., Tiko, T. J., Milin, S. 1996. Field evaluation of the potential of *Trichoderma* spp and *Gliocladium virens* and *Beauveria brassiana* (Bals.-Criv) Vuill.

(Cordycipitaceae) for the control of black pod disease and cocoa pod borer of *Theobroma cacao* L. 13th Conf. Intern. Rech .Cacaoyère (Malaisie), pp. 481- 487.

Bong, C.L., Stephen, M. 1999. *In vitro* assessment of sensitivity of cocoa clones to Phytophthora isolates. In: Sidek, Z., Bong, C.L., Vijaya, S.K. Ong, C.A. et Hussan,A.K. eds, Sustainable crop protection pratices in the next millennium, MCB-MAPPS plant protection conference 99, Kota kinbalu, Sahah, Malaysia, pp.32-37

Bowers J.H., Sanogo S., Tondje P.R., Hebbar P. K., Lumsden R.D. 2001b. Developing strategies for biological control of black pod, monilia pod rot, and witches'broom on cacao. *In : Proceedings of the 1st West and Central Africa training workshop on biocontrol of plant diseases, with special reference to cacao black pod diseases.* 25-29 June, 2001. Douala (Cameroon). pp. 10-16.

Burbage D.A., Sasser M. 1982. A medium selective for *Pseudomonas cepacia*. Phytopathology. 72: 706.

Carroll G. C.1998. Fungal endophytes in stems and leaves: from latent pathogen to mutualistic symbiont. Ecol. 69: 2-9.

Chen C, Bélanger RR., Benhamou N., Paulitz TC. 1998. Induced systemic resistance (ISR) by *Pseudomonas* spp. Impairs pre-and postinfection development of *Pythium aphanidermatum* on cucumber roots. European Journal of Plant Pathology 104: 877-886.

Chet, I., Inbar, J., Hadar, I. 1997. Fungal antagonists and mycoparasites. In : Wicklow, D.T. , Söderström, B (eds) The Mycota IV : Environmental and microbial relationships. Springer- verlag, Berlin, pp. 165 – 184

Davet, P., Rouxel, F. 1997. Détection et isolement des champignons du sol. Eds. INRA, Paris. France.194 p.

De Cal A, Garcia-Lepe R., Pascual S., Melgarejo P. 1999. Effects of timing and method of application of *Penicillium oxalicum* on efficacy and duration of control of Fusarium wilt of tomato. Plant pathology 48: 206-266.

Druzhinina I., Kubicek C.P. 2005. Species concepts and biodiversity in *Trichoderma* and *Hypocrea* : form aggregate species to species clusters. J zhejiang Univ. Sci. B.6 : 100-112.

Duijff BJ., Gianinazzi-Pearson V., Lemanceau P. 1997. Involvement of the outer membrane lipopolysaccharides in the endophytic colonization of tomato roots by biocontrol *Pseudomonas fluorescens* strain WCS417r. New Phytologist 135: 325-334.

Evans H.C., Holmes K. A.,Thomas S. E. 2003. Endophytes and mycoparasites associated with an indigenous forest tree, *Theobroma gileri*, in Ecuador and preliminary assessment of their potential as biocontrol agents of cocoa diseases. Mycological Progress 2 (2): 149-160.

Flood J. 2006. The threat from global spread of cocoa pests and diseases : hypothetical scenario or clear and present danger. *In*: proceedings of the 15th International Cocoa Research Conference. 9-10 October 2006. San Jose (Costa Rica). pp.857-872.

Gerlagh M, Goossen-van de Geijn H.M., Fokkema N.J., Vereijken P.F.G. 1999. Long-term biosanitation by application of *Coniothrium minitans* on *Scleroritinia sclerotiorum* – infected crops. Phytopathology 89: 141-147.

Guizzardi M. M., Pratella G. C. 1996. Biological control of gray molt in pears by antagonistic bacteria. Biological Control 7: 30-37

Harman E.G., Howell R. C., Viterbo A., Chet I., Lorito M. 2004. *Trichoderma* species - opportunistic, avirulent, plant symbionts. Nature reviews, Microbiology. 2 : 43-56.

Hibar K. Daami-Remadi M. Khiaaareddine H., El Mahjoub M. 2005. Effet inhibiteur *in vitro* et *in vivo* de *Trichoderma harzianum* sur *Fusarium oxysporum* f. sp *radicis-lycopersici Biotechnol.* Agron. Soc Environ. 9 (3) 163-171.

Hebber K.P., Martel M.N., Heulin T. 1998. Suppression of pre-and post-emergence damping-off in corn by Burkholderia cepacia. European journal of plant pathology. 104 : 29-36

Howell, C.R. 2003. Mechanisms employed by *Trichoderma* spp. in the biological control of plant diseases : the history and evolution of current concepts., Plant Dis. 87 : 4 –10.

Howell C. R., Hanson L. E., Stipanovic R.D., Puckhaber L.S. 2000. Induct ion of terpenoid synthesis in cotton roots and control of *Rhizoctonia solani* by seed treatment wi th *Tr ichoderma virens*. Phytopathology 90: 248-252

International Cocoa and Chocolate Organisation (ICCO). 2000. Célébration du cacao, 115 p.

Kébé I. B., N'goran J. A.K., Tahi, G.M., Paulin, D., Clément, D., Eskes, A.B. 1996. Pathology and breeding for resistance to black pod in Côte d'Ivoire.*In :* proceedings of the International Workshop on the Contribution of disease Resistance to cocoa Variety Improvement. 24th- 26th November, 1996. Salvador, Bahia (Brazil). pp. 135-139.

Koné Y. R., 1999. Etude de la structure actuelle des populations de *Phytophthora* spp., agents de la pourriture brune des cabosses du cacaoyer (*Theobroma cacao* L.) en Côte d'Ivoire. Mémoire de Diplôme d'Agronomie Approfondie, Option Défense des cultures, Ecole Supérieure d'Agronomie, Yamoussoukro. 111 p.

Kouamé K. D. 2006. Structure et dynamique des populations de *Phytophthora* spp., agents de la pourriture brune des cabosses du cacaoyer (*Theobroma cacao* L.) en Côte d'Ivoire. Mémoire de DEA. UFR Biosciences. Université de Cocody, Abidjan, Côte d'Ivoire. 74 p.

Krauss U., Soberanis W. 2001. Biocontrol of cocoa pod diseases with Mycoparasite mixtures. Biological control 22: 149-158.

Krauss U., Soberanis W. 2002. Effect of fertilization and biocontrol application frequency on cocoa pod diseases. Biological Control 24: 82-89.

Krauss U., Martijn ten Hoopen G., Hidalgo E., Martinez A., Arroyo C., Garcia J., Portuguez A., Palacios M. 2003. Biocontrol of moniliais (*Moniliophthora roreri*) and Black pod (*Phytophthora spp*) in Panama with mycoparasites in two formulations. In: Akrofi A.Y., Achonor J.B., & Ollennu L.A.A. (eds). Proceedings of INCOPED 4th International Seminar on Cocoa pests and diseases. 19- 21 October 2003. Accra (Ghana). pp 53-58

Lass R. A. 1985 Disease. In Cocoa.. Wood G.A.R., & Lass R. A. Eds. 4th edition, Longman, Longman, 265-365.

Maurhofer M., Hase C., Meuwly Ph., Métraux J-P., Défago G. 1994. Induction of systemic resistance of tobacco to tobacco necrosis virus by the root colonizing *Pseudomonas fluorescens* strain CHAO: influence of the gac A gene and of pyoverdine production. Phytopathology 84: 139-146.

Mpika J. 2002. Lutte contre la pourriture des cabosses du cacaoyer en Côte d'Ivoire : isolement et identification des antagonistes naturels de *Phytophthora* sp.

Mémoire de DEA. UFR de Biosciences Université de Cocody, Abidjan, Côte d'Ivoire. 74 pp.

Nyassé S., Cilas C., Hérail C., Blaha G. 1995. Leaf inoculation as an early screening test for cocoa (*Theobroma cacao* L.) resistance to *Phytophthora* black pod disease. *Crop Protection* 14 (8): 657-663.

Papavizas, G.C., Lumsden, R. D. 1982. Improved medium for isolation of *Trichoderma* spp. from soil. Plant. Dis., 66 : 1019-1020.

Paulitz T.C., Linderman R.G. 1991. Lack of antagonism between the biocontrol agent *Gliocladium virens* and *vesicular arbuscular* mycorrhizal fungi. New physiologist. 117 : 303-308

Peixoto-Neto P.A.S. 2002. Microorganismos endofiticos. *Biotecnologia Ciência & Desenvolvimento.* 2002: 62-76.

Rapilly F. 1968. Les techniques de mycologie en pathologie végétale. Annales des Epiphyties Volume 19. N° hors-série. INRA. Paris. pp 25-39.

Rubini M. R., Silva-Ribeiro R. T., Pomella, A. W. V., Maki C. S., Araujo W. L., dos Santos D. R., Azevedo J. L. 2005. Diversity of endophytic fungal community of cacao (*Theobroma cacao* L.) and biological control of *Crinipellis perniciosa*, causal agent of Witches'Broom Disease. Int. J Biol Sci. 1: 24-33.

Samuels G.J. 1996. *Trichoderma*: a review of biology and systematic of the genus. Mycological Research 100: 923 - 935.

Sanogo S, Pomella A., Hebbar P.K., Bailey B. Costa J.C.B., Samuels G.J., Lumsden R.D. 2002. Production and germination of conidia of *Trichoderma stromaticum*, a Mycoparasite of *Crinipellis perniciosa* on Cacao. Phytopathology 92: 1032-1037.

Shari Fuddin S. 2000. Studies on cocoa rhizosphere Bacteria for biological control of *Phytophthora* species. *In:* proceedings of the 13th International Cocoa Research Conference. 9-14 October 2000. Kota Kinabalu, Sabah (Malaysia). pp.489-493.

Sid Ahmed A., Sanchez P.C., Candela E.M. 2000. Evaluation of induction systemic resistance in pepper plants (*Capsicum annuum*) to *Phytophthora capsici* using *Trichoderma harzianum* and its relation with capsidiol accumulation. European journal of plant pathology 106 : 824-824.

Singh P.P.S., Shin Y.C., Park C.S., Chung Y.R. 1999. Biological control of Fusarium wilt of cucumber by chitinolytic bacteria. Phytopathology 89 : 92-99.

Tahi M., Kebe, I. Eskes A.B., Ouattara S., Sangare A., Mondeil F. 2000. Rapid screening of cacao genotypes for field resistance to *Phytophthora palmivora* using leaves, twigs and roots. European Journal of Plant Pathology 106: 87-94.

Tim, A., Gearge, A., Martijn ten Hoopen, G., Krauss U. 2003. Rhizosphere populations of antagonistic fungi of cocoa (*Theobroma cacao* L.) clones tolerant or susceptible to Rosellinia root rot. In: Akrofi AY, Achonor J.B, Ollennu, LAA (eds): Proceedings of INCOPED 4th International Seminar on Cocoa pests and diseases at Accra, Ghana, pp. 104-115

Tondje, P.R., Hebbar, P.K., Samuels, G., Bowers, J.H., Weise, S., Nyemb, E., Begoude, D., Foko, J., Fontem, D. 2006a. Biossay of *Geniculosporium* species for *Phytophthora megakarya* biological control on cacao pod husk pieces. African journal of biotechnology 5 (8) : 648 - 652.

Tondje, P.R., Mbarga, J.B., Atangan, J.B., Tchana, T., Begoude, D., Nyemb T. E., Deberdt, P., Bon M. C., Samuels G. J, Hebbar P.K., Fontem D. 2006b. Evaluation of

Trichoderma asperellum (PR10, PR11, PR12, 659-7) as biological control agents of *Phytophthora megakarya,* the causative agent of black pod disease on cacao (*Theobroma cacao* L.). In: Proc. 15th Int. Cocoa Res. Conf. San Jose, Costa Rica, pp. 981-993.

Tondje P. R., Roberts D.P., Bon M.C., Widmer T., Samuels G.J., Ismaiel A., Begoude A.D., Tchana T., Nyemb-Tshomb E., Ndoumbe- Nkeng M., Bateman R, Fontem D., Hebbar K.P. 2007. Isolation and identification of mycoparasitic isolates of *Trichoderma asperellum* with potential for suppression of black pod disease of cacao in Cameroon. Biological control. 43: 202-212.

Vidhyasekaran P., Muthamilan M. 1999. Evaluation of a powder formulation of *Pseudomonas fluorescens* Pf1 for control of rice sheath bligh. Biocontrol Science and Technology 9: 67-74